The Cradle of Words

The Cradle of Words

Language and Knowledge in the Spanish Empire

Valeria López Fadul

Johns Hopkins University Press
Baltimore

9 8 7 6 5 4 3 2 1

Johns Hopkins University Press
2715 North Charles Street
Baltimore, Maryland 21218
www.press.jhu.edu

Library of Congress Cataloging-in-Publication Data

Names: López Fadul, Valeria, 1986– author.
Title: The cradle of words : language and knowledge in the Spanish
    empire / Valeria López Fadul.
Description: Baltimore : Johns Hopkins University Press, 2025. |
    Series: Information cultures | Includes bibliographical references and
    index.
Identifiers: LCCN 2024013761 | ISBN 9781421450216 (hardcover ; acid-free
    paper) | ISBN 9781421450223 (ebook)
Subjects: LCSH: Spain—Languages—History—16th century. |
    New Spain—Languages—History—16th century. | Spain—Intellectual
    life—1516-1700. | New Spain—Intellectual life—16th century. |
    Knowledge, Theory of—Spain. | Knowledge, Theory of—New Spain.
Classification: LCC P381.S6 L57 2024 | DDC 462.09171/24609031—dc23
    /eng/20240805
LC record available at https://lccn.loc.gov/2024013761

A catalog record for this book is available from the British Library.

*Special discounts are available for bulk purchases of this book. For more
information, please contact Special Sales at specialsales@jh.edu.*

Contents

Figures

Note on Transcriptions and Translations

This book reconstructs how early modern writers in the Iberian world understood words as sources. It charts the itineraries through which names circulated orally, in manuscript materials, and in print. I have maintained original spelling throughout because the way orthographic conventions changed over time, especially for words borrowed from other languages, tells a story in itself. Given the availability of numerous early modern printed books in digital format, my endnotes prioritize the transcription of manuscript materials. Whenever available, I compared critical editions of primary sources with facsimiles, manuscripts, and early printed editions. Unless otherwise noted, all translations are my own.

Acknowledgments

Several early modern compendia of Castilian sayings include the maxim "De hombre agradecido, todo bien creído" (Of a grateful man, all is well believed). This ancient adage argues that gratitude to others bespeaks credibility, since no one can achieve anything alone. After many years of researching and writing this book, I too can vouch for this fact.

Writing this book would have been impossible without the financial support of numerous institutions. The Princeton History Department, the Center for the Study of Religion, the Princeton International Institute for International and Regional Studies, the Princeton Program for Latin American Studies, and the John Carter Brown Library supported my research in its earliest stages. I am also grateful to the Department of History at the University of Chicago, where I held an Andrew W. Mellon Provost's Postdoctoral Career Enhancement Fellowship. At Wesleyan University I have received assistance from the History Department, including a generous Meigs publication grant, and the Center for the Humanities. The Folger Shakespeare Library funded a timely writing fellowship. An Andrew W. Mellon Foundation Fellowship for Assistant Professors in the School of Historical Studies at the Institute for Advanced Study enabled an essential year of leave during which the project took its current form.

This book began as a dissertation at Princeton University. I am profoundly grateful to my advisor Anthony Grafton for his teaching, kindness, and unwavering generosity over many years. (His enthusiasm for etymologies helped, too.) Adam Beaver's wisdom and guidance in the early stages shaped my approach to early modern Spanish history. Mauricio Tenorio Trillo's commitment to recovering Latin American philology has served as a model. Francesca Trivellato has remained an unfailingly kind mentor since my undergraduate days; her sharp suggestions strengthened my arguments. I also thank my teachers in Colombia and in the United States: Jeremy Adelman, Vera Candiani, Arcadio Díaz Quiñones, William C. Jordan, María Portuondo, Rob Karl, Stuart Schwartz, and Moisés Álvarez

Marín. Valerie Hansen encouraged me to pursue history from the beginning of my college career and has become a valued friend.

I have enjoyed the good fortune of pursuing research in libraries and archives in Spain, the United States, and Latin America. I would like to express my appreciation for the librarians and staff of the Firestone Library at Princeton University, the Beinecke Rare Books & Manuscripts Library and the Sterling Memorial Library at Yale University, the John Carter Brown Library at Brown University, the Hispanic Society of America, the Joseph Regenstein Library at the University of Chicago, the Newberry Library, the Huntington Library, the Benson Latin American Collection at the University of Texas at Austin, the Olin Library at Wesleyan University, and the Historical Studies-Social Science Library at the Institute for Advanced Study in Princeton. In Spain and Latin America, I am thankful for the assistance of archivists and librarians in the Biblioteca Nacional de España, the Real Academia de la Historia, the Biblioteca Real del Palacio, the Real Biblioteca del Monasterio de San Lorenzo de El Escorial, the Biblioteca Capitular Colombina, the Archivo General de Indias, the Archivo Histórico Nacional, and the Biblioteca Nacional de Colombia.

Wesleyan University has become a true intellectual home; my colleagues in the History Department and in the Program for Latin American Studies have offered constant support and sustained innumerable exchanges. I owe special thanks to Gary Shaw, who read the manuscript and provided feedback, encouragement, and mentorship. My students' enthusiasm for early Latin American history heartened me while I completed this book; their questions and comments led me in new directions. The members of the Indigenous Studies Research Network read and commented on several chapter drafts, as did the Wesleyan Center for the Humanities Spring 2020 cohort and its director Natasha Korda. At the Institute for Advanced Study, the members of the early modern seminar (2020–2021) and of the School of Historical Studies created a warm virtual community during the times of social distancing. Thanks to Suzanne Akbari, Eleanor Hubbard, Samantha Kelly, Pamela O. Long, Arnaud Orain, and Jonathan Sheehan. My colleagues at *History & Theory* have enriched my work through many conversations; Ethan Kleinberg and Courtney Weiss Smith have offered steadfast encouragement.

Along the way, numerous interlocutors have read chapter drafts or commented on presentations. I thank Ann M. Blair, Daniela Bleichmar, Dain Borges, Larissa Brewer-García, Sabrina Carletti, Mackenzie Cooley, Thomas B. F. Cummins, Claire Gilbert, Jessica Delgado, Caroline Engelmayer, Brodwyn Fischer, Antonio Feros, Tamar Herzog, Adam Jasienski, Richard Kagan, Seth Kimmel,

Emilio Kourí, Domingo Ledezma, Julia Madajczak, César Manrique Figueroa, Stuart McManus, Hannah Marcus, Miguel Martínez, Ann Moyer, Katrina Olds, María Ospina, Stefania Pastore, Jesús de Prado Plumed, Ronny Regev, Melissa Teixeira, Fidel Tavárez, and Pier Mattia Tommasino. Early on Mercedes García Arenal provided encouragement and incisive comments; her exemplary scholarship continues to inspire me. I am immensely grateful to Alexander Bevilacqua for his feedback at every stage of this project and for his friendship.

An earlier, more concise version of chapter 2 appeared as "Language as Archive: Etymologies and the Remote History of Spain," in *After Conversion: Iberia and the Emergence of Modernity*, ed. Mercedes García-Arenal (Leiden: Brill, 2016), 95–125. Some themes of chapter 1 were touched on in "Juan Páez de Castro and the Project of a Universal Library," in "Wars of Knowledge: Imperial Hegemony and the Assembling of Libraries Forum," *Pacific Coast Philology* 52, no. 2 (October 2017): 173–83. I am grateful to Brill and to the Penn State University Press for allowing me to reproduce these materials here. I also thank Earle A. Havens, Matthew R. McAdam, and everyone at the Johns Hopkins University Press for their investment in and commitment to this project.

While working on this project I have enjoyed the solidarity of many wonderful friends: Marco Aresu, Juanita Aristizábal, Catherine Abou-Nemeh, Abigail Boggs, Frederic Clark, Heidi Hausse, Justene Hill Edwards, Reut Harari, April Hughes, Jennifer Jones, Radha Kumar, Madeline McMahon, Amanda Nelson, Gabrielle Ponce, Laurie Nussdorfer, Paula Park, Andrei Pesic, Jenna Phillips, Helen Pfeifer, Margaret Schotte, Diana Schwalowski, Paris Spies-Gans, Victoria Smolkin, Ying Jia Tan, Laura Ann Twagira, and Stephanie Weiner. I am likewise indebted to Marta Albalá Pelegrín and Javier Patiño for stimulating conversations and collaborative projects about early modern Spanish history and literature. I would also like to record my appreciation for the many modern scholars, grammarians, and lexicographers whose works I cite in the pages that follow. They made an investigation about sixteenth-century etymological practice in a vast multilingual world imaginable.

Finalmente, pero no con menos afecto ni devoción, dedico este libro a mi familia, que ha apoyado y protegido todos mis proyectos desde los Estados Unidos hasta el gran Caribe. Especialmente, a Jorge López Morales (E.P.D.), a Jorge López Fadul, a Emilia Fadul Rosa, y sobre todo, a Conrad Vahlsing.

The Cradle of Words

Introduction

In December 1579, in the Mexican Diocese of Guadalajara, the Spanish administrator of the town of Ameca appeared before a scribe to certify the answers submitted to a royal census. The questionnaire required the official to summon both Spaniards of "faith and credit" and the town's "old and leading" Indigenous authorities to describe the settlement's geographical features, population, history, natural resources, and languages. They were to be discussed in a succession of fifty prescribed chapters. The first question asked respondents to state not only the town's name but also "the meaning of the name in the indigenous language and the reason it is so named."[1] To convey the etymology, or origin, of Ameca, the Crown official explained that this name of a "town of Indians" could be translated into Castilian from the "Cazcan language, which is the one spoken in this town," as "above the water." The name was derived from the town's position above a river. The official also clarified that two additional settlements, Huitzquilic and Jayamitla, were subject to Ameca. Huitzquilic, or "town of thorny herbs," was thus named on account of a spiny plant that grew in the area. Jayamitla signified "apiary," since it was constructed with a series of boulders that resembled beehouses (fig. I.1). Later chapters of the census allowed the Spanish official to briefly relate the town's history as he had understood it from local authorities, piecing together Indigenous testimonies, words, names of founders, and noteworthy landmarks into a narrative that both began and culminated by asserting the town's incorporation into the domains of the Spanish king.[2]

That same month in the kingdom of Castile, near the city of Madrid, the authorities of the town of Barajas appeared before a scribe to answer a similar

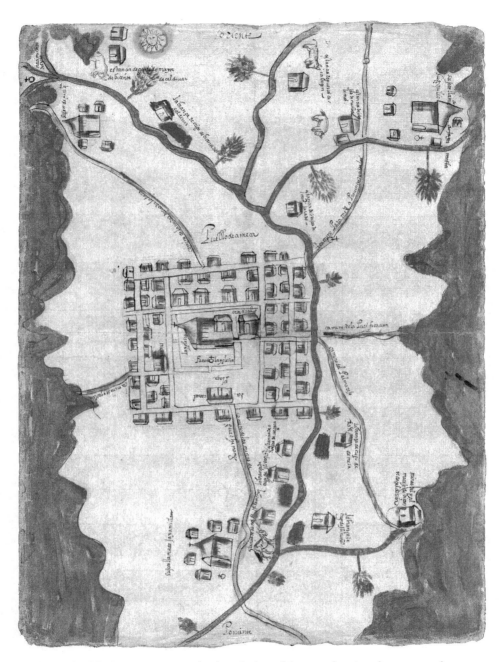

Figure I.1. Map accompanying the description of Ameca, showing the town settlement beside the river with the two subject settlements of Jayamitlan (*left of the main settlement*) and Huitzquilic (*bottom right*). Courtesy of the Nettie Lee Benson Latin American Collection, University of Texas Libraries, University of Texas at Austin.

questionnaire. Regarding the place-name's origin, the officials declared that Barajas had both Hebrew and Arabic parts. The many "ancient and curious people of the village" held as "common opinion and truthful thing" that the two parts of the name, Bar and Aja, referred to a former lord of the town, from the time "when the Moors occupied Castile," a "Moor named Baraja." This lord was the son of a Moorish woman called Aja, and "thus named the town Baraja." In response to subsequent questions, the elders of Barajas related how ancient stones, inscriptions, and coins revealed the town's antiquity. When the "Romans possessed Spain," a townswoman erected a temple to Jupiter, as a surviving stone indicated, and coins with the figure of the emperor Trajan betokened the local circulation of Roman currency. The town was eventually "won from the power of the Moors" during the reign of the Castilian king Alfonso VI, who conquered nearby Toledo and Alcalá de Henares in 1085. While the authorities of Barajas did not establish the settlement's pre-Islamic name, their account sought to demonstrate the town's ancient origins, its persistence in the face of conquest, and finally its pious incorporation into the Christian kingdom of Castile.[3]

Thus two inquiries in different hemispheres began with the same request for an etymological explanation of each town's name. This query inaugurated the questionnaires distributed both in the transatlantic viceroyalties of the Americas and in the Iberian kingdoms beginning in 1569.[4] The censuses display the multiple layers of translation that characterized Spanish imperial knowledge making. They sought to harness local knowledge while also making it fit into categories meaningful to Spanish administrators. The censuses strove to make a diversity of local traditions, which were often orally transmitted, both collectible by and legible to a distant monarch who increasingly relied on such compendia to make statements about his power and omniscience.[5]

Why did the inquiries begin by asking locals to unravel the history of their settlement's name? What could be gained from such a seemingly arcane request? In the mid-sixteenth century, etymology, as this book shows, became a meaningful method to study the natural and human history of the Spanish Empire. Early modern European scholars believed that languages changed relentlessly. They mutated according to the will, practices, material, and social conditions of those who spoke them, which I will call speech communities.[6] Languages therefore had histories and could be treated as archives. They recorded the origins of peoples and the collective choices that speech communities transmitted from generation to generation. These choices, or customs, were the foundation of local history and, hence, also of customary law. Thus, the Crown officials and informants of both

Ameca and Barajas, regardless of their location on the globe, the specific languages that their communities spoke, or the modes of inscription (alphabetic or otherwise) they used, began their reports by relating the origins of their settlements as archived in their town's names. To the administrators, missionaries, and scholars charged with learning about the peoples and places under the jurisdiction of the Spanish king, local authoritative sources held the greatest evidentiary value. Linguistic patrimony was intrinsic to each speech community that adapted their words to convey specific—and often ancient—knowledge of place.

This book investigates how and why chains of information contained in language, such as those embedded in the names of places, became significant to knowledge production of all kinds in the middle decades of the sixteenth century. Early chronicles about the ancient history of the Americas, works on Spanish antiquities, writings on the medicinal virtues of plants and animals, and attempts to discern the arcane meanings of the Bible all relied on modes of linguistic analysis, like etymology, and on the formulation of detailed linguistic genealogies. In turn, insights drawn from the history of language were used in support of historical, political, and even legal arguments that were by no means linguistic in nature.

Since early modern Europeans understood language to encode various kinds of historical information, studying language was the starting point for all inquiries, whether scientific, religious, legal, or historical. This book contends that Spanish scholars, missionaries, and administrators treated the empire's multiple tongues, at home and abroad, as rich archives of local knowledge that could be exploited alongside other resources like the Americas' vast mineral and natural wealth and the labor of its Indigenous inhabitants. To make sense of their newly broadened world, Spanish scholars actively drew on a long European tradition of linguistic thought as well as on the Iberian Peninsula's unique centuries-long experience with multilingualism—not just with Castilian, Catalan, Portuguese, and Basque but also with Latin, Arabic, Hebrew, and even ancient Greek.

Soon after his coronation in 1556, Philip II (r. 1556–1598) initiated a variety of projects at great expense to his perennially depleted coffers.[7] They included the expansion of the Crown's archival system, the preparation of a royal edition of all the works of the late antique scholar Isidore of Seville (c. 560–636), the founding of an all-encompassing royal library in San Lorenzo de El Escorial, the gathering of a collection of relics, the edition of an authoritative Bible in Antwerp, a botanical expedition in the American provinces of the Spanish Empire, and the censuses of towns that eventually included Ameca and Barajas. Linking these initiatives was the overarching desire to compile specialized knowledge, whether religious, natural, historical, or commercial, in central repositories that the king's administrators could easily access.[8]

The present study demonstrates that linguistic studies bound these knowledge-making pursuits together in ways that have come to seem incompatible from the perspective of our modern disciplinary divisions. Today we would distinguish among "science," "history," "religious indoctrination," and "statecraft," but in the eyes of King Philip II, these disciplines were united as a single project that would constitute the cornerstone of his reign. The new Renaissance emphasis on first-hand observations in botany or medicine, moreover, dovetailed with the search for linguistic information. Direct experiences of the natural or human world of the Americas did not displace the search for authoritative local knowledge recorded in language. Instead, linguistic research was an integral part of gaining firsthand knowledge of the so-called New World.

Multilingualism in the Early Modern Iberian World

In the late sixteenth century Philip II governed peoples who inhabited radically distinct environments, possessed disparate political traditions, and spoke mutually unintelligible languages. From Castilian Spanish, Arabic, Basque, and Catalan in Iberia to Dutch in the Low Countries, Italian in Milan and Naples, Tagalog in the Philippines, and Quechua, Nahuatl, Aymara, Tupi, and Guaraní in the Americas, the empire encompassed an impressive multiplicity of languages. Spain was the first European polity to possess a transatlantic empire. Since the outset, the Spanish conquest of the Americas introduced foreign diseases, brutal labor regimes, and violent social dislocations that brought about a dramatic decline of the continent's Indigenous population. Despite the onslaught of plagues and violence, and the disappearance of numerous societies and their languages, Spanish administrators and missionaries nonetheless sought to govern a region that remained fundamentally multilingual.

To the Spanish elite, their empire's linguistic variety reflected its universal aspirations. Yet the empire's polyglossia also represented an enormous challenge to their sovereignty. How to unite such a vast and disparate political body under the rule and spiritual guidance of a single king? Modern scholarship long insisted that the Spanish Crown sought to impose Castilian Spanish in its transatlantic realms immediately after 1492. More recent accounts have revised this view, showing that the monarchy handled its multilingual subjects in varied and even contradictory ways in order to placate vying corporate groups, including the regular (monastic) orders as well as the secular clergy. The Crown at times sought to establish the supremacy of Castilian and at others pursued pragmatic strategies, including employing interpreters and missionaries to learn local languages. The common goal was to ensure the flow of revenue, the dispensation of justice, and the advance of Christianization. As a result, Spanish administrators generated a haphazard set of linguistic decrees and seemingly incompatible approaches to linguistic diversity.[9]

In medieval and early modern Europe, as in most premodern societies, linguistic pluralism was commonplace. In royal courts, official documents, religious ritual, and everyday interactions, different languages, dialects, and registers of the same tongue coexisted as a fact of life. As Peter Burke has argued, it is inaccurate to label the royal pronouncements on language of this period "language policies" because this phrase connotes an intent on the part of the monarch to shape a linguistic landscape for the future. Instead, rulings on language were merely meant to remedy immediate difficulties that arose from the everyday exercise of governance.[10]

In line with this function, royal pronouncements on language instruction in the Americas varied substantially over the course of the early modern period.[11] In 1524, with the arrival of the first missionaries in New Spain (modern-day Mexico), Spanish officials began to emphasize teaching in local languages. In 1550, Emperor Charles V issued a royal decree to ensure that Indigenous people were taught Castilian. In 1574, however, another royal decree required prelates and other ecclesiastical authorities to select candidates to catechize Indigenous communities who knew the languages in which they were to propagate the Catholic faith.[12] By 1585, when Archbishop Pedro Moya de Contreras (d. 1591) convened the Third Mexican Provincial Council, the use of Indigenous languages for religious conversion was preferred. The training of clergy and practical difficulties associated with learning languages, however, sometimes led to conflicting positions between the religious orders and the secular clergy. Attitudes toward Indigenous languages remained highly contextual and dependent on regional circumstances, and even on the inclinations of specific parishes and missionaries.[13]

In the 1580s the Third Council of Lima, likewise, decided to continue teaching Indigenous subjects in their own languages and to advance the use of what it identified, problematically, as "general languages."[14] These were language varieties that the Spaniards believed to be widespread in regions like the Andes and Mesoamerica, which large empires had ruled at the time of the conquest. The early modern term *lengua general* encompassed a variety of types, grouping together different languages, varieties, and dialects under the same moniker.[15] Sometimes sixteenth-century scholars perceived an entire language family to be part of the same general language.

To remedy variations in the teaching of religious doctrine and to standardize the materials that missionaries had at their disposal, the Third Council of Lima commissioned a group of scholars to compose grammars, catechisms, and other liturgical materials in these "general languages." Alan Durston's study of Standard Colonial Quechua in the former Inca domains and what would become the vice-royalty of Peru demonstrated how through the crafting of grammars, catechisms, and dictionaries, the missionaries not only relied on what they perceived to be widely spoken languages but actually "created their own vehicular varieties through complex processes of selection, codification, and standardization, rather than working with what was 'already there.'"[16] These standards, Durston further argues, anticipated similar strategies carried out by nineteenth- and twentieth-century colonial empires in Asia and Africa. Standard Colonial Quechua, was not, however, a "lingua franca," in the sense that it facilitated communication between people, but rather was a written standard created for pastoral purposes that never "acquire[d] significant administrative or legal roles." It was, additionally, overall a restrictive medium; Indigenous pupils were supposed to memorize and perform texts in it, rather than debate or even converse with their instructors.[17] The approach of relying on these so-called general languages was also adopted elsewhere, as in New Granada (modern-day northeastern South America), albeit with mixed results in a region with no clearly predominant languages.[18]

The tradition of using Indigenous languages to catechize remained more or less in place until 1770, when the Bourbon monarch Charles III ordered the imposition of Castilian and the eradication of Indigenous languages as part of his reform program to remake the Habsburg kingdoms into colonies.[19] This royal decree signaled a significant break with the past.[20] Until the eighteenth century, no universal solution, no overarching decree regulated administrators' interaction with the linguistic diversity of the Americas and the Philippines (which became part of the viceroyalty of New Spain in 1565).

Demographic and social forces like Iberian emigration to the Americas and the demands of the colonial economy did facilitate the spread of Castilian as a language of administration and communication. But this was not a uniform process. The Taíno language ceased to be spoken in Hispaniola as early as 1550, a casualty of extreme Indigenous demographic collapse. However, the forceful congregation of Indigenous communities (a process called *congregación* or *reducción*) into Spanish-style towns in some instances allowed these communities to preserve their languages. In the Andean regions, the mining economy and labor tribute impositions encouraged the spread of Southern Peruvian Quechua and Castilian to the detriment of other varieties of Quechua and Aymara. The mobilization of tributary laborers from various regions, urban growth, and commercial integration all stimulated the use of Southern Peruvian Quechua and Castilian as common languages.[21]

Increasingly, alphabetic writing was used to record Indigenous languages throughout the colonial period, yet it did not displace other forms of inscription. In Mesoamerica, still one of the most linguistically diverse regions of the world, Indigenous intellectuals adopted alphabetic writing to compose histories in Indigenous languages for their own communities.[22] Indigenous-language writings in Mesoamerica also became significant to municipal administrators and litigants.[23] In the Andean world, where fewer Indigenous-language documents in alphabetic writing survive, it was commonplace to use interpreters to communicate with Spanish judges.[24] In this region, various forms of literacy also coexisted as different communities adopted visual and alphabetic forms of inscription.[25] In sum, Indigenous languages did not decline linearly after the Spanish conquest.[26]

In contrast to the dozens of Indigenous-language grammars written and printed between the 1540s and the late eighteenth century in Spanish America, few comparable materials exist in African languages.[27] Missionaries generally relied on interpreters to teach Christianity to enslaved African laborers. Moreover, the African diaspora was multilingual, no single language predominated, and the Black men and women who were subjected to the transatlantic slave trade were forced to speak Spanish, Portuguese, or even "improvised creole languages" when they arrived in American cities.[28]

In the Iberian Peninsula, multilingualism persisted throughout the sixteenth century, with the important exception of Arabic. In medieval Spain, multilingualism had been commonplace, and speakers of various Romance languages and Arabic communicated with each other. After Ferdinand and Isabel conquered the kingdom of Granada (1482–1492), Muslims and Jews (until 1492) could live "in their own law and languages" as vassals of Spanish kings. Beginning in 1502, however, the Catholic monarchs began to institute policies to encourage the teaching of Christian doctrine in Castilian and Latin. This was followed by forced conversion and progressively repressive legislation against the speaking of Arabic. Language, among other features like dress, became central to discussions about the integration of moriscos (former Muslims living under Christian rule) into Spanish society as external markers of religious identity. The Crown's policy shift from a "regime of legal pluralism" to one of "religious and legal orthodoxy" inaugurated a century of debates about the moriscos that culminated with this group's expulsion from the Spanish domains between 1609 and 1614.[29]

The reality of linguistic multiplicity at home and overseas forced a broad range of Iberian scholars to consider whether language could be disaggregated from a society's other cultural or religious features. This question became an acute theological problem that concerned converts from Judaism and Islam and the Indigenous people of the Americas. Debates about speech communities, then, were often grounded in broader, often polemical debates about religion and religious belonging.[30]

Recent scholarship on multilingualism in the Spanish Empire has explored the strategies that missionaries and administrators employed to deal with the exigencies of everyday governance and conversion. Furthermore, by turning to language as a source, contemporary scholars have relied on philological methods to transform, sometimes radically, our understanding of the dynamics of social and cultural change under Spanish colonial rule. They have shown the many ways in which Indigenous societies preserved their historical traditions and adapted to colonial society.[31] This book builds on these insights but argues that writers and scholars in the premodern past likewise treated language as an object of historical inquiry. The pages that follow reconstruct their specific scholarly practices. The goal is to reconstruct these early modern scholars' understanding of language and to show how their understanding informed their investigations of linguistic pluralism and social change in the Spanish Empire.

Words Unraveled: Language Studies and the Early Modern Uses of Etymology

In early modern Europe, intellectuals made what Burke has called the "discovery of language."[32] Language education began to focus on other tongues than Latin.[33] Curiosity, commerce, and missionary interests brought about the proliferation of grammars, dictionaries, conversation manuals, and treatises on orthography. Scholars compared languages and established genealogical filiations or searched for common principles.[34] This curiosity extended to Castilian and to other Iberian languages. In 1492 the humanist grammarian Antonio de Nebrija (1444?–1522) published a *Grammar of the Castilian Language* (*Gramática de la lengua castellana*).

Admittedly, and despite the book's posthumous and enduring fame, the *Grammar of Castilian* enjoyed limited circulation and was not published again until the eighteenth century.[35] Nebrija's works on Latin, especially his Latin-Castilian dictionary, were more influential among early modern scholars.[36] Nebrija's effort to reduce spoken Castilian to rules, however, was part of the broader scholarly interest in vernacular languages. The "discovery of language" entailed, as in Nebrija's case, treating the spoken tongue as object of scholarly interest, describing its features, and establishing sources to study its history.

Renaissance writers sought to establish the Castilian tongue's history, orthography, and literary canon. In 1546, the antiquarian Ambrosio de Morales (1513–1591), for instance, urged fellow scholars to cultivate their spoken language. Morales believed that eloquence in one's native tongue was a sign of prudence and the "main ornament with which a wise man adorns his person."[37] Even though eloquence was possible in all languages, one should especially strive for it in one's own tongue.[38] It was because the Romans treated their language with reverence that Latin had acquired such nobility and honor. By contrast, too few scholars chose to write in Castilian. "The bad luck of our Castilian language," he lamented, "always pains me, that being equal to all the good ones in abundance, character, variety, and beauty it is so forgotten, and little considered." On account of its speakers' negligence, he complained, "it had lost much of its value."[39] He asked, "Who could point to many books in Castilian that they have read, and with confidence claim that they have improved the use of our language?"[40] Morales's pleas inadvertently remind us that "Castilian" was still being created in the sixteenth and seventeenth centuries.[41]

The strategy of gathering words and then mining them for their etymologies reflected a long-standing European understanding of the relationship between language and knowledge acquisition. By treating words as archives, Spanish chroniclers implemented an approach to language study that had served European scholars since antiquity. Names and their meanings could serve learning about any topic. Writers of all genres, including Pliny the Elder and Isidore of Seville (c. 560–636)—two prominent models for sixteenth-century chroniclers, believed that words contained clues about the histories of cities, land formations, deities, rituals, items of clothing, and even social roles, because they revealed the insights of their coiners. This field of study was often described using a well-known Stoic phrase that Augustine had quoted: etymology contained the "cradle of words" (verborum cunabula), their roots or very beginnings.[42] For ancients and for moderns, language and its transformations represented a way to access the unrecorded past, gain insight into the experiences of previous generations, and learn about people's origins and the histories of their regions.

Consequently, the study of words underlies a wide spectrum of intellectual pursuits. In medieval and early modern mystical, exegetical, and historical writings, the study of names played a prominent role. Isidore of Seville defined etymology as "the origin of words when the force of a verb or noun is inferred through interpretation." Knowing the origin of a word made it easier to understand its force, and "one's insight into anything is clearer when its etymology is known."[43] An etymology was "an ornament of brevity," and its force resulted from gathering in the mind associated words and concepts, serving as the beginning of a "memory chain."[44]

In the early modern period, the question of how languages signified, and whether they did so naturally or conventionally, acquired a renewed vigor.[45] The rediscovery of Plato's *Cratylus*, the revival of Neoplatonism, the translation of Hermetic texts, the Renaissance interest in magic and Kabbalah, and debates about biblical translation all contributed new approaches to etymological studies.[46] Despite long-standing debates about the possibilities (and failures) of etymological knowledge and the proper way to mine words, scholars generally agreed that this method had the potential to reveal great amounts of information about the objects that words signified.

Early modern scholars approached the study of words in several interrelated ways that can be subsumed under the following types: Place-names contained traces of their founders, so they could be etymologized to establish origins. The names of plants and animals revealed clues about their function or characteristics, that is, the knowledge that their name makers had discovered about their properties through experience. Names of objects and rituals betrayed hints about their purposes; thus, they could encapsulate collective, social understandings. In these two cases, words could be treated exegetically to uncover historical information. In works devoted to sacred texts, but not exclusively, etymologies could also be employed to access the arcane meaning of words and thus unveil prophetic knowledge. Finally, word analysis could also help determine the qualities of a language and how suitable its sounds were in conveying the properties of the objects signified.

Depending on the preferences and linguistic principles espoused by each writer, a work might emphasize one approach to the study of words over another or combine several types. In historical works interested in resolving questions about the origins of settlements, for instance, scholars could etymologize place-names to link them to illustrious founders. In natural historical investigations, etymology could be deployed to capture what name makers knew about plants, animals, and minerals. Some works might combine all approaches to make an argument about a language's antiquity, copiousness, and capacity to signify. Etymologies in these works move across what now seem to us the borders between philology and fantasy—at times they appear like modern attempts to use concepts to recreate social history; at times they resemble Isidore's outlandish claims based on phonetic similarity (*onomatopoeia*) alone. Isidore asserted, for instance that "the walking stick [*baculus*] was invented by Bacchus, the discoverer of the grape vine, so that people affected by wine might be supported by it."[47]

What is more, early modern scholars believed that the method of mining words to extract the knowledge of their name makers could be applied to any language and any society. In the *Cratylus*, for instance, Socrates reminds his interlocutor that the lawgiver, "whether he be here or in a foreign land, so long as he gives to each thing the proper form of the name, in whatsoever syllables, to be no worse lawgiver, whether here or anywhere else."[48] Students of words increasingly turned to etymologizing vernacular languages, which in the sixteenth century were beginning to acquire grammars and histories of their own.[49] In principle, etymology was a universally applicable tool.

Although these ideas and commitments were not unique to Spain, Iberian researchers were the first to apply etymology and other methods of word study to the languages of the Americas. Chronicles about the human and natural history of the so-called Indies employ all the approaches to the study of words outlined above. However, the study of words in these works is most often devoted to introducing readers to previously unknown objects and thus seek to convey the historical experiences of Indigenous name makers. The choice to include etymologized or glossed Indigenous words in the chronicles betrays the assumption that these lexical units transmitted specific knowledge that challenged the limits of translation. Thus, chroniclers collected lexical objects and established Latin or Castilian equivalents as best as they could.[50] Pietro Martire d'Anghiera's (c. 1456–1526) account of early Spanish incursions in the Americas included a list of foreign words, including Taíno and Antillean nouns like *canoa* (canoe), *iucca* (yuca), and *manati* (manatee). These appear in alphabetical order, with Latin equivalents, appended in the 1516 edition of the chronicle, which Nebrija edited. The "vocabula barbara" was to aid the reader's comprehension of Martire's account but also to encourage them to "learn new words and new names together with new and admirable things."[51]

All royally sponsored information-gathering projects, whether scientific expeditions, censuses, libraries, or histories, had one thing in common: in every domain, Spanish scholars sought to understand the Americas by studying its languages. Going beyond the formal study of grammar or language acquisition, they did so in ways that have become largely unfamiliar and even strange to us. For instance, in a set of reflections on the importance of nomenclature for the study of the natural world, the Neapolitan physician Nardo Antonio Recchi (1540–1594) noted that the first impression to reach a reader of an unknown plant is its name.[52] Extant nomenclature was a necessary entry point to the study of natural history.[53] As in the case of the names of plants, not only etymology but also other forms of lexicographical fieldwork, in addition to broader reflections about the histories and transformations of languages, could yield insight into the origins of peoples and the natural and human histories of the places they inhabited.

Assembling Information: Uniting and Preserving Variety

New words could circulate in manuscripts, printed books, letters, and through the oral accounts of those who had been in the Americas or corresponded with those who settled there. The Crown's ambitious surveys depended on geographically expansive networks of information gathering that linked Iberia's most distinguished scholars, members of the clergy, and various types local of experts across cities like Lima, Mexico, and Santo Domingo and the publishing centers of Antwerp, Rome, and Seville. A revelatory example of how information moved across the Iberian Republic of Letters, where manuscript circulation was as significant as the dissemination of printed works, is the production of a royal edition of all the works of Isidore of Seville. Philip II commanded many of the scholars in his service to pursue the traces of this Iberian author.[54] The king's instructions included tracking surviving manuscripts, comparing them, and eventually using them to produce new and improved royal editions of the works of this Christian authority.[55] The decades-long effort eventually culminated in the publication, first in 1597 and then in 1599, of Isidore's complete works.[56] Significantly, the *Etymologies* inaugurated the edition, offering a model of how to gather multiple types of information in a single work.[57] More than that, this late antique summation of universal knowledge had as its main organizational pillar the tradition of name study.

As capacious as the parameters of early modern research appear, whether in the form of a physical library, a work of natural history, a census, or a lexicon, they invariably reflected the compilers' worldviews. These collectors established the limits of what was knowable and how information ought to be organized. As John Henderson has argued about Isidore's *Etymologies*, that "monumental text" replicates its "author's belief system," which was deeply rooted in an Iberian, Catholic medieval worldview.[58] So, too, sixteenth-century Spanish knowledge-gathering projects reflected the views of their time. They both adapted ancient and medieval compilatory practices and innovated on long-standing institutional and scholarly models, from the medieval encyclopedia to the historical account known as the *relación*, to the collection of proverbs. As much as scholars experimented with new modes of collection and transformed old practices, the general parameters of inherited models proved to be resilient and flexible.[59] Moreover, inherent to their compilatory goals was the desire to capture, within the broad meta-categories afforded by these models, information about particulars. They sought to bring together information while also preserving its local specificity.

Overview

The chapters that follow reconstruct concrete processes of knowledge making across the Iberian world. They highlight crucial moments in the translation and circulation of words, books, objects, methods, and ideas among different kinds of scholars and local experts in Spain and in Spanish America. Using a wide variety of sources, from treatises to archival government documents, scholarly correspondence, literary works, scientific images, and maps, the pages ahead recreate the intellectual networks that crisscrossed Spain's overseas possessions and that informed those at the imperial court who wrote about the American domains. Writers in the Americas, despite the publication barriers that relegated many of their works to manuscript circulation in the mid-sixteenth century, after all, imagined a global audience.[60]

This study brings together the scholarship on linguistic studies in early modern Europe with that of intellectual production in colonial Latin America to explore sixteenth-century categories of interpreting information. It contends that oral traditions were a fundamental and valued source of knowledge for sixteenth-century scholars both in the Americas and the Iberian Peninsula, as they were for European scholarship more broadly. Although modern historians have generally held that sixteenth-century writers considered information not encoded in alphabetic writing to be inherently inferior, this book demonstrates that early modern scholars routinely relied on oral traditions, from etymologies to adages, to reconstruct local histories on both sides of the Atlantic.[61]

To these scholars, oral knowledge was as valid as written evidence. Expanding on the works of Fernando Bouza and Daniel Woolf on early modern notions of orality and authority, I argue that oral knowledge either from Iberian peasants or Indigenous American informants, in conjunction with the examination of various kinds of pictorial and material remains, was central to historical and scientific works.[62] Scholars eclectically drew from an ample set of approaches, including antiquarianism, natural history, linguistic studies, and the collation of legal testimony to make sense of oral accounts and thus complement their study of written documents.

This book concentrates mainly on intellectual production in the sixteenth century that resulted from Philip II's patronage or that responded to works produced by scholars connected to the royal court. By 1586 the royal seal of King Philip II contained a globe and a legend adopted from the *Satires* of Juvenal: *non sufficit orbis*, "the world is not enough."[63] This phrase sought to exalt the monarch's vast territories, but also describes the Spanish Crown's aspiration to possess information. Even so, the historical and linguistic works about the Americas studied in this book focus mainly on the islands of the Caribbean and the viceroyalties of New Spain, established in 1535 (modern-day Mexico), and Peru, established in 1542 (modern-day South America except Venezuela and the Portuguese territories until the eighteenth century when it was reduced). New Spain and Peru were considered the core centers of Spanish power in the Americas until the eighteenth century.

The book follows a chronological arc, though also sometimes moving back and forth in time to consider the precedents and repercussions of specific linguistic polemics. By maintaining this chronological organization, it also surveys the ways in which the conceptual possibilities of etymologies exploded in the middle decades of the sixteenth century and how scholars eventually came to doubt the reliability of this form of evidence to write history. The first chapter examines the intellectual production of chronicler of the king Juan Páez de Castro (?–1570) between 1550 and 1570 and the response in his writings to the experimental methods of history writing developed in the early published chronicles on the Americas. The second and third chapters explore the interconnected rise of polemical etymological histories to answer questions about the identity of the first inhabitants of ancient Iberia and the Americas. The fourth chapter examines the uses of etymologies in procuring natural historical information by focusing on the writings of the doctor Francisco Hernández (c. 1515–1587). The final chapter explores the works of scholars who believed that Hebrew contained the rudiments of all languages, among them Benito Arias Montano (d. 1598) and Luis de León (d. 1591), and the ways writers based in the Americas read and debated their works in manuscript and print. Together the chapters demonstrate the transformative effect of the new linguistic data that Spanish scholars amassed. Most significantly, the comparison afforded by the Indigenous traditions and languages of the Americas gradually altered how Spanish scholars understood cultural, linguistic, and historical change at work in their own, Iberian society.

Spanish scholars observed not only new languages but also the impact of Indigenous demographic collapse caused by the Spanish introduction of diseases and ruthless labor regimes, religious conversion, and the general deep social restructuring and violence of the conquest. This led them to formulate theories about the transmission of oral knowledge, the genealogies of historical commemoration, and the flourishing and decay of polities. Some even compared the recent experience of the Americas to Spain's ancient and medieval past. Through their historical research, scholars sought to defend elaborate hierarchies of languages and peoples within the Spanish realms. Their goal was to legitimize projects to administer and catechize multilingual subjects in an increasingly repressive and persecutory religious society. Language was at the center of their often polemical historical reconstructions. Purportedly ancient knowledge from the Americas helped to shape reconstructions of Iberia's remote past. In this way, intellectual lessons learned overseas served to reinterpret the past back home and even to argue for different possible futures.

CHAPTER ONE

The World in the Library

In various unpublished manuscripts from the 1550s, the king's chronicler Juan Páez de Castro (d. 1570) meditated on the difficulties of historical writing as well as on its possibilities. History could bring lost worlds back to life, and knowledge of the arts and sciences could be transmitted and augmented by each successive generation who would improve on the experiences of their forebears. Yet history was also fragile, as humanity's fragmented records attested. Ancient and more recent sources told of the decline and rise of nations, empires, and lineages, and of the destructive force of floods, pestilences, and fire.[1] Many venerable authors mourned knowledge forever lost or that could be reconstructed only partially. The largely irretrievable history of the Greeks, the Romans, and their descendants in the Iberian Peninsula was a reminder of the fragility of knowledge. Páez found traces of catastrophic loss not only in the records of Europe but also in "those of the Indians [of the New World], if we wish to read their histories."[2] From his study of early sixteenth-century chroniclers of the Americas, Páez concluded that genealogies and knowledge had been ruptured and lost in both Europe and the Americas. This made historical preservation even more universally necessary: the preservation of records ultimately allowed for the "invention of the arts and political life [*el vivir politico*]."[3] The lack of surviving sources hindered many important inquiries, many of which, Páez argued, were crucial to royal governance.

To remedy this decay Páez addressed to the king of Spain, Philip II, a proposal to make a "royal public library in some place safe like Spain."[4] He returned to this topic in several other writings. Much like the "continuous and perpetual" history that he endeavored to write, Páez suggested the creation of a multichambered library, all encompassing in its contents, replete with rare, ancient books and objects, carefully organized and cataloged, and constantly expanding to include all the knowledge available in the world.[5] Like other humanists of his time, Páez emphasized the role that libraries could serve in the long-term preservation of knowledge.[6] Emulating the imperial writers of ancient Rome, Páez proposed not only the foundation of a public library worthy of modern Spain's geographical extent but also the production of works that synthesized knowledge of its domains.[7] The ever-expanding compendia would mobilize the Spanish king's transoceanic networks of administrators, scholars, and missionaries.

Páez's library would not only contain records of the past; it would also gather and store new information about the present. Given the range of peoples and places under the rule of Spain, Páez called for investigating oral traditions, consulting the accounts of local experts and informants, relying on the insights of cosmographical studies, and the examination of ancient remains. To catalog everything and everyone under the rule of the Spanish monarch, Páez imagined projects that relied on forms of knowledge gathering common to his royal patron and to his contemporaries. These methods of collecting information mirrored Spain's well-known preference for a decentralized, collaborative, and "composite" administrative model. The goal was to unite knowledge while also preserving its context and its specificity.[8]

In particular, Páez's reflections exhibit a desire to acquire all the information available about the Americas.[9] Traditionally, historians of the Spanish Empire have tended to separate the investigations and historical interests of scholars in the Iberian Peninsula from those concerned with the affairs of the Americas.[10] Páez's example helps us see the anachronism of these divisions: in fact, the intellectual networks that crisscrossed Spain's domains were polycentric. Literary works, scientific images, manuscripts, maps, and even words from distant realms circulated through official and personal channels, informing the writings of scholars at the heart of the royal court, who were as interested in the Americas as they were in Iberia. Páez's compilatory ambition was consistent with that of a humanist culture that aspired to universal knowledge.

Páez, who produced his reflections on historical writing at a transitional political moment and served as royal chronicler to both Charles V and Philip II,[11] serves as an illuminating case study to explain the relationships between knowledge gathering, language, and history writing in the mid-sixteenth-century Spanish Empire.[12] This chapter explores how Páez conceived of a set of knowledge-gathering projects to meet the needs of a newly global Spanish monarchy. His proposals included a method of assembling documents, linguistic and oral evidence, and complementary visual evidence to reconstruct the natural and human history of the Spanish realms. He envisioned the creation of a royal library, the writing of chronicles, and the distribution of formal questionnaires to harness geographically expansive networks of administrators.

Páez's manuscripts, correspondence, and annotated books demonstrate his interest in Indigenous American words. His writings are representative of how Spanish curiosity about the Americas' Indigenous societies, and their languages, exploded in conjunction with simultaneous innovations in the study of the Greco-Roman tradition, European vernaculars, and biblical exegesis. While the grammars (*artes*) of Indigenous American languages did not circulate far outside of missionary orders, their dissemination is not the only indication of European scholarly interest in the tongues of the Americas. Linguistic studies in the early modern world encompassed a much broader range of approaches than the formal study of grammar or attempts at language acquisition.[13] They included such genres as etymological studies, that is, investigations into the origin of words and theoretical discussions of the history of languages and the causes underlying their transformation. The polyglossia of the Americas, combined with the territories' vastness, opened new challenges for historical writing and for thinking about linguistic diversity. Iberian chroniclers of the Americas applied established European methods, like the collection and study of local words, to mine, harness, and then exploit Indigenous knowledge. Chroniclers, missionaries, cosmographers, and natural historians routinely collected and glossed words in the many languages that they encountered, including the Taíno language of the Caribbean islands and the Quechua spoken in the Andean region.[14] Their works abound with what they claimed to be Indigenous names for natural phenomena, customs, and deities, as well as with their attempts to extract meaning from these words by questioning local authorities. The scholars who undertook these historical projects, whether in Spain or in the Americas, approached their tasks with similar scholarly methods that were far from set.

By examining the transmission, translation, explanation, and misinterpretation of representative Indigenous American words, like the song-dances known in the Taíno language of Haiti as *areito*, and how their conceptual underpinnings came to be transformed in their oceanic crossings, this chapter argues that in both the Iberian Peninsula and in the Americas, scholars were willing to reach across various cultural and temporal boundaries to explain the changes taking place in their world. Charting the movement of words through a geographically expansive network demonstrates how writers across the Atlantic were bound by similar questions, methods, and concerns. Spanish scholars writing in a diversity of genres harnessed linguistic evidence from ancient classical sources as eagerly as they used the words originating from Indigenous informants. Most significantly, the emphasis on Indigenous words and on creating repositories of knowledge that aimed to preserve local categories of authority shaped Spanish scholarship as well. As early as the 1550s, scholars connected to the royal court concluded that oral knowledge, which was crucial to reconstructing the history of the Americas, was significant to understanding the Iberian Peninsula's and Europe's both remote and more recent past.

Páez de Castro's Vision of Universal Knowledge between Old and New Worlds

None of Páez de Castro's works were printed during his lifetime, but his activities can be recreated from his manuscripts and correspondence. These writings attest to his influence as a scholar who connected networks of intellectual exchange and books and manuscripts circulating among different parts of Spain, Italy, and the Low Countries.[15] His correspondence with the chronicler of Aragón Jerónimo Zurita (1512–1580), one of his oldest and most intimate friends, provides insight into the overarching goals of his scholarship.[16]

Páez's early years and family background are almost unknown. He attained his knowledge of Greek, Hebrew, Latin, and Chaldean at the University of Alcalá de Henares in the early to mid-1530s.[17] Páez later studied canon and civil law, perhaps at the University of Salamanca. In 1545, he accepted a position in the service of the jurist and theologian Francisco Vargas (d. 1566). Páez then traveled to the Council of Trent as part of Vargas's entourage. In Italy, he established a close relationship with Diego Hurtado de Mendoza (d. 1575), whose famous collections of books and manuscripts he constantly cited in his letters to Zurita. He then moved to Rome for a short time and in 1555 was appointed Latin chronicler and royal chaplain, first of Charles V and later of Philip II, a position that he held until his death in 1570.[18]

In his *Memorandum to the King on the Formation of a Library* (1556), Páez hoped to definitively convince the recently crowned Philip II to found a public royal library.[19] He imagined a collection in which new and ancient knowledge could be physically and intellectually linked. "Chained" libraries, with valuable books, reference works, and frequently consulted books affixed to shelves by chains, had existed at least since the fourteenth century.[20] Páez relied on the image of the chain to expound on the intellectual, material, and political potentials of the library he envisioned. "Everything is chained together," he explained. "Wise men follow books and these are in turn followed by those who desire to be their disciples, and these require scribes and printing presses, and these necessitate materials such as paper and parchment."[21] For Páez, libraries not only ennobled nations but also brought economic prosperity. In the case of Spain, Páez emphasized, a library of these dimensions would allow for a better understanding of distant crowns, their geographies and tributes, and the histories of their peoples and natures. Its construction would be a great achievement for the king, as those who saw its rooms "could think that they had visited the main parts of the universe."[22]

Páez's scheme possessed features of forms of collecting already present in Spain and Italy since the early sixteenth century. His outline for the royal library stands out from contemporaneous assemblages, however, in its attempt to theorize the collection, and subsequently the integration, of these objects and knowledge from the Americas into its hierarchy of topics. The imagined library would become a space to think about the relationship between general categories and particular phenomena in light of new knowledge.

Páez imagined in detail how scholars would consult his repository's rare and unique materials. He held Valladolid to be an ideal location for the library because the king resided there, as did the royal *audiencia* (court).[23] Páez emphasized preservation as a central pillar of his project. Successful preservation was a feat that had eluded so many great kings in both ancient and modern times. The preservation of knowledge was also central to Páez's treatise on history writing. Accurate and cheap printed editions were an ideal way of assuring the widespread transmission of information. They provided a safeguard against the ravages of war, natural disasters, and the mortality of book collectors. Reliable editions would emanate from the originals that were safely preserved in the royal library. Páez offered the example of France, where the Royal Library then in Fontainebleau produced so-called royal editions that the king approved through royal privilege.[24]

For the Spanish library, Páez envisioned three successive spaces with separate identities, similar to the original Vatican Library he frequented during his Roman stay. The first room would house sacred texts: the Holy Scriptures in their original languages, the Septuagint and the Vulgate, the works of the Church Fathers, records of church councils, and texts concerning law and philosophy.[25]

The second room would be reserved for cosmographies, globes, and histories, especially those arriving from the Indies. Adorned with extensive maps, navigational charts, and drawings, this reading space would display "the portraits of Hernán Cortés, Christopher Columbus, and Magellan," alongside the Spanish king's ancestors. It would also include portraits of Ptolemy mapping the world "and Aristotle composing the books on animals with many hunters and fishermen before him, who on Alexander's command brought him many things to consider."[26] Artificial objects such as wonderful machines, weapons, and inventions like clocks and mirrors would complement the paintings. The room would also pair Greek and Roman antiquities, such as vases, medals, and coins, with parts of rare animals and "trees, herbs, and fruits made out of metal, and given their very own colors; of foreign things that will not cause less admiration than everything else," since "many Persian kings, and even those of the Indies, as barbarous as they were, had trees and herbs of gold." The ministers of the Indies would oversee providing these rare objects, and "through the Portuguese navigations many others would be collected."[27]

Páez's ideal library would culminate in a third, secret room. This space would resemble an archive: it would house the papers of the state and government, papal privileges, treaties between kingdoms, royal decrees and wills, and other royal documents.[28] In addition, the archive room would preserve the original "books of laws, the *fueros* of Spain and its crown," in case it were necessary to consult them "to resolve a doubt."[29] The information in this room would serve the needs of government in the ways that Arndt Brendecke has outlined, not through an anachronistic "maximization of political rationality through a high degree of empirical reference" but rather by allowing the king to know enough to fulfill his duty of rewarding and punishing his subjects adequately.[30] The room would also include a special section for the unedited memoirs of the king to pass down to his heir and to be continued by the new monarch.[31] The third room would be appropriately adorned with the portraits of famous rulers commanding their information-gathering projects: for instance, Julius Caesar ordering Marcus Varro to establish a public library and Vespasian holding a book that contained the most significant information about the Roman Empire. Finally, it would also include the portrait of Charles V transferring all his kingdoms to his son Philip II.

Páez's interest in the Americas can be documented as far back as 1548, when he was traveling in Italy. In numerous letters addressed to Zurita, he requested the books of Gonzalo Fernández de Oviedo and news of the most recent publications about the Indies, including the writings of Hernán Cortés.[32] In a letter of 1546, he expressed how much he longed to be an Aristotle of the Indies. He "desired greatly to see that world," he confessed, so that he might observe the sky and "illustrate the plants, terrestrial and aquatic animals, and birds and insects that are so different from those of our world."[33] In this and other letters, Páez also begged Zurita to write to their common friend Agustín de Zárate (c. 1514–1585), who had been sent to Peru in 1544 as a *contador de mercedes* (general accountant), to request "a letter of all the wonders that he had seen."[34]

Páez's yearning to collect information and objects from the Indies, both East and West, was accompanied by an intensified study of Greek and Latin ancient works. His studies were facilitated by access to the manuscripts of Hurtado de Mendoza and by his encounters in Trent, and later in Rome, with distinguished physicians such as Girolamo Fracastoro (d. 1553). While in Italy, Páez developed a strong interest not only in Aristotelian philosophy but also in the writings of the ancient botanists Dioscorides and Theophrastus. He complemented these studies by observing dissections by Paduan anatomists. He was also instrumental in obtaining and then forwarding to Spain the Greek manuscript of Dioscorides that Andrés Laguna (d. 1559) used for his translation of the work into Castilian.[35]

In a 1547 letter, Páez detailed how his studies of Aristotle required not only philological and mathematical abilities but also engagement with plants, minerals, and landscapes. Writing from Trent, he declared that he was "immersed in Aristotle with the greatest zeal" that "any Christian had ever applied to [the Stagirite]." Páez further explained that he had great "expertise with the language" and knew even more mathematics "than what is necessary for Aristotle." He also possessed "a great knowledge of anatomy," which he confessed had been effortful to obtain. For the moral writings he could draw on his legal studies, and for other subjects on his knowledge of rhetoric and poetry.[36] To approach the natural writings, he claimed to have investigated collections of minerals in Trent and its vicinity in the company of the doctor Giulio Alessandrini (d. 1590).[37]

Páez's library would provide one single space where the joint study of text and objects that he described in his letters could be systematically undertaken. Additionally, by insisting on the collection of objects, maps, and parts of animals alongside books, Páez presented an insight that he would explore more thoroughly in his *Memorandum on the Things Necessary to Write History*. For him, a comprehensive chronicle of the human and natural history of the Spanish global realms should draw not only on books and documents but also on oral testimonies, antiquities, and other visual and cartographic materials. This principle applied to both Spain and the Indies but was particularly important in the Americas, where the lack of alphabetic writing made the reliance on oral interview and the study of material remains all the more necessary.

Páez's imagined library strove, like other collections before and after him, to create a microcosm of the universe for the contemplation and education of the beholder.[38] The plan also possessed two characteristics that would become increasingly common in Iberian and Italian collections: the blending of a library and a portrait gallery, and the inclusion, alongside books and manuscripts, of scientific instruments and natural and artificial objects. The collection of objects from the Americas was commonplace in elite and ecclesiastical Spanish collections of the early and mid-sixteenth century: items like precious and semiprecious stones, coconuts, *jícaras* (vessels) for chocolate drinking, featherwork, figures of Indigenous American deities, pearls, and crosses made out of corn husks are listed in inventories, wills, and literary and philosophical works.[39]

The image of the chain, of the concatenation of successive types of knowledge, is prominent in the *Memorandum to the King on the Formation of a Library* and in several of Páez's other writings. In a passage on history writing, Páez highlighted how the natural and human diversity of the Indies inevitably brought to his mind how the world's immeasurable multiplicity—now more obvious than ever with the discovery of the Americas—was in fact linked by "the chaining together that God, our lord, put in everything."[40] In Páez's library, the connections between nature, human history, and sacred writings would be constantly highlighted by the proximity of different kinds of information in one physical space. The library's layout would also allow scholars to think about broader categories and questions, and of the possibility of integrating the Old and New Worlds into a single comprehensive and chained narrative.[41]

Páez's bibliographic recommendations reflect his own interests as a book collector. He possessed a broad-ranging humanistic library that included a significant number of volumes in Greek.[42] Among the books that would have been stored in the imagined second room's *armaria*, Páez owned and annotated a 1554 edition of Francisco López de Gómara's *General History of the Indies* (*La historia general de las Indias, y todo lo acaecido en ellas dende que se ganaron hasta agora y la conquista de Mexico de la nueva España*) and the only published part of Pedro Cieza de León's *First Part of the Chronicle of Peru* (1553, reprinted in 1554, *Parte primera de la crónica del Perú*).[43] In his 1548 letter to Zurita requesting publications about the Americas, Páez asked whether the "other two volumes that Oviedo promised of the *History of Indies* had been printed," signaling that he had read the *General History of the Indies* (1535, reprinted in 1547, *La historia general de las Indias*) and was eager to obtain its sequels. Some scholars have contended that at some point after Cieza's death in 1554, Páez had a copy of the second part of the chronicle that dealt with the history of the Incas.[44] A passage in the treatise on history writing also indicates that Páez, like other scholars in the royal court, came across American illustrated codices.[45]

Lexicography in Early European Chronicles of the Americas

The histories of the Americas that appeared in print in the first half of the sixteenth century were in themselves compendia of cosmographical descriptions, accounts of people previously unknown to Europeans, distant natural words, and violent struggles for conquest, royal concessions, and evangelization.[46] Compelled by their own agendas and interests, authors experimented with techniques to bring together information from a variety of sources into narrative form. These early chronicles collated and integrated oral accounts acquired through direct experience, letters, formal testimonies rendered before the king and his counselors, and the questioning of Indigenous and Spanish experts.

The chroniclers' compilatory efforts extended to lexical objects.[47] Pages upon pages of what they claimed to be Indigenous words, phonetically transcribed into Latin or Castilian, glossed, and even etymologized, abound in these early histories of the Americas.[48] Words in these histories appear woven into a broader narrative and not exclusively in list form, as lexicons. Despite this, their authors' sustained attention to nomenclature as a primary source of historical information resembles the approaches of early modern lexicographical fieldwork.[49] Their inquiries resulted in word assemblages that then circulated in manuscript and printed form embedded in historical narratives. These word collections, in turn, sparked the reflection and writings of readers such as Páez, as his careful interest in and annotations about Indigenous words in his books about the Americas reveal.[50] While grammars and lexicons of Indigenous American languages did not circulate widely in Spain or in the rest of Europe beyond the missionary orders or among networks of administrators, words embedded in early manuscripts and published letters, chronicles, and natural histories did reach a broad readership.[51] From these materials, noteworthy words made their way into dictionaries or lexicons. Antonio de Nebrija's *Vocabulario de romance en latín* (c. 1495) already incorporated the Taíno word *canoa*, known to Christopher Columbus since the first voyage, as the equivalent of a *linter* or *monoxylon*, a ship carved from the trunk of a single tree.[52] The attention devoted to Indigenous words in these early chronicles is representative of European linguistic studies in the sixteenth and seventeenth centuries, which, as Toon Van Hal has emphasized, remained largely concerned with words as the "basic units of grammar."[53]

The words that appeared in these histories passed through chains of translation and, invariably, of interpretation and misinterpretation. The links in these chains extended from the Indigenous informants who conveyed the meanings of words to their Spanish interlocutors, to the Spanish writers who claimed to understand them and phonetically transcribed them into their own languages, to the chroniclers who recorded them and provided their explanations in printed works. Eventually, they were even rendered into many other languages. From their inception in Spanish chronicles, letters, and oral testimony, these words were already heavily mediated objects.

Through the etymological method, word collectors approached their objects of study both synchronously and also diachronically. Etymologies could reveal the origins of names and the transformations that words had undergone through time to reach their present form. This knowledge was historical because it pertained to particulars. These word collectors also believed that words possessed antiquity, as the names of things were passed down from one generation to the next. Not only that, but words encapsulated a certain social, collective knowledge since their use was the result of the will of various speech communities. In the *Memorandum*, Páez, for instance, compared language to money. Like the "coin that circulates, in the languages that are changing every day by the will of usage, the words that are spoken the most are the best."[54] The historian John Considine has argued that lexicographers understood words as "an inheritance from the past."[55] The early Castilian chroniclers of the Americas approached words as vestiges. Wary of the disfigurations that previously unknown Indigenous proper names might undergo in translation, López de Gómara, for instance, begged translators in a dedication in his *General History of the Indies* to both "observe the properties of our *romance*" and to not "take away nor add nor change any letters from the proper names of the Indians, nor the last names of Spaniards," so that information about lineages would be properly preserved.[56]

The writings of the first official chronicler of the New World, Pietro Martire d'Anghiera, demonstrate the varied uses of local nomenclature to write history. A native of Milan, Martire served in the Council of Indies. He never traveled to the Americas but obtained his information through interviews, letters, and other kinds of testimony at court. His extensive correspondence with diplomats, members of the clergy, princes, and scholars spread some of the earliest news about

Spain's overseas activities to Europeans and was published in Latin between 1493 and 1525.[57]

Martire collected names of social roles, ceremonies, musical instruments, and plants and animals. Whenever more information was available, he devoted particular attention to the etymologies of toponyms. In the *Third Decade* (1514–1516) of his *Decades of the New World* (*De orbe novo*), Martire contended, for instance, that the first inhabitants of Hispaniola had named the island Quizqueia and then Haiti. He distinguished between the Indigenous inhabitants' ancient language and their current one and attempted to explain the reasoning behind the place-names. "These denominations," he warned, "were not the children of fickleness, but of the meaning that they had." Quizqueia is "a big thing that does not have an equal." This word signified "vastness, universe, everything." Martire compared this name to the word *pan* (all) that the Greeks used to refer to their islands, since, like the Caribbean Indians, they believed that their "magnitude was the whole orb, and that the sun does not warm anything else [but theirs] and the neighboring islands." Haiti was equally meaningful. It signified roughness. The islanders applied it to the whole territory through metonymy. In many parts of Haiti, "expanses of steep mounts can be found, thick and terrifying jungles, and fearful and dark valleys given the altitude of its mounts, and in other parts it is very pleasant."[58] In their use of metonymy, the Indigenous inhabitants of the island resembled the poets who sometimes referred to all of Italy as Lazio (Latium), which was formally the name of one central region. For Martire, the island's name contained important traces of ancient knowledge that were worth relating: Quizqueia provided clues to the Indigenous inhabitant's worldview, and Haiti information about the island's landscape (fig. 1.1).[59] The etymological discussion was accurate because the Indigenous inhabitants of what was now Hispaniola preserved such valuable information about their antiquities in their historical poems, which Martire identified as *areitos*.[60]

Oviedo produced some of the most detailed and widely read accounts about the first decades of Spanish rule in the Americas. In all of his writings, Oviedo positioned himself, in contrast to Martire and other Europeans who wrote about the Americas, as an authoritative witness, reporting on the basis of his own direct observations and interviews with local experts.[61] Oviedo arrived in the Indies in 1514 as a member of an expedition to conquer the Darién region in modern-day Colombia and Panama. Throughout his long career in the Americas, he occupied several positions. In 1532, the king appointed him general chronicler of the Indies and governor of the fortress of the city of Santo Domingo on the island of

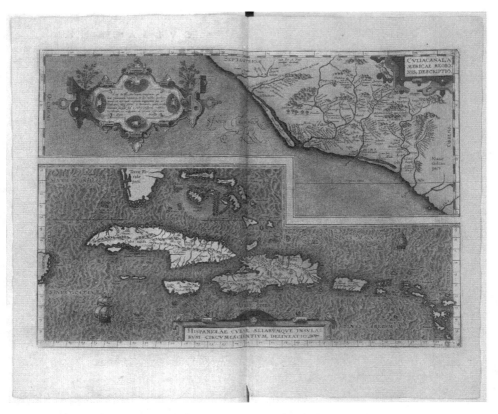

Figure 1.1. Cartographers like Abraham Ortelius (1527–1598) were careful to preserve both the Spanish and the Indigenous toponyms in their representation of the Caribbean islands, attesting to the endurance of the Indigenous names. "Ayti, sive Spaniola," in Abraham Ortelius, "Hispaniolae, Cubae, aliarumque insularum circumiacientium, delineatio," in *Theatre de l'univers: contenant les cartes de tout le monde. Avec une brieve declaration d'icelles* [translation of *Theatrum orbis terrarum*, 1595] (Antwerp: Plantin Press, 1598). Courtesy of the John Carter Brown Library at Brown University.

Hispaniola. In 1549, he became perpetual keeper (*alcaide*) of the city's fortress and harbor. He occupied these posts until his death in 1557.[62]

Throughout his extensive writings on the Americas, Oviedo sought to compile and explain local nomenclature. This was a central aspect of his efforts to chronicle the natural and human history of the Americas. In the prologue of the 1547 edition of the *History*, he pointed out the difficulties of apprehending the "diversity of languages and customs of the men of these Indies" and anticipated objections over the quantity of new, Indigenous words in the text. "If some strange and barbarous words are here found," Oviedo explained, "the cause is the novelty of what is treated" and not the quality of his *romance*, that is, his own language. These new lexical items were names or words collected and "set down to make clear the things that the Indians signified by them."[63]

The study of names furnished Oviedo with further evidence to bolster his authoritative eyewitness status when interpreting and describing plants, animals, polities, and ceremonies. For example, the iguanas that lived throughout the islands and on the mainland were not, as Martire had claimed, like the crocodiles of the Egyptian Nile. Oviedo believed that Martire was susceptible to "notorious errors" because his writings were based on hearsay. He explained how these American "dragons" were much smaller than their African counterparts and did not possess the saffron color (*croceus*) that the "glorious doctor Isidore of Seville" had attributed to the crocodile in his *Etymologies* and that supposedly gave the creature its name (*crocodillus*). Oviedo, too, had his limits: it was unclear to him whether iguanas were aquatic or terrestrial and, consequently, whether they could be consumed during Lent. He left it to each reader to decide. Despite his uncertainties, Oviedo could attest to their size (fig. 1.2). He had seen many of them and even kept a few in his home. The "*yuana*," Oviedo explained meticulously, "is pronounced *y* and after a short interval *u* and then the three subsequent letters *ana* are uttered together, and in this way in the whole name [one] makes two stops in the manner in which it has been described." Oviedo sought to replicate the sounds of the animal's Indigenous name, as if to transmit to his readers the word's materiality. Oviedo collected the animal's name and rendered it akin to an object. In this way, too, he sought to convey his authority: he had been in Hispaniola, he had possessed an iguana, and he knew in detail its Indigenous name.[64]

Scholars of Oviedo have noted the importance of his direct experience of the Americas in both determining his scholarly credentials and shaping his chronicle. It is worth noting that the concept of experience in Oviedo's work was more capacious than knowledge derived empirically through his own senses.[65] Direct experience included the information gleaned not only from trustworthy sources, like written testimonies, but also from the knowledge of informants and from various oral traditions, including spoken language. Oviedo sought to access firsthand accounts of Spanish actions in the Americas and the historical knowledge of Indigenous societies through a variety of techniques that he then strove to

lagartos que he dicho crã cocodrilos. Pe
ro enla vdad la/y v ana/ muy diferente ani
mal es del cocodrilo: y en ninguna cosa a el
semejãte. Esta que aqui yo pinte quiere al

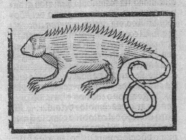

go semejar ala/y v ana/ y a la verdad esta
forma tienen/cõla protestacion o auiso que
al principio deste capitulo se toco:que aque
ste animal es como neutral/i q no se deter
minan los hombres si es terrestre o d agua
porq enla vna y enla otra estimaciõ se pue
de sospechar que tiene mucha parte.

¶ Capitu. iiij. del pexe lla
mado Gihuela i de sus armas.

L pexe o pescado llamado vi
huela es grande animal / i la
mandibula o hocico elto/o su
perior õl es vna espada orlada
de vnos colmillos/o nauajas
de vna parte i de otra/tan luenga como vn
braço de vn hombre i algunos mayores i
menores segun la grandeza i cuerpo deste
animal/que tales armas tiene. Y ole he vi
sto enel darien en la tierra firme tan grande
q vn carro con vn par de bueyes tenia har
ta carga i peso que traer enel desde el agua
hasta el pueblo. Estas espadas que digo es
tan llenas de vnas puntas de huesso maci
ças i rezias i muy agudas/ o punçantes d
vna parte i otra dela espada conla qual no
sele para pescado delante sin que le mate. y
tambien ay estos pescados enlas costas de
sta i delas otras yslas dstas partes. Estos
pescados me dizen ami los hombres de la

mar que los ay en españa pero sin estas pũ
tas o puas enlas espadas/no se si lo crea por
q en algunos templos en españa las he vi
sto colgadas pero o no se de donde las han lle
uado/o si las ay enel mar de españa assi fie
ras:mas aca enestas mares delas indias i
tierra firme muchas destas he visto d la ma
nera q tengo dicho. Son buenos pescados
de comer/ po no tales como los pequeños
dellos mismos/i de otros de los menores
ó otras especies / porque por la mayor par
te los pescados muy grandes no son sanos
aca alo que yo he entendido i las mas ve
zes se comẽ por necessidad/excepto el Ma
nati que aun que son muy grãdes son muy
buenos i sanos : del qual Manati se dira
mas adelante en su lugar.

¶ Capitulo. v. Delos pe
res boladores que se hallan en el grande
golpho del mar oceano/ viniendo de Espa
ña a estas indias.

Lguno preguntara la causa
porque digo que estos pesca
dos boladores se hallan a la
venida a estas pres enel grã
de mar i golpho del oceano:
i no dire su buelta desde aquestas indias
a españa/o Europa. Y por sacar desta dub
da al lector/digo que aun que ala buelta se
hallan los mismos pescados / assi como ala
venida no son tantos en mucha manera/ ni
los nauios buelven por el mismo rumbo/o
derrota que aca vinieron/i ala vanda del
norte no ay tantos como por estotra via ha
zia el sur o parte dla tierra firme. Hallan se
desde tan pequeños como vn Abejoncico
hasta tamaños como grandes Sardinas.
Estos quando las naues van corriendo en
su viage i ala vela se leuantan de vna par
te i de otra a manadas grandes i peque
ñas/pero enellos es grandissimo i incon
table el numero destos peces boladores:y
de vn buelo acaesce y a caer espacio de tre
zientos passos/i mas/i menos: i acaece al
gunas vezes caer dẽtro elas naos: i yo los
he tenido biuos en las manos i los he co
mido . Y son muy buen pescado al sabor/

make visible, as proof of the reliable chains of transmission through which he had secured information. This led him, for instance, to include transcriptions of whole interviews in the form of dialogues.

Oviedo's explanation of Indigenous words represents an effort to access, translate, and exploit local information as he learned it from a reliable informant. An instance occurs in book 33, dedicated to New Spain, or modern-day Mexico, where Oviedo sought to clarify Cortés's actions in the city of Tenochtitlan, and their aftermath, by questioning Juan Cano, a veteran of the conquest who was then married to Doña Isabel, one of Moctezuma's "legitimate daughters." Cano happened to be on the island of Hispaniola on September 8, 1544.[66] The interview transcription starts with Oviedo presenting his reasons for summoning Cano, whom he considered both "an intelligent man" and an "eyewitness." Oviedo clarified that it was his "custom to give the context and the names of the witnesses." He conceded that often witnesses remembered events differently, "because understanding is sometimes better in some men than in others, and for this reason it not surprising that they disagree in their accounts, specially about similar, things." To correctly render events that "he had not seen" and to make sense of a "diversity of information," Oviedo aimed to listen to many testimonies and to compare and collate them. Thus, he would become aware of "certain errors" and emend that which was dubious. He thus laid out his method for comparing oral testimonies.

One of Oviedo's first questions for Cano was how the Indigenous nobility of Tenochtitlan determined the "legitimate children" of rulers, since he had heard from other sources that Moctezuma, the supreme ruler of the Mexica, had many wives and many more children. Cano clarified that the custom was quite clear and consistent, and that "legitimate unions" were celebrated when the families of both partners were present and the union was consummated during an "*areito,* or dance, which was called in the Mexican language *mitote.*" Only the children conceived during this ceremony were considered legitimate.[67] Oviedo's transcription of Cano's responses highlighted the local terms of Indigenous ceremonies. It also showed, in practice, how Spanish settlers translated New World concepts using other Indigenous terms that they believed to be equivalent.

Likewise, Oviedo recorded the names that he believed were used to signify the same object in different languages. For instance, in book 6 of the *General and Natural History,* which is dedicated to miscellaneous topics, Oviedo introduced the dwellings of the Indigenous inhabitants of Hispaniola as *buhio,* "which in the tongue of the Indians means house." In a subsequent expanded version of this chapter, which remained unpublished, Oviedo added that these homes were called *buhios* in "all the islands," but that in the tongue of Haiti they were actually called *eracra*. Linguistic specificity mattered.[68] He went on to explain how the inhabitants of the island built these dwellings, of which there were two kinds: a smaller type called *caney* (fig. 1.3) and bigger and more elaborate ones where *caciques*, or rulers, lived. He described the bonds that held the wooden beams together: *bexucos*, "which are some veins" that "grow around trees" to bond the separate pieces in place of nails. *Bexucos* comprised many types and even possessed medicinal qualities. A type of wood called *corbana* was ideal for its durability. Some of the thatched roofs were covered with the leaves of the plant called *bihao*.[69]

For Oviedo, local nomenclature was among the most authoritative ways of learning about people and places. Firsthand knowledge of the Americas, as Oviedo's extensive chronicle shows, did not displace the search for authoritative information about the region's nature and about its past and present inhabitants. European chroniclers like him used locally obtained information to connect geographies and societies according to European categories, though local knowledge remained a crucial interpretative framework for understanding the Americas.

This was also true of Pedro Cieza de León's widely read and translated *First Part of the Chronicle of Peru*, which introduced generations of European readers to the Andean region and the linguistically diverse territories that had once comprised the Inca state. Cieza remained in the Americas between 1535 and 1550.[70] He engaged in numerous expeditions and military activities and made extensive notes about his experiences. Throughout his writings, he often cited his own observations but also detailed the importance of Indigenous informants and translators in securing his information. In the *First Part*, Cieza acknowledged the crucial assistance of the Dominican missionary Domingo de Santo Tomás (d. 1570), who composed the first grammar and lexicon of the Quechua language.[71] Even before reaching Peru in 1547, in the company of Pedro de la Gasca (1493–1567), the acting viceroy of that territory, Cieza was aware of the centrality of interpreters for conquering and for obtaining any historical information. For instance, when the Spaniards founded the town of Anzerma, upon a settlement that the Indigenous inhabitants called Umbra, Cieza lamented that the lack of interpreters (*lenguas*) made it impossible to "discover any of the provinces' secrets."[72]

Páez owned and annotated a copy of Cieza's *First Part*, revealing how a scholar in the royal court engaged with these early published histories. His annotations range from underlining significant passages to providing complementary information from other texts, summarizing whole sections, noting lexical variations of what the reader presumed to be the same object signified, and finally to analyzing

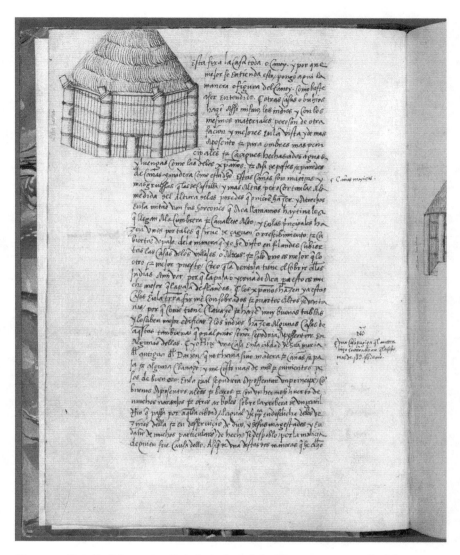

Figure 1.3. Detail of the *caney.* Oviedo included the "figure or manner" of the *caney* "so that it can be better understood." Gonzalo Fernández de Oviedo, *Natural y general hystoria de las Indias,* 1539–1548, HL, HM 177, vol. 1, fol. 3v. Courtesy of the Huntington Library, San Marino, CA.

etymologies. For instance, in chapter 36 Cieza offered a preliminary description of what the Spaniards would come to know as the viceroyalty of Peru, emphasizing that the mountainous topography had facilitated the Spanish conquest. Páez added a disagreement in the margin: "They were not conquered because of this, but because of the division of the two brothers, and the passion of Atabalipa."[73] Páez referred here to the wars between the sons of the Inca ruler Huayna Capac, Huascar and Atahuallpa, which had resulted in Atahuallpa's victory. Francisco Pizarro and his companions arrived in 1532 to the Andean region when Atahuallpa was in the process of consolidating his power.[74] Páez might have learned about this in Agustín de Zárate's *History of the Discovery and Conquest of Peru* (1555, *Historia del descubrimiento y conquista del Peru, con las cosas naturales que señaladamente alli se hallan y los successos que ha avido*), one of the first chronicles of the conquest of the Incas and the many civil wars that Pizarro and his brothers fought against their former allies in Peru. Zárate explicitly linked the swift success of the Spanish conquest of Peru to the instability caused by these wars, since "Guascar's army was undone and [had fled] and Atabaliba's army was for the most part dismissed because of the new victory."[75]

In other chapters of Cieza's text, Páez took care to record new words in the margins, like "quinua," a grain eaten in the Andean world, and to add the etymologies of important names, like those of the noble families who lived in the city of Cuzco.[76] Cieza narrates how Huayna Capac set out from Cuzco with members of the "two most famous lineages" of the city, "the Hanancuzcos and the Orencuzcos," to set up a fort in the port of Túmbez, and from there conquer the island of Puná. These events would culminate in the foundation of Guayaquil, on the Guayas River of modern-day Ecuador. Páez's annotations clarified that "the city of Cuzco is on one part on a hillside, the other on a plain. Those above are called Hanancuzco. Hana means above. The others are called Huricuzco because Hura means below. But those above were more distinguished."[77] At the end of the book, he also included a short note about the meaning of the name Huayna Capac, which he had learned meant "rich boy."[78] Other annotations indicate his interest in natural history, in particular collecting the names of plants and animals and their equivalents, if they existed, in other languages of the Americas. He proposed, for example, that *guavas* are the same as the *pacay*, and that the appellation *aguacate* is equivalent to a fruit known in Peru as the *palta* (fig. 1.4).[79] From words for plants and animals to the names of rulers, Páez actively engaged with these lexical objects on the printed page.

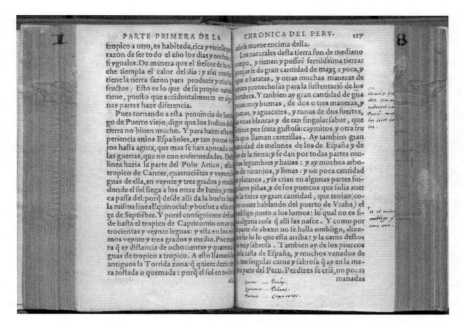

Figure 1.4. Pages from Páez de Castro's copy of Pedro Cieza de León's *First Part of the Chronicle of Peru* (1554) showing annotations about New World plant names and their equivalents in other New World languages. Pedro Cieza de León, *Primera parte de la cronica del Peru* (Antwerp: En casa de Juan Steelsio, 1554), 126v–127r. Real Biblioteca del Monasterio de El Escorial, Esc. 20.VI.17. Courtesy of Patrimonio Nacional.

Words in Motion: The Example of the Areito

Páez's interest in Indigenous American words also emerges from his annotations to López de Gómara's *General History*. Unlike Cieza, López de Gómara never traveled to the Americas; he composed his works as Hernán Cortés's secretary. Páez had become aware of López de Gómara through Zurita. In various letters, Páez and Zurita exchanged news about Gómara's financial and health woes.[80] While reading Gómara's *General History*, Páez wrote down new toponyms in the margins and the names of edible and medicinal plants like the "yuca" and the "bixa" that, according to Gómara, the women of Hispaniola used to paint their bodies when they "dance their areitos, and because they tighten the flesh." Páez noted the words "AREITO" and "CANTARES," or songs, in the margin of the chapter dedicated to the religion of Hispaniola. "Areito," Gómara explained, "is like the Zambra of the Moors, who dance singing ballads [*romances*] in praise of their idols, and their kings, and in memory of their victories, and notable and ancient events, and they have no other histories."[81] Páez would lift the *areito* from these printed books into his writings about the origins of historical commemoration.

The information that Páez gathered allowed him to pursue questions that required a consideration of New World nature and humanity alongside what he knew about Eurasia and Africa. As we saw, in the *Memorandum of the Things Necessary to Write History*, for instance, he contemplated the origins of historical remembrance, and later of historical writing. He believed that they could be traced to the most primitive days of humanity, holding the impetus to "leave memory of their deeds" to be a natural inclination of humankind. In this, people imitated the model set forth by God in nature, particularly in their basic ability to reproduce. Ancient peoples created different ways of memorializing the past that eventually evolved into the chronicles and, subsequently, the carefully researched and constructed accounts of present times.[82] This was an arduous process. Despite human beings' natural dispositions, historical writing did not emerge spontaneously. "Before they found letters," people had to rely on other ways of preserving the accomplishments and tribulations of their ancestors.

Páez was as interested in oral memory as in written records. He related that before the invention of writing, people "composed songs of their feats to keep them in their memory better and they painted them on skins and fabrics as best as they could." He recalled that this manner of conserving the past could be found in the Americas. Indeed, American codices that were reportedly filled with meaningful figures frequently arrived at the Spanish court.[83] In addition to the *Memorandum*, Páez elaborated on the nature of ancient and New World songs in at least two more writings.[84] In the first set of notes, Páez discussed how "before people invented letters and their uses, men entrusted their feats and the memorable things of their time to the memory of people arranging them in the genre of *trova* that was used in their lands." A sufficient example of this form of commemoration, Páez claimed, "without counting the antiquities of our world," is "that which the western Indians did in the New World [and] in their dances singing the things of their past kings without having a more truthful history."[85] Here as elsewhere, Páez directly compared Indigenous forms of remembrance with European ones. In a subsequent set of remarks, Páez further refined his analogies by summoning examples from the Hebrew Bible. In the beginning, wise men were satisfied with having "memorable things sung publicly with musical instruments as Moses sang the victory of the Red Sea and other women the feats of David."[86] He then once more shifted to the Americas, referring specifically to "those songs that they called areytos," but editing the phrase in favor of a more general descriptor "those songs that were said to last for many days." He concluded by claiming that "ancient poetry existed to celebrate such things and even our old ballads [*romances*] originated in this. So, in this manner it is believable that after letters were founded, by the great majesty of God, that the first things that were written down were those rudely composed songs [*cantares*] and histories, as usually happens with human inventions which are perfected little by little."[87] In other words, Páez imagined the use of musical accompaniment, referred specifically to the Caribbean songs of remembrance known as *areitos*, and compared the poems with old Iberian *romances*. Moreover, although he believed these early poems to be "rudely composed songs," he stressed the coherence of oral knowledge and its historical content even when it was not committed to writing.[88] He emphasized that oral knowledge could be transformed seamlessly into written, granting it greater ease and permanence.

Páez's adoption of the word *areito* exemplifies a concrete instance of knowledge making tied to the circulation of a Taíno word in manuscripts and printed books across Spain's transatlantic domains in the first half of the sixteenth century. It allows us to trace a crucial moment in the translation—both in the early modern sense of movement and also of the rendering from one language to another—of a lexical object and how in the process of transmission many of the word's conceptual underpinnings came to be made and reconstituted in the works of various readers and authors, with important consequences both for their modes of argumentation and their subsequent reception. Spanish writers shaped the meanings that New World words acquired, carrying with them the presuppositions, prejudices, and linguistic principles that their authors used to understand the many languages of the Americas. The *areito*, a term Indigenous to the Caribbean islands, would eventually encompass a broad range of American types of historical remembrance that were primarily oral and often sung. Peninsular readers and chroniclers, like Páez, but also antiquarians and even poets, apprehended the word and the various meanings that it now contained to make sense of Spain's oral traditions and to argue for their historical value. In doing so, these writers applied first their own categories to interpret American practices, then reconstructed their vision of ancient Iberia by analogy with the Americas' Indigenous traditions.

In his rendition of the historical practices of the New World, Páez summoned the descriptions of songs and ballads already available in the manuscript and printed works of early chroniclers like the Hieronymite friar Ramón Pané (d. sixteenth century), who traveled with Columbus on the second voyage, and Martire.[89] In the *Third Decade* (1514–1516) of his *Decades* (*De orbe novo*), Martire described the *areitos* as poems that were passed down by wise members of a society to preserve history. He distinguished between two kinds of *areitos* that existed on the island of Haiti, or Hispaniola.[90] The first were more general and recounted events in succession. The second told of the noteworthy deeds of the *cacique*, or ruler, relating the accomplishments of their fathers, grandfathers, great-grandfathers, and all of their ancestors. The inhabitants of the island would recite these poems accompanied by drums, which Martire believed to be called *maguay*. Choirs and dancers participated in the recitations and modified their performances based on the contents of the poem since some *areitos* were songs of war, some were elegies, and others recorded love stories. Whatever their themes, Martire believed that they were structured oral traditions that were passed down from one generation to the next. He attributed to the *areitos* the capacity to preserve the past faithfully, down to the most remote origin story of the island's first inhabitants and the meanings of the names that they assigned to their lands.[91]

No printed description of the *areito* was more extensive than that of Oviedo, who claimed to have witnessed several of these performances in the Caribbean islands, as well as even more lavish ones on the mainland.[92] Oviedo maintained that the inhabitants of the West Indies did not possess written records. He searched for them in vain but could find only "their songs, which they call *areitos*," which served as "their books or memories."[93] He characterized the songs as "a good and gentle manner of remembering ancient and past things." The song-dances reminded him of the passages where Roman historian Titus Livy discussed the first dancers who came to Rome from Etruria. These dancers coordinated their voices with their movements "to forget the labors of the pestilence" in the year when the early Roman statesman Camillus died in 364 BCE.[94]

Oviedo believed that the *areito* as form of "dancing and singing was very common in all the Indies."[95] In the fifth book of the *General and Natural History*, Oviedo provided a long discussion of the general characteristics of the *areito* based mainly on the songs that he had heard and seen on Hispaniola. He described them as carefully regimented performances and as highly structured forms of oral remembrance. The island's Indigenous inhabitants would celebrate an *areito* to commemorate "some notable feast" or to pass the time. Many men and women, or sometimes only the men and other times only the women, would get together to recall the "vanquishing of an enemy, or the marriage of a *cacique* or king of the province." To do so, they "would sometimes take each other's hands, or they would conjoin themselves arm upon arm," as though threaded, and one of them, either a man or a woman, would guide the others by demonstrating the steps of an ordered choreography. In this manner, all would follow singing, while they moved, in a low or high tone depending on the directions of the guide. Oviedo observed that the dancers' steps were in concert with the verses or words that they sang, like the dancers described by Livy. And this was carried out in the most orderly fashion. Indeed, when the guide spoke and moved, the others would answer with "the same steps, words, and order." The dancing and singing would last for three or four hours, until the guide finished the story. But sometimes the *areito* would extend for much longer, even from one day to the next.[96]

From the dancing to the sounds of drums, Oviedo then turned to explain the various purposes that these sung poems fulfilled. With this "bad instrument or without it," in their songs they "relate their memories and past histories, and in these songs, they narrate the way their past caciques died, and how many and who they were, and other things that they do not want forgotten." This manner of singing, Oviedo emphasized, was nothing less than an "image of history" or a "remembrance of things past." Through the *areito*, the Indigenous inhabitants of Hispaniola would remember war and peace, good and bad times, genealogies of their rulers, and whatever they wished for "the young and old to communicate," so that their deeds were "well sculpted into the memory." In this way, their history would endure from one generation to the next in "place of books."[97] It is noteworthy that while in this section Oviedo described the purposes of the *areito*, he did not provide concrete examples of histories or genealogies that he learned by listening to the songs.

Areitos, Oviedo declared, could serve other purposes besides recording history. They could also be performed to communicate battle plans before war, to bestow justice upon a member of the community, to mourn the dead, or even simply for sheer enjoyment. Oviedo believed *areitos* sometimes involved excessive drinking and the veneration of demonic deities, which he condemned with disdain. For instance, he recounted having been present at an *areito* in 1529 in Nicoya, in the *gobernación* of Nicaragua, in which the sole purpose seemed to have been the enjoyment of alcohol. At the time, Oviedo was accompanied by only one priest and several other Spaniards in the settlement of a recently baptized *cacique* known as Don Alonso to the Spaniards, and Nambi in the Chorotega language. Characteristically, Oviedo pointed out in this passage that the name Nambi signifies "dog." Having heard a report of an *areito* that preceded a retaliatory attack against Spaniards on the island of San Juan (Puerto Rico), Oviedo feared the worst.[98] His party was far from any other Christians who could assist them if their hosts were to suddenly turn hostile. Oviedo described the locals as "people so bestial and idolatrous and so full of vices." He also expected them to "truly have no kind disposition towards Christians, because by becoming [their] lords they have enslaved them." While the Spaniards remained alert, the *areito* began. The locals drank a cacao beverage and *chicha* (an alcoholic drink made of maize) and smoked "what they called yapoquete," which in "Haiti or Hispaniola is called tobacco." Many of the people, Oviedo reports, "became drunk," fell on the ground, and "remained where they fell until the wine wore off the next day." When the *areito* concluded, Oviedo claimed to have chastised Don Alonso for the celebration's excesses. The *cacique* replied that he knew "that the drunkenness was bad but that that was the custom of his ancestors" and claimed that if he prohibited the practice, his people would leave.[99]

While in some instances Oviedo praised the artistry of the *areito* and its capacity to preserve history through generations, as his assessment of the *areito* in Nicoya shows, he also disapproved of significant parts of local behavior, characterizing the people of Nicoya as "idolatrous" and "full of vices." As disparate as these judgments may seem, they were part of the same worldview that allowed Oviedo to seek out local knowledge, while simultaneously disparaging the Americas' Indigenous inhabitants. As Nicolás Wey Gómez has demonstrated, Oviedo and his contemporaries inherited an ancient cosmological paradigm that posited the tropics as a place of extremes where humidity and heat engendered a lush vegetation but made people morally inferior to those of the temperate European latitudes. By determining that the "Indies" were a cohesive region located in the "torrid zone" of the globe, Oviedo could praise the region's wondrous natural characteristics, its people's ingenuity, their aptitude for the mechanical arts, and their practical knowledge of nature, while also condemning their moral qualities. By emphasizing these shortcomings, he sought to justify the Spanish conquest.[100]

While Oviedo presented the *areito* as common to the entire New World, he acknowledged that the songs and dances varied across regions. He declared the intention to discuss these variations as his chronicle unfolded outward from the Caribbean toward the mainland as it followed the itinerary of the Spanish conquerors.[101] Even so, he retained the Taíno word *areito* to designate, classify, and interpret numerous other kinds of performances in places distant from the islands, even when he learned their specific names in local languages. For instance, also in the *gobernación* of Nicaragua, Oviedo described witnessing an *areito* that the locals called a *mitote* (fig. 1.5).[102] Oviedo's choice to classify the *mitote* as a kind of *areito* would contribute to the dissemination of this Caribbean word in print and manuscript throughout Europe and the Americas as a meta-category.[103] Paul

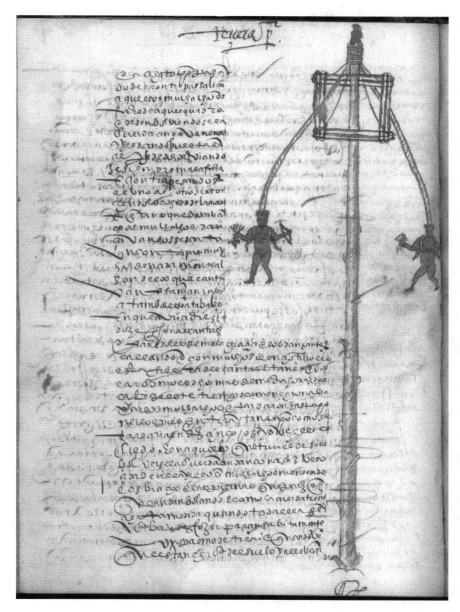

Figure 1.5. Detail from Gonzalo Fernández de Oviedo's *Historia natural y general de las Indias*, showing the *areito* or *mitote* in Nicaragua. This manuscript is a sixteenth-century copy of Oviedo's original. PR II/3042, fol. 84v. Real Biblioteca, Madrid. Courtesy of the Real Biblioteca del Palacio, Madrid.

Scolieri has argued that early chroniclers like Oviedo "'invented' the term *areito* to represent an evolving early modern concept of performance as an embodied way of knowing and transmitting knowledge that is distinct from writing."[104] But "inventing" does not capture the range of Oviedo's uses of local knowledge to formulate his own categories. While linguistically specific to the Caribbean, the term *areito* ended up subsuming numerous kinds of dances throughout the Americas.[105]

Oviedo's efforts to describe the *areito* using various Castilian signifiers irretrievably shaped its meaning for European readers.[106] His decision to maintain the term *areito* as opposed to adopting a Castilian translation indicates that, far from purely "inventing" the category, Oviedo placed value on what he had learned in the Caribbean. Historians have long recognized that Spaniards' early experiences there were crucial to the creation of the colonial institutions that they would later use to govern the mainland.[107] The words and knowledge of the people of the Caribbean also allowed chroniclers like Oviedo, who aspired to use local knowledge, to make sense of phenomena that they believed to be widespread in the West Indies because cosmological commonalities underpinned this part of the terrestrial globe.

The custom of using Taíno or Arawak words to designate social and natural phenomena beyond the Caribbean was not lost on Spanish scholars who routinely commented, sometimes critically, on this choice. In his chronicle of Peru, for example, Zárate claimed that in all of the provinces of that Andean region, lords were called *curacas*. Yet the Spaniards referred to them by the Caribbean word *cacique*. When reading Cieza's chronicle, Páez too noted in the margin of a passage where Cieza referred to Andean rulers as *caciques* that "in Peru" the lords are called "Curacas."[108] The reason for this lexical projection, according to Zárate, was that "the Spaniards that set out to conquer Peru went there accustomed, in all words and general and common things, to the names with which they called [things] of the islands of Santo Domingo, San Juan and Cuba and the mainland where they had lived." Since the Spaniards did not know what things were called in Peru, they named them not with Castilian equivalents but with the words of the islands. Words like *cacique* were soon adopted by Indigenous Peruvians themselves when interacting with the Spanish. The same occurred when designating everyday objects like bread or alcoholic drinks.[109] As Zárate's analysis shows, Spanish scholars pondered lexical borrowing, the use of Indigenous words far beyond their communities of origin, and linguistic mutations more broadly as they contemplated the social and cultural transformations wrought by the Spanish conquest.

To convey his experience of the *areito*, Oviedo did more than use Indigenous terminology. He also employed a strategy common to early modern cosmographical accounts: the use of comparisons and analogies.[110] He proposed several analogies between objects familiar to European readers and their American equivalents, all the while highlighting the insufficiency of his comparisons and more generally the limits of narrative description. Pineapples were a little bit like artichokes but required an additional five pages of description, and an image, to better convey their properties; manatees had heads like oxen but were otherwise utterly dissimilar; and the temples of Nicaragua were like mosques in some ways but not others.[111] Analogies could illuminate aspects of an object, yet they could also obscure its central features, mislead the reader, and misrepresent the object of analysis. They most often betrayed fundamental misunderstandings of the comparers.

For Oviedo, *areitos* resembled the dances and songs of the ancient Etruscans that Livy described but also the contemporary dances of Spanish, Flemish, and Italian peasants, who with tambourines performed songs for solace in the heat of the summer. In other words, *areitos* were an effective way of remembering the past among common people who could neither read nor write. They coexisted with written records and were another form of recalling, preserving, and socially circulating historical knowledge. Oviedo explicitly compared the *areito* to the popular ballads recited and sung in Spain by both Christians and Muslims. They were the main way of remembering the past, "at least, amongst those who do not read."[112] The *romances* were narrative poems sung mainly in Castile and later diffused to other parts of the Iberian Peninsula.[113] They were "chronicles or rhymed novels" that told the stories of kings, knights, and their military deeds and often also narrated the feats of simple men.[114] *Romances* exist wherever Castilian is spoken. Many are anonymous and have been transmitted orally. The earliest written ballads can be traced to the fifteenth century. *Romanceros*, or collections devoted to these poetic compositions, began to be published in the sixteenth century. Illiterate, semiliterate, and erudite authors composed *romances*.[115]

To demonstrate that *areitos* were similar to *romances*, Oviedo first dispelled the idea that *romances* were merely entertaining poems and showed that they could transmit history faithfully—not only general events but also specific details. From the *Summary* of 1526 to the first part of *General and Natural History* (published in 1535, 1547), Oviedo substantially increased his discussion of the *areito*. While in the *Summary* Oviedo compared the *areito* to the songs of peasants (*labradores*),

in the *General and Natural History* he introduced its direct correlation with *romances*.[116] In an in-depth discussion of various *romances* and their capacity to
preserve the past, Oviedo cited a small collection of Iberian poems. To explain
these examples, Oviedo interchangeably referred to the Castilian medieval *romances* as both songs and *areitos*, thus reversing the formula to present the *areito*
as evidence of the *romance*'s historicity. *Romances* allowed subsequent generations to learn, for example, that on March 28, 1344, the Castilian king Alfonso XI
had been in Seville when he decided to set out and conquer the southern Taifa of
Algeciras in southern Spain. Oviedo claimed that in the year he composed it, 1548,
his account, "or *areito*," had already endured for two-hundred and four years.

Likewise, in an account of French and Neapolitan princes, Oviedo wrote that
common people also lauded Charles V's military victories in song so that "neither
children nor the old will ever cease to sing this *areito*, for as long as the world lasts."
In this manner, he emphasized, "many more ancient and modern memories circulate among people," even when "those who sing them and recite them do not
know how to read." Oviedo concluded by bringing the discussion full circle to the
Indies, whence he wrote, reinforcing to his reader that "the Indians do well in this
part to have the same care, since they lack letters, which they substitute with their
areito, which supplement their memory and fame, since with these said songs they
know about things that happened many centuries ago."[117]

Oviedo's decision to repeatedly refer to the Spanish poems that he cited interchangeably as *romances* and as *areitos* demonstrates his integration of Indigenous
American practices into his view of the global past. Another example helps to
clarify this heuristic technique. As his description of the *mitote* in Nicaragua
shows, Oviedo was fascinated by the Indigenous custom of painting the body. The
inhabitants of Hispaniola, for instance, used the seeds of a plant known to the
locals as the *bixa or xagua* (*bixa orellana*) to produce a bright red paint. They did
so with great "industry," since they harvested the plant year-round and prepared
the paint from its seeds by mixing them with "certain gums." After concocting
the paint, they would cover their faces and bodies. The Indigenous painters
claimed that the tincture protected them by relieving their wounds during
battles.[118] To explain the process, Oviedo coined (or at least used for one of the
first times in print) a new verb in Castilian, *embixar*, for "to paint with the *bixa*."
He warned his reader to not judge this painting as something savage or exclusive
to the Indies. After their triumphs in war, the Romans would ride in chariots
with their faces painted red to imitate fire. Julius Caesar also wrote of ancient
Britons who would paint their faces with an ointment that Oviedo claimed to be

of "*bixio* color or red" to gain "a more horrible appearance for battle."[119] Oviedo used the adjective *bixio* to describe the ancient Briton tincture, thus establishing that the Americas could also shed light on Roman times.[120]

The connection Oviedo drew left a powerful impact on his readers and sometimes led them to more drastic conclusions. For example, the natural historian Francisco Hernández (c. 1515–1587), writing two decades later, used Oviedo's verb *embixar* to describe the painting of the body before the celebration of an *areito* in Mexico. By then, both the noun *areito* and the verb *embixar* had been incorporated as commonplace terms into the lexicon of those who wrote about the Americas.[121] Hernández related how "it was a custom of many nations, and it is still today, to paint themselves with many colors taken from many natural things. This was done in New Spain before the Spaniards came to it, when the people were to perform in battles or in *areitos*, they colored their bodies with *bixa* [*embixavan*], believing that they would seem more beautiful or more fearsome and daunting to their enemies." Hernández embraced Oviedo's comparisons between American practices and European traditions not only to illuminate the past but also to challenge his readers' assumptions. He posed the question, "Why do we censure barbarous and bestial nations when we have at hand so many familiar and domestic examples?" Hernández further inquired, "Do our women not paint their faces white and their lips red and dye . . . their hair blond? Do they not print upon themselves, where they like it the most, moles and paint on their arms diverse forms and characters?"[122] Thinking about the *bixa* and other Indigenous American forms of beautifying the body prompted Hernández to estrange himself from his own context and to treat Spanish customs with detachment and even disdain.

Oviedo's *Summary* and *General History* were some of the most widely read, quoted, and discussed texts about the Americas in sixteenth-century Europe. The first part of the *General History* was the only section published in the author's lifetime. It was translated into French in 1555, and the Venetian compiler (and a correspondent of Oviedo), Giovanni Battista Ramusio, included it in the third part of his Italian *Raccolta de navigationi et viaggi* (Venice, 1550–1556).[123] Given this extensive circulation, Oviedo's account of the *areito* shaped how writers in both the Americas and Europe used the term. For instance, the French lawyer François Baudouin included the *areito* in his *Prolegomena* of 1561 to discuss the role of oral traditions in the historical practices of ancient Romans and Germans and to propose new possibilities for history writing.[124] The English writer Sir Philip Sidney also explicitly incorporated the *areito* in his *Apology for Poetry* (c. 1583) to intervene in a debate about the proper use of rhyme in English compositions.[125]

Three central features that Oviedo attributed to the *areito* recurred in sub-sequent Spanish writings that used the term. First, *areitos* were structured oral traditions that could be historical, and they were stably passed down from one generation to the next. Second, this form of memorializing the past was wide-spread in the Americas, where there was no alphabetic writing. And third, *areitos* resembled the songs and *romances* of illiterate peasants, past and present. This final claim contributed to forging new understandings of the *romance* tradition. This point is significant because it highlights how the Castilian canon itself was being created in the mid-sixteenth century. It also demonstrates that Spanish scholars often interpreted the customs and knowledge of the Americas' Indige-nous inhabitants by relying on their understanding of European peasants. How-ever, as Tamar Herzog has shown in the realm of the law, this "growing identifi-cation," at least by the middle decades of the sixteenth century, also included the use of Americas' Indigenous people to illuminate the behavior of European peas-ants.[126] Subsequent scholars would rely on these convergences to develop increas-ingly elaborate tools to assess all kinds of Iberian and New World oral traditions like sayings, local histories, the vernacular knowledge of plants and animals, and the agricultural practices of various regions. The dissemination of Oviedo's chronicle and its description of how the inhabitants of the New World used song and dance to memorialize the past contributed to a concurrent process of both validating, at least in theory, the use of *romances* as historical evidence and of treating the *romance* itself as an object of antiquarian inquiry and research, a substantial departure from previous assessments of these compositions as merely entertaining.

Oviedo's descriptions made a powerful impact on not only Páez but on poetry collectors of the mid-sixteenth century who sought to gather *romances* from vari-ous sources and reproduce them in printed compendia. The Sevillian writer Alonso de Fuentes (d. 1550) used the New World *areito* in his *Forty Songs of Diverse and Wondrous Histories* (1550, *Quarenta cantos de diversas y peregrinas historias*) to elevate the role that *romances* played in the remote and even more recent his-tory of Iberian society. His collection included songs and "histories" extracted from sacred scriptures, stories from antiquity, songs about people from various nations, and tales of Christians and Spaniards. Fuentes believed that these songs had preserved the memory of events that would otherwise have been lost to poster-ity. These compositions also captured an older language that had slowly changed into its present form. As such, they contained now-archaic words that had fallen out of use but whose meanings could be conjectured.[127] In the epistle that inaugurates his work, Fuentes noted that the practice of arranging events into verse was imagined not only by the "men of our hemisphere who have some light of reason, but [also by] the barbarous nations of Indians that are governed almost by natural instinct even until our times." Fuentes claimed that "these [Indians] have for their feasts and *areitos*, certain chants that are sung there; in these they celebrate and represent the famous past doings of their *caciques* and previous lords, for the example of those present."[128] By comparing the *areito* and the *romance*, Fuentes presented a currently attested example of memorializing the past in song. He argued for the fundamental role that orally transmitted poetry fulfilled in a variety of human societies, in both hemispheres, before the existence of writing. By using examples from a vast geographical expanse, Fuentes incorporated the Americas into a history of poetry and historical remembrance that, despite the epistle's brevity, had universal pretensions.

References to the *areito* increasingly appear in poetic and historical treatises of the second half of the sixteenth century, especially in works published in Seville or composed by members of its learned elite. Its incidence in works connected to Sevillian circles demonstrates how that city had become a central node in the transmission of books, knowledge, practices, and even words from the Americas. Twenty years after Fuentes, the Sevillian scholar and former soldier Gonzalo Argote de Molina (d. 1596) discussed the *areito* in a treatise about Castilian poetry. In 1575, Argote published the first edition of Don Juan Manuel's (d. 1348) *Tales of the Count Lucanor*, evidencing his profound interest in the history of the Castilian language. A grandson of the Castilian king Ferdinand III and nephew of King Alfonso X, Don Juan Manuel had produced a compilation of *enxiemplos*, or stories with moral teachings, around the year 1335. Some of the stories can be traced back to the *Calila and Dimna*, while others were locally transmitted.[129] Argote reveals in the opening to the work that he came across the manuscript in the court of Philip II. He was first drawn to the book on account of the author's illustrious genealogy. However, what compelled him to edit and publish the collection was Don Juan Manuel's "enjoyment and antiquity of the Castilian language."[130] At the end of his edition of the *Count Lucanor*, Argote appended an original "discourse on the Castilian poetry contained in this book," followed by an index of "ancient Castilian words," with their present-day equivalents. The glossary mined *Count Lucanor* for specific words to trace the mutations that Castilian had undergone from the fourteenth century until Argote's time.[131] The *Discourse*'s main intent, however, was to pinpoint Castilian poetry's principal features, like its various meters, and the possible genealogies of some of its forms.[132]

In thinking about the relationship between poetry and history, Argote echoed Fuentes's assertion that verse represented the primary way all human societies had preserved and transmitted their memories not only before the invention of writing but also after. Argote, however, widened his references substantially: alongside Gothic poetry, he included fragments of a poem in Basque.[133] Using language similar to that of Fuentes, Argote contended that Castilian romances attested to the practice of recording history in song.[134] Argote believed that this manner of remembering and singing history publicly in Spain originated with the Goths, "among whom it was accustomed, as Ablabius and John of Uppsala write, to celebrate their accomplishments in songs, if it was not understood that it was a custom of all peoples, and these must have been the rhapsodies of the Greeks, the *areitos* of the Indians, the Moor's *zambras,* and the songs of the Ethiopians, who nowadays we see get together on feast days with their *atabalejos* (drums) and *vihuelas roncas* to sing the praises of their ancestors, for whom there was no other meaning than this."[135] Argote's passage, which included mentions of Gothic songs, Moorish *zambras,* and African songs, perhaps performed in Seville, and Indigenous *areitos,* sought to convey both the universality and the dignity of this type of composition. By the 1570s, the *areito* had become yet another genre of historical poetry, alongside those of the Greeks, Goths, and Castilian peasants.

In his 1588 work *The Nobility of Andalucia (Nobleza de Andaluzia),* Argote revisited this theme and reflected more broadly on the challenges of history writing. In a dedication to the king, he described how in writing about the noble families of the region, he relied on a variety of materials, from manuscripts to all of the most up-to-date published histories. In some instances, however, "the lack of writing" had compelled him to seek out other sources. He claimed that the "proper ancient names of farmhouses [*cortijos*] and inheritances" could also serve to reconstruct old lineages. In addition to names, Argote believed that "romances and old songs, and ancient sayings [*refranes*] that have remained from our parents," were also a valuable source. Once more, Argote cited the precedent of the history of King Alfonso X, "who made use of these in his [history.]"[136] He then explicitly reproduced the passage on the origins of historical remembrance that he had introduced in the *Discourse* of 1575 in its entirety but included an additional category in his list of types of poems. He added the "*Endechas* [elegiac poems] of the Canary Islanders."[137] The poems of the Indigenous inhabitants of those islands served as further proof of the universality of poetry to memorialize the past.

By the 1580s, although the word *areito* continued to circulate in Europe as a form of oral historical commemoration common to the Indies, its semantic field had narrowed significantly for writers in the Americas. As missionaries and scholars learned of forms of historical memory specific to the Andean and Meso-american regions, the *areito* came to be associated mainly with singing and dancing, and sometimes the consumption of alcoholic beverages, to commemorate victories or important events.[138] The Sevillian dramatist and poet Juan de la Cueva (1543–1612), for example, who spent three years in Mexico City, succinctly claimed in his *Second Epistle*,

> Of no less glory and excellency are
> the ancient *Romances* where we see the same number of correspondences,
> The antiquity and character
> of our language we have preserved in them, and
> because of them we know antiquity.
> Singing was a custom of the Goths, (their glorious feats)
> and from them in us it is reproduced. The Rhapsodies the famous Greeks used
> were without a doubt of this kind
> and the mournful Indian *Areytos*.[139]

For de la Cueva, the *areito* was a "mournful song," celebrated to commemorate events. His example demonstrates that *areitos* were no longer regarded as the dominant form of historical transmission in the Americas.

The Great Year

From the information available in his book collections, Páez also drew conclusions about the emergence of history writing and the rise and decline of various societies. He sought to reconcile ancient European accounts with what he learned from his American sources to demonstrate both the difficulties inherent in producing historical accounts and their universal importance. The similarities that Páez perceived between the *areitos* and the songs, ballads, and *romances* of Europe inspired him to announce his intention to write a comparative and chronologically ambitious treatise. Its tentative title would be *Treaty of the Conformity That There Is between the Customs and Religions of These Western Indians with the Ancient [Ones] That Historians Write [about] in These Parts That We Inhabit*, and it would establish the parallels between the religious beliefs and customs of Indigenous Americans and ancient Iberians and Europeans.[140] No copy of this work has been accounted for; perhaps it was never completed. However, some of Paéz's annotations and notes point to this comparative interest. Gómara had stated that the "priests" of Cumaná on the mainland were known in the local language as *piaches*. The *piaches* were adept at using herbs to heal the body, but they also practiced the divinatory arts. In the margin, Páez noted that these priests were "like druids."[141] In his writings about Homer, Páez was already developing ideas about the role of druids and their counterparts in ancient societies. He listed various types of religious men who advised rulers, healed, and also practiced divination.[142] Like druids, *piaches* were to Páez a recognizable social type that had existed across various eras and societies.

Elsewhere in his *Memorandum*, Páez continued to describe the human progression toward historical writing. The use of only *trovas* and ballads ended "after God found it good to reveal the truly celestial gift of letters." However, this transition did not come about immediately, so that poems coexisted with written forms of remembering the past, since "many years passed in which history was not written with the greatness and majesty required." The Greeks, "the most political and careful people we know," were the first ancients to write histories, led by Herodotus and Thucydides. Reflecting on this early tradition, Páez claimed that even though the ancients were rustic and even uninformed "of doctrine and art, they still understood that the principal foundation of history was to not dare to say anything false and to only speak the truth, and not write anything to please or burden others, and to always display a free intention."[143] As Cicero had claimed in *De oratore*, truth was a fundamental component of historical writing. So, too, was eloquence, which was necessary to properly convey not only the actions and physical expressions of past actors "but also the conditions, inclinations, and passions of the spirit," which explained their behavior. In this, he asserted, history resembled poetry, "as in many other things."[144]

From this brief disquisition, Páez turned to praise the permanence of the written word. No system of recording information could measure up to writing because God willed it "to be the memory of memories."[145] Arts and sciences declined with the demise of the Romans. "If all the good Greek and Latin authors" had "been completely lost men would have necessarily returned to being savages, and very slowly, over many thousands of years, the arts would be discovered" once more. Libraries allowed for only a fragmentary preservation of knowledge, but this was enough, Páez believed, to rescue some knowledge from oblivion. He offered as an example his perception of the inhabitants of the Canaries. They "must have had commerce with Africans and Romans, according to the particular signs in those islands, but since they lost all their letters they [had] become," he wrote, "like savages" when the Spaniards first encountered them. Páez then asserted that the

Indigenous inhabitants of the Americas, though still barbarous, were "improving themselves."[146] Their current state, he averred, was the result of a long process of decline. "To this miserable state," he proclaimed, "they must have arrived after many years since they lost their letters and memories." He contrasted this with the inhabitants of China, who did have law enforcement and industry "when they were discovered, since they had not lost their letters."[147]

Páez was not alone in believing that the inhabitants of the Americas had lost their letters and memories. Various Spanish authors sought to explain how the Americas were inhabited by people genealogically connected to the lineages that appeared in the Hebrew Bible or in the writings of ancient Greek and Roman authors like Plato. Gómara, for instance, found the Atlantis explanation plausible because "in Mexico they call water *atl*," which was a word similar to that of the name of Plato's Island.[148]

The same year that Páez composed the first iteration of his treatise on history writing, Zárate published his *History of the Discovery and Conquest of Peru* (1555), which described the conquest of the Incas and the many civil wars that the Pizarros fought against their former allies in Peru. These protracted conflicts culminated in the assassination of the viceroy Blasco Núñez de Vela in 1546. Given his position as a royal administrator, Zárate sided with the viceroy and was briefly imprisoned before returning to Spain. The *History* narrated these struggles, recounted the events of the conquest, and provided an overview of the natural history of the region and the ancient customs of the Inca.[149] Other chroniclers of Peru, like Cieza de León, praised Zárate for his humanist erudition, which he demonstrated through his many citations of Plato in the Latin translations of Marsilio Ficino (1443–1499). These references were particularly prominent when Zárate speculated about the origins of the inhabitants of the New World and the similarities between the Americas and the Atlantis of Plato's *Timaeus*.

Zárate believed that the customs of the people of Peru resembled those described in Plato's account of Atlantis, especially in the description of a noteworthy temple covered in silver and gold.[150] Zárate also made many comparisons of the Incas to the Romans.[151] Like other scholars at the time, Zárate believed that he had found traces of long-gone populations of giants in Peru, which had vanished from the earth by the will of God like the inhabitants of Sodom.[152] Despite his assessment, he admitted that knowledge of Peru's ancient past was difficult for a foreigner to access because "the natives have no letters nor know or use writing, not even the paintings that serve in place of books in New Spain, but only the memories that they preserve in each other." Yet, in addition to these oral testimonies, the

Incas also recorded their accounts and tributes in unique devices: "some cotton ropes, that the Indians call *Quippos*, denoting the numbers by knots of different makings, ascending through spaces on the rope that denote tenths, and in this manner, and from above, and choosing the rope of color [depending on] what they want to show." These ropes were efficient at storing information because, as Zárate explained, "in each province there are people in charge of memorizing, through these ropes, general things," whom they called "Quippo Camayos." The Inca emperors carefully preserved these devices in public houses that were "filled with these ropes, which are understood with great ease by whoever is in charge, even if they are much more ancient than he is."[153] From Zárate's account it was clear that the Indigenous inhabitants of the Andes had concerned themselves with historical preservation.[154]

Evidence of the awareness of the unity and diversity that bound the world survived in the writings of ancient philosophers who believed in the Great Year. Plato and his followers defined the Great Year as the time during which all the planets and stars complete a revolution in the heavens. Ficino, Plato's most influential early modern commentator, highlighted how the conclusion of the cycle brought about floods and fire as forms of "reshaping powers used by Providence."[155] Páez wrote about the Great Year in at least two texts. In this time span of perhaps forty thousand years, "large empires are made and are passed from one people to another. Nobility and lineages are lost and the memories of men are obliterated by deluges, pestilences, wars, and great fire[s], of this the ancients tell of that of Phaeton and the Indians . . . that of Viracocha," of the Incas.[156] Páez believed that notions about destruction and renewal in ancient Greece and in the Andean world were commensurate. They served as proof of the importance of history and its physical preservation. History was challenging because it demanded that the historian consider both "unity and diversity."[157] Unity necessitated the "chaining of all the subject matter from the beginning until the end."[158] But the world, although finite because of its corporal nature, appeared infinite to the senses.

Conclusion: Chained Histories

According to Páez, Spain's newly acquired position in the history of humankind required a new chronicle researched in "Europe, Asia, and Africa, where the arms and standards" of the king had reached. He believed that this work had to extend to the "the newly discovered worlds, not known by the ancients, at least to be able to travel to them." This work would compare the ancient and the modern, and for this, "great wit and wisdom" was required. Optimistically, Páez concluded his treatise on history writing by declaring that the histories written under the patronage of the Spanish king would "paint a new sky never seen by our ancestors, a newly imaged land, with the rarity that it had, where we will not find things that resemble ours, new trees, herbs, beasts, birds, and fish, new men, customs and religion, great deeds in the conquest and possession of that which was conquered."[159]

Influenced by Oviedo and other chroniclers of the Americas, scholars of Páez's and later generations used information about oral traditions in the Americas to reinterpret European ballads and *romances* as historical sources. Their writings represent an early effort to contend with the variety of oral commemoration, some of which was believed to represent a stable transmission across time and to preserve memories that could then be translated into written histories. Lexical objects became a central source in these works. Chroniclers collected them as vestiges of the Americas. Páez writings show that the progress of the printing press, archives, and textual documents did not yet diminish the authority of oral discourse or of visual images. These were perceived, as Fernando Bouza has emphasized, at least until the seventeenth century, as also dependable, though they were thought to appeal to different aspects of human cognition.[160] However, oral testimony was much more difficult to preserve. Finally, at a time when the history of the Castilian language was coming under greater scrutiny, these collectors also realized that the history of Castilian necessitated an examination of older poetic compositions, some of which were now preserved in writing but had originally been transmitted orally.

In Páez's library, knowledge of particulars would first and foremost be preserved with great specificity. Yet their proximity to each other would make it possible to unite this information, to trace chains of continuity and similarities, to understand the natural and moral forces that bound the world, and the universe that it was a part of, into a great machine. In the library it would thus be possible to recreate the diversity and unity of the Spanish monarchy, and even of the world. Although Páez's proposals were not realized according to his designs, Philip II's information-gathering projects did mirror Páez's desire for a universal history of the Spanish realms. In the middle decades of the sixteenth century, both official chroniclers and archives proliferated.[161] They were meant to facilitate the king's ability to rule the massive corporate body of the Spanish Empire. When Jerónimo Zurita was proclaimed the first chronicler of Aragón, for instance, the official documents declared that since the majority of those "kingdoms in great part are governed by custom," it was fundamental to observe the succession of events, because it was through deeds that custom, not always recorded in writing, was created.[162]

The library of the Escorial, which would be part of a monastery complex and royal residence that Philip II began building in 1563, was located not in Valladolid, as Páez had suggested, but in the less accessible foothills of the Guadarrama Mountains, away from commercial routes. One of the central components of Páez's scheme was the inclusion of a printing press, a feature that was never realized at the Escorial, to the frustration of many scholars.[163] Thus, the library that Páez imagined was never completed. Many of the activities he thought it would house, however, such as the study of history and cross-cultural comparison, took place anyway, as this chapter has shown, not only in his own pages but in of those of his peers and successors, who within a few decades absorbed the historical lessons of the Americas, even those that existed merely as oral narration, and used them to interpret not just the history of the Americas but of antiquity and of Spain itself.

CHAPTER TWO

The Search for Spain's Most Ancient Language

The first eight books of Esteban de Garibay Zamalloa's *Historical Compendium* (1571) narrate the history of all of Spain's kingdoms, from the time of the creation of the world to the demise of the Visigoth kings. Determining the chronology of remote times, before the Roman arrival in the Iberian Peninsula, required a variety of methods to overcome the lack of surviving historical accounts. Garibay (1533–1599) believed that the study of place-names was a helpful tool to excavate this early history. Since the first age of the world, the founders of a region would confer their own name upon their territories. For instance, the second king of Spain, Ibero, named several settlements after himself. The Ebro River also took the name of this monarch, as did all of Spain, which "was called for some time Iberia."[1] This naming practice, which was amply attested in ancient authorities, allowed Garibay to use toponyms to trace the origins of people.[2] Garibay thus believed that names, like antiquities, were vestiges of past times.

To further explain the logic behind ancient name choices, Garibay compared them with contemporary processes. For him, "the Spanish who pass to the Western Indies do the same thing every day by imposing [upon] lands discovered once more, and cities by them built, the names of cities and regions of Spain . . . since just as Spaniards now have this custom in the Indies, in that manner our forefathers did the same when they came from Armenia and began to populate Spain."[3] Besides giving their own names to the places they established, settlers also used appellations from their places of origin. Hernán Cortés, for instance, conferred upon the lands that he conquered the name of "New Spain."

The Spanish conquest of the Americas was also commensurate with other aspects of Spain's ancient past. Spain was "such an abundant land," Garibay claimed, that it had been "the Indies of the Romans."[4] In several passages, he presented his arguments for establishing these equivalences. Spain possessed an abundance of precious stones and metals, especially silver and iron, so that Carthaginians, Romans, and other "foreign nations came to earn riches in Spain, like the Spaniards themselves now do in the Indies."[5]

Speculation about the "memory" of the flood in Indigenous historical reckonings inspired Garibay, like Páez de Castro before him, to establish chronological parallels between the history of the Americas and Europe. The Americas offered further proof that the flood described in the Hebrew Bible was universal. It had been a scourge of such proportions, Garibay asserted, that "these terrible waters and perishing of peoples were so fixed in peoples' understanding that even the barbarous people of the Western Indies preserve [it] in their memory, even after so many centuries, inheriting the memory from fathers to sons."[6] In other words, the knowledge of the Americas bolstered Garibay's explanations. An understanding of Indigenous American societies, together with the many demographic, political, and economic transformations brought about by decades of Spanish rule, could serve scholars like him in their effort to imagine Spain's ancient past. In particular, it could help unveil the dynamics that underlay the establishment of place-names as well as the transformations that languages underwent in response to the profound disruptions brought about by conquest. For Garibay and his contemporaries, language and its mutations represented a way to access the unrecorded past.

In the absence of other kinds of records about Spain's primordial habitation, scholars routinely turned to place-names. Etymologies were a fiercely contested source of knowledge regarding the history and culture of the early inhabitants of the Iberian Peninsula and, after the middle decades of the sixteenth century, increasingly of the broader Spanish Empire as well. Their study was a serious pursuit that struck at the core of some of the most controversial aspects of Spanish history and self-understanding. In the sixteenth century, a renewed interest in Neoplatonism, biblical studies, and Hebrew, coupled with increased attention to vernaculars and the history of classical languages, injected etymological studies with a renewed, almost frenzied energy. Politically fraught histories that sought to explain the relationship among origins, genealogy, and the religious piety of the various communities that inhabited the territories under the rule of the Spanish king often had recourse to etymologies.

Scholars in various genres used etymological derivations especially when no histories were extant or when the testimonies recorded in surviving accounts were incomplete or dubious. The etymological approach was common, to some degree, throughout Europe; among the most prolific of early modern etymologists was the French scholar Jean Bodin (1530–1596). Yet Spanish writers used etymology to overcome a challenge particular to their history, a problematic historical and religious heritage that threatened to relegate Spain to a lesser status among its more purely Christian rivals.[7] The history and languages of early modern Spain bore the indelible imprint of centuries of Arabic and Hebrew habitation. Iberian scholars' obligation to grapple with the history of populations of non-Christian origin, and to weigh their future integration into what was becoming an ever more repressive and persecutory society that privileged those of long-standing Christian descent (the "Old Christians"), invested Iberian etymological histories, and the theories of origins in whose support they were marshaled, with an unparalleled polemical urgency.

Responding to concerns over the status of Castile within the Iberian Peninsula, the loyalty of recent Muslim and Jewish converts to the Crown, and Spain's place within Europe and Christendom, Spanish writers increasingly placed language at the center of a politically fraught project to "convert" the history and geography of their homeland. This metaphorical conversion was intimately connected with the contemporary conversion of Jewish, Muslim, and Indigenous American communities to Christianity. Whether working directly under the patronage of Philip II or beyond the royal court, Spanish scholars engaged in linguistic debates to reclaim the Iberian Peninsula's Basque or Hebraic origins, assert its Greek and Roman heritage, and neutralize and assimilate its Arabic legacy.

To do so, Spanish scholars usually resorted to two related methodologies for converting linguistic patrimony into proof of antiquity. The analysis of place-names, as we have seen, required identifying languages of origin and how they changed over time to reach their present form. The second method involved establishing the relationship between the ancient languages spoken in an area and Hebrew, Greek, and Latin, the three languages of the Catholic Bible. Both of these strategies demanded that scholars implicitly or explicitly subscribe to, or in some instances even develop, accounts of the linguistic transformations that they sought to describe. The theories on which they drew to do so ranged from the story of the Tower of Babel in Genesis, to theories about the cyclical rise and decline of languages, to interpretations that emphasized climate and the effects of trade and political ruptures, and, in some cases, to a combination of all of the above.[8] After the mid-sixteenth century, moreover, scholars increasingly came to rely on examples from their readings about the Americas to validate their linguistic explanations, as this chapter demonstrates. Like Garibay, they tended to hold that the contemporary Americas offered living examples of what Spain's ancient past must have resembled. In doing so, they turned to Indigenous languages as a site to understand historical change and to make arguments about the possible futures of various Iberian linguistic communities.

Early modern Spanish scholars, among them chroniclers like Garibay and Ambrosio de Morales (1513–1591), the biblical scholar Benito Arias Montano (1527–1598), the Basque lawyer Andrés de Poza (d. 1595), and the Arabic lexicographer and Franciscan Diego de Guadix (d. 1615), attempted to extract historical knowledge from Iberia's etymological landscape. This chapter focuses on their efforts to account for the seeming ubiquity of Hebrew and Arabic place-names and loanwords in Spain's many territories and languages. The resolution of two problems became central to deciphering Spain's ancient history and its ensuing linguistic diversity: First, what had been the Iberian Peninsula's primordial tongue and, consequently, the identity of its first inhabitants before the advent of successive foreign conquests? Second, what was Castilian's relationship to Arabic? Had Castilian and its speakers changed irretrievably during the Islamic centuries?

Despite the increasingly influential association between language and religious identity that ultimately resulted in the prohibition of Arabic and the expulsion of the moriscos between 1609 and 1614, many scholars concluded in response to these questions that the learned traditions of Spain's "Oriental languages" could not be ignored without great loss. Determining how to incorporate their linguistic arguments into a historical narrative without any Islamic religious elements became a challenge that concerned lexicographers, antiquarians, and even forgers. Etymologies became an effective way through which scholars in an increasingly intolerant society could acknowledge no longer acceptable religious minorities in their understanding of the Iberian Peninsula's past.

The commitment to prove that certain toponyms derived from Hebrew or Arabic, or neither, however, would also help to bring about an unintended consequence. An accumulation of ever more detailed explanations as to how and why certain toponyms related to each other, and the conflicting interpretations upon which these convoluted linguistic genealogies rested, would eventually prompt scholars like Jerónimo Román Zamora (1535–c. 1597), Juan de Mariana (1536–1624), and Bernardo de Aldrete (d. 1641), to question whether the etymological method, in any of its iterations, could serve to reconstruct ancient history. The comparative perspective offered by Indigenous American place-names, which had bolstered earlier etymological conjectures, would also contribute to undermining the search for origins through linguistic vestiges.

Etymologies, Origins, and the History of Languages

While etymologies abounded in works of all genres, determining the proper way to execute etymological analysis and establishing the value of the evidence it offered had always been a matter of controversy.[9] In his *Commentary on the Common Principles of Languages and Letters* (1548, *De ratione communi omnium linguarum et literarum commentarius*), the Swiss scholar Theodor Bibliander (1505–1564) presented some of the most-contested aspects of etymologies and their uses since antiquity. Although Bibliander appeared on the Spanish Index of Prohibited Books as a banned author, his work synthesized the same passages from ancient and Christian authors also available to Spanish writers.[10] Bibliander believed that etymology, which studied the "origin and inflexion of words," was necessary for the knowledge of languages. Understanding a word's etymology "opens the way to unravelling its power of signifying." Etymologies could point to where words came from, whether they were adopted from other languages, and, consequently, the principles underlaying their creation or else their "migration."

Etymologies could also mislead. For Bibliander, however, "this should be attributed to individuals, rather than to the discipline itself." To present the possibilities and limitations of etymological derivation, he cited Quintilian, Varro, and Augustine. Augustine expressed a fourfold apprehension about etymologies. As Bibliander wrote, they were, first, that it was "unnecessary and too curious business; [second,] that pursuing it is an infinite occupation; [third,] that, like the interpretation of dreams, people interpret the origins of words each according to their own whims and talent"; and, finally, that the origins of many words can never be truly ascertained. Nonetheless, even Augustine conceded that the origins of words should nonetheless be investigated.

In sum, as Bibliander's synthesis conveys, the Western tradition could be summoned to argue both for and against etymology, making it a fairly flexible, and more often than not polemical, method that could be applied to various forms of argumentation. One could establish etymological analogies through phonetic similarities (*onomatopoeia*), for instance. The student of linguistic origins could also trace how words are derived from other words by making phonetic connections, or they could conjecture about how and why a language borrows words from other languages.[11]

Building on the writings of Josephus, Jerome, Isidore of Seville, as well as on the forged genealogies of Annius of Viterbo (1432–1502), sixteenth-century scholars further refined the historical and exegetical uses of etymology. Annius's genealogies were published in his *Antiquities* (1498). Working under the patronage of Pope Alexander VI (Rodrigo de Borja), Annius included a section that dealt specifically with the antiquities of Spain and its first twenty-four kings, rulers who preceded the arrival of Romans conquerors on the Iberian Peninsula.[12] The book's first edition was dedicated to the Catholic monarchs of Spain. By relying on forged ancient authors and manuscripts, Annius claimed that the first inhabitants of Spain descended from Tubal, the fifth son of Japhet and a grandson of Noah. He thus conferred upon the earliest dwellers of Spain biblical origins—an extremely desirable ancestry in sixteenth-century Europe.

Annius's works were widely read and extremely influential in Spain and elsewhere. As Anthony Grafton has argued, they offered a model to imagine an ancient pre-Hellenic and pre-Roman past.[13] Scholars in Spain repurposed some of Annius's modes of argumentation, mainly the recourse to etymology and its connection to biblical genealogies, to exalt the antiquity of cities and people not attested in other ancient sources. Charles V's royal chronicler Florián de Ocampo (d. 1558), for instance, infamously incorporated Annius's writings into his own *General Chronicle of Spain* (*Crónica general de España*, 1543). Ocampo was aware that Annius was a controversial authority to rely on, especially his citations of the alleged lost works of Berosus. However, Ocampo justified his choice by acknowledging that he could find no other source for "such ancient times" and that Annius had dedicated his works to the Catholic monarchs. This imbued them with a certain respectability.[14] To support his family trees of the founders of cities, Ocampo often offered linguistic proofs by studying the origins of the names of said cities and towns and establishing a direct filiation between their specific toponym and their corresponding biblical or mythological founders. For example, he asserted that the region of Tudela in the kingdom of Navarra had been established by Tubal. Ocampo believed that the passage of time degraded or "corrupted" languages, so that years of use had corrupted the word Tuballa into Tudela.[15]

The chronicler Pedro de Alcocer also began his laudatory account of the city of Toledo (1554) by tracing Noah's ancestry, recounting the events of the Flood, and the linguistic confusion that ensued at the Tower of Babel. He then assumed the view, widespread in the sixteenth century, that the biblical Tubal, was the first man to populate Spain.[16] Annius's model, as Don Cameron Allen and Giuseppe Marcocci have shown, would also inspire missionaries and other scholars who wrote the history of the Americas. These authors attempted to integrate Indigenous Americans into a monogenetic understanding of humanity by deploying etymologies to connect American societies with the events and characters of the Hebrew Bible.[17] The source material that Annius created also allowed scholars like Ocampo and Garibay, who sought to write "universal" histories of Spain from the beginning of time until the present, to gain information about the otherwise opaque first ages of the world and how the regions of Spain had been first settled.

This commitment to etymology was by no means an Iberian peculiarity. In his *Method for the Easy Comprehension of History* (1566, *Methodus ad facilem historiarum cognitionem*), one of the most influential early modern treatises on how to read and assess historical writing, Jean Bodin exalted the value of the etymology and explained how historians could employ this type of proof.[18] In the ninth chapter of the *Method*, he declared that no other question had "exercised the writers of history more" than the origins of peoples.[19] Historians expended countless hours researching and recounting the rise and decline of polities or the minutia of civil wars only in order to prove "the fame and splendor of their [own] race."[20] This was an error in which, according to Bodin, even the most famous of the ancient authorities had indulged, and some contemporaries indefensibly continued to do so. The reason such a pursuit was wrongheaded could be found in the Holy Scriptures, which taught that all people "were of the same blood" and "ultimately allied by the same bond of race." From all corners of the world, then, looking backward would inevitably lead to only one point of origination, the Chaldeans, "the most ancient of all people," as the reliable testimonies of Moses, Metasthenes, Herodotus, Ctesias, and Moses Maimonides, among many others, proved.[21] Metasthenes was among Annius's invented authors. From the country of the Chaldeans, humans spread and propagated into diverse kinds until modern days.[22]

Bodin's emphasis on the unity of human origins did not make him doubt the value of etymology. In the movements and intermingling of populations, Bodin argued, the scholar could trace patterns that would serve as interpretative keys useful for deciphering the origins of peoples. This process would ultimately lead to an understanding of the genealogical bonds that united seemingly Indigenous nations to other branches of humanity. Three types of proofs, based on an awareness of the principles that governed nature and change over time, could be used to reach back toward origins. For Bodin, chief among the three was the study of etymologies, followed by "the situation and character of the region," and finally, the reliability of the historian relating the conditions of previous times.[23] It is in the "linguistic traces," Bodin wrote, that "the proof of origin chiefly lies."[24]

For Bodin, the mechanisms that governed linguistic change were of utmost importance to the historian.[25] Bodin synthesized the main causes that sixteenth-century scholars believed brought about linguistic change. Scholars configured them in a variety of arrangements. Changes in languages, he argued, resulted from three major causes. The first was the passing of time, which causes everything to be inevitably transformed. The second cause was "the fusion of colonies and people." When different peoples came into contact, their languages altered one another through the borrowing of words or the mutation of sounds. This was a common phenomenon.[26] The third cause dictating linguistic change was, in accordance with Bodin's more general theories about the importance of geography and climate in determining the differences among peoples, the "nature of the area."[27] By this logic, the form and sounds of a language contained clues about the moral and intellectual aptitudes of its speakers.[28] Despite the myriad changes to which languages were constantly subjected, Bodin argued that traces of their original form always managed to endure the passing of time, the corruption prompted by the contact with other languages, and the changes brought about by climate and geography. If names preserved ancient local traditions, then scholars willing to investigate these traditions and collate their contents with etymological derivations could make use of the information contained in topographical names to promote their accounts of how places developed and who settled them at first. Names had the capacity, as Angus Vine has argued, to "continuously transmit origins" and "communicate between the past and the present."[29]

Bodin's linguistic reasoning and reliance on the etymology as a road toward one primordial, pre-Babelic, tongue was characteristic of Renaissance linguistics. As Marie-Luce Demonet and Claude-Gilbert Dubois have shown, sixteenth-century scholars simultaneously sought to retrieve the original language, which the majority believed to be Hebrew, and its inherent wisdom but also wished to explain the diversity of contemporary idioms.[30] According to Genesis, seventy-two languages appeared after the Tower of Babel, but many more were known in the modern world. As chapter 11 of Genesis states, before the confusion of Babylon, all peoples used the same language. This language, the most excellent of all, would prevail among those who did not "conspire in the deviated edification of the tower," and "one of the just [ones] was Heber, from whom the Hebrew tongue takes its name," as Juan Luis Vives explained.[31] Besides collecting linguistic information, numerous scholars attempted to account for how and why languages had mutated.[32]

Although early modern scholars of languages generally limited their investigations to languages, peoples, and events that appeared in the Bible, the boundaries of their inquiries were porous. The biblical text provided few "unambiguous statements on language," and scholars could "speculate freely on linguistic problems despite a background of traditional Biblical lore."[33] Some writers, like Bodin, the Italian linguist Celso Cittadini (1553–1627), and the Spanish scholar Bernardo de Aldrete (more on whom below) attempted to explain linguistic change through broader systems of causation combining natural and political factors.[34] Etymological derivations could be used to advance specific agendas, as they were by the French Hebraist Guillaume Postel (1510–1581) or the Dutch physician and linguist Johannes Goropius Becanus (1519–1572), who promoted a system of beliefs or a particular theory of history that exalted the antiquity or primordial characteristics of their communities of origin. For instance, Becanus argued through etymologies that Dutch had not only been the language of Adam but that it had remained unchanged even after the events at Babel and into his own time.[35]

Explaining how and why languages change, and how this understanding could be used to account for the history of Castilian, and of Spain more generally, lay at the core of Juan de Valdés's (c.1509–1541) well-known *Dialogue on Language* (1535, *Diálogo de le lengua*). A religious reformer from the city of Cuenca and a follower of Erasmus, Valdés fled Spain for Italy in 1530 after the publication of his *Dialogue on the Christian Doctrine* (1529, *Diálogo de la doctrina Cristiana*) to escape the Inquisition.[36] After a brief stay in Rome as a political agent of Charles V, he then settled in Naples, where he composed the *Dialogue on Language* and continued to write until his death in 1541.[37] In the sixteenth century, the *Dialogue* circulated in manuscript form—Jerónimo Zurita, for instance, possessed a copy of it.[38] The *Dialogue* remained unpublished until the eighteenth century, when the Valencian scholar Gregorio Mayans y Siscar (1699–1781) rediscovered it, though he did not know the identity of its author. Valdés's *Dialogue* was one of the first works to discuss the history of Castilian and its relationship with the other languages extant in the Iberian Peninsula. It introduced important questions, themes, and exegetical tools that subsequent scholars would reaffirm or challenge in their search for the origins and history of Spain's multiple languages and their speakers.[39]

Covering what we would now designate as lexicography, phonetics, and grammar, Valdés articulated several hypotheses about how different types of linguistic alterations took place and how they related to political and social events. For him, commerce and conquest were the most significant factors triggering linguistic change. He argued that the transformation of languages could occur suddenly, due to violent political ruptures, but also gradually, as a result of quotidian interactions with neighboring societies. Spain possessed such a great number of languages, Valdés argued, because it had been subject to so many different overlords across its ancient and modern history. Further linguistic mutations ensued through trade with neighboring societies.

Valdés approached the controversy surrounding ancient Spaniards' original, pre-Roman tongue with ambivalence, conceding that it belonged more to the realm of history than to that of grammar. However, he believed that the question could be answered by considering the evidence provided by grammar alongside that rendered by historical study. Although many claimed that Basque was the most ancient language of the Iberian Peninsula, given its unique characteristics and the fact that foreign invaders had never managed to conquer the Basque territories, a closer reading of ancient historians led Valdés to claim that the most widespread tongue in ancient Spain was not Basque but Greek. The reign of Greek lasted only until the Romans conquered Iberia, at which time Latin gradually "banished Greek from Spain." This was so until the Goths invaded Spain and brought their tongue, which did not replace Latin but merely corrupted the Roman language. The use of a hybrid Latin, with elements of Greek and Gothic, lasted until the arrival of the Moorish kings.[40]

In his proposition that Castilian emerged from decadent Latin, Valdés expressed a similar position to that of Nebrija. Nebrija summarized the origins of Castilian succinctly in his *Grammar of the Castilian Language* (1492): "When the Roman Empire began to decline, the Latin language declined with it, until it arrived at the state in which we received it from our parents. Truthfully, in comparison to the Latin of those times, it has as little to do with it as it has to do with Arabic."[41] This mixed language, Valdés believed, subsequently also incorporated numerous Arabic words, because even though "the kingdoms were regained, there still remained in them many Moors as dwellers that maintained their language."[42] Castilian's lexical borrowing from Arabic was extremely broad and the product of extended social and political relations with speakers of that language.

Reflecting on the specific mechanics of linguistic borrowing, Valdés asserted that by establishing the categories of words that speakers of one language borrowed from the speakers of another, an observer could determine the nature of the social, political, or commercial relations that had once obtained between these societies. Thus, one might work backward from lists of the Arabic words in the Castilian lexicon to determine what kinds of objects the Arabs had introduced into the Peninsula. He remarked that, "even though many things that they name with Arabic words have equivalents in Latin," usage made Castilians prefer the Arabic word to the Latinate one. For example, people called rugs *alfombras* rather than *tapetes*, opting for the Arabic derivative over the Latin.

Even so, Valdés maintained that although Castilian had appropriated numerous words from Arabic, Latin "was still the main pillar of the Castilian language."[43] Linguistic borrowing, he emphasized, regardless of its particular dynamics or motivations, was inevitable, governed by the ebb and flow of usage, and devoid of moral or religious connotations. Valdés's selection of Greek as Iberia's primordial language, his omission of Hebrew, and his choice to set the arrival of Arabic late in the history of Spain, derived at least in part to his reading of ancient historians like Pliny and Strabo, who described the Greek colonies scattered on the Peninsula's eastern shores.[44] But it can also be attributed to his humanist preferences. Valdés sought to emphasize Spain's connection to the Greek tongue, since this language possessed expressive capabilities and qualities that exalted the genealogy of Castilian and Spain. His linguistic hierarchy located Spain's origins in the classical world already centuries before the Roman conquests.

From the history of Castilian to the particularities of lexical borrowing, Valdés next led his two interlocutors in the *Dialogue* to the problem of what should be the standard for elegant linguistic usage. Like Desiderius Erasmus and Conrad Gessner, Valdés judged that "vulgar" languages could not be reduced to proper arts, or grammatical rules, since these spoken tongues were always in flux. One of the *Dialogue*'s interlocutors reminded the other that "vulgar tongues in no way can be reduced to rules that can be used for learning."[45] The problem that arose from this proposition was how to set a standard for a language that was always changing. Early in the *Dialogue*, Valdés alluded to Pietro Bembo's *Prose della volgar lingua* (1525). Yet he disagreed with Bembo's formalistic understanding of linguistic usage and proposed instead to set a style that was based not on artificial literary models but on the sayings (*refranes*) "born amongst the common people" and the speech of the courtiers of the city of Toledo.[46] The sayings that he wished to recover were not the literary *sententiae* of the ancients: the Castilian ones were to be "taken from vulgar sayings, many of them born and raised amongst old women, behind the fire spinning their wheels." This proposition might have also originated in debates over the correct standard for Italian, a literature with which Valdés was very likely familiar.[47] In contrast to Castilian sayings, "the Greek and Latin ones . . . are born amongst learned people and celebrated in books of much doctrine."[48] But to consider the properties of the Castilian language, "the best proverbs have to be born amongst the common people."[49] In the process of proposing this standard, Valdés criticized Nebrija's Castilian dictionary and his tendency to include only words that possessed clear Latin roots in the compendium. Valdés's conclusions reveal a fundamental assumption about the causes of linguistic change, one he shared with many contemporaries: the history of a spoken language parallels in fundamental ways the lived experiences of its speakers and, as such, could be used to reconstruct that society's history. Languages change according to the will and collective wisdom of speech communities, reflecting the needs of the communities and even their material history. For Valdés, the standard for correct linguistic usage should be based on the tried and tested choices of speakers, which were then passed down through generations in the form of words and sayings. In this way languages were always also archives.

Hebrew, Basque, and the Search for Spain's Most Ancient Language

As we have seen, scholars disagreed about the languages first spoken in Spain. While Valdés argued for Greek, several of his contemporaries took the side of Basque. In his *Historical Compendium* (1571, *Los XL libros d'el compendio historial de las chronicas y vniuersal historia de todos los reynos de España*), Garibay explained that after the Flood the patriarch Tubal arrived in Spain in the region of Cantabria, where, compelled by natural affection for their original homeland, Tubal and his followers named their new settlements after the mountains and rivers of Armenia. For instance, following Ptolemy, some geographers believed that the "mountain where Noah's Ark stopped [upon] the flood in Armenia," was called Gordeya and "between the province of Alava and Biscay," Garibay offered, there was a very tall mountain that "changing only the D to B, they call Gorbeya."[50] Garibay provided an ancient proverb as supporting evidence: "If anyone wants to note their affection for their fatherland [*patria*] . . . they say: the cow of Gorbeya always desires Gorbeya."[51] If others, following the Holy Scriptures, "wish to say that the mountains of Armenia where Noah's Ark stopped, was called Ararath," indeed a mountain in Cantabria that had once been called by the same name was now known as Aralar. Close to the town of Mondragón, where Garibay wrote his history, there was also a town of mills called Babylonia, which had been its name since ancient times. To Garibay, all these names proved that Tubal and his companions first disembarked in the region of Cantabria.

What language had Tubal brought to Spain? Garibay conceded that "there [was] also much debate among our chroniclers." Some held it to be the Chaldean language because of some supposedly Chaldean toponyms in Andalusia. However, as Garibay would explain, these came with Nebuchadnezzar, "Prince of the Chaldeans of Babylonia," over "one thousand five hundred and seventy years" since Tubal settled Cantabria. For Garibay, the first language of Spain was Basque, the same one that "is spoken in most parts of Cantabria, especially in the provinces of Gipuzkoa, Alava, Biscay, and in most of the kingdom of Navarra," as well as in parts of France. Garibay asserted that Basque was one of the seventy-two languages that originated after the confusion of the Tower of Babel.

Besides the evidence of names, Garibay offered other considerations to support his claim. "It is of great consideration and mystery . . . that at least in Spain all children from their birth bring this language to their lips, because the first words that they speak are *tayta*, that is how they name the father, and mama, that is how they name the mother." These names, Garibay explained, "are from the Cantabrian language in which the father is called *ayta*, and the mother, *ama*, in the manner that children because of their youth or for other reasons corrupt [these words]: the small difference is in one letter in the beginning that the Cantabrians take away or that they add: because to the name of *ayta* they add the T, and to that of *ama*, the M."[52] In conclusion, Garibay explained that Basque was preserved in those mountainous regions where no other language was introduced, so that it did not mix "with foreign nations outside of its law, either because of the strength of its lands, or the strength of its people, or one and the other."[53]

Other scholars were skeptical as to whether Spain's first language could ever be discovered. For the antiquarian Ambrosio de Morales, the first language of Spain was impossible to determine because in the time of the Romans, when the earliest surviving histories were written, the natives of those kingdoms already spoke very different languages. He noted that when writing about the languages of the Iberian Peninsula, Seneca, Cornelius Tacitus, and Strabo asserted that Iberians did not use one single language but rather each community possessed its own natural tongue. The antiquarian and archbishop of Tarragona Antonio Agustín (1517–1586) agreed, regarding the Basque hypothesis as untenable given that neither books nor any other written memories survived in that language. It was, at the very least, impossible to prove its antiquity.[54] Agustín was well known for his skepticism about how many of his contemporaries grounded comprehensive theories on what he considered to be slender foundations of evidence.[55] The Jesuit historian Juan de Mariana would articulate the same arguments against the antiquity of Basque in his *Historia general de España* (Lat. 1592, Castilian 1601).[56]

Morales's skepticism did not preclude him, however, from participating in discussions on the origins of the label Spain, or Hispania, itself, a topic that had been in dispute ever since the seventh century, when Isidore of Seville had traced it to the word Hispanus, a corruption of the name Hispalus, the legendary founder of the city of Hispalis (now known as Seville).[57] The majority of authors who endeavored to decipher the identity of the Peninsula's first settlers contended with this issue. In many instances, their interpretations served as a cornerstone to their broader theory of origins and as a pivotal explanation of Spain's subsequent historical development.

Morales's favorite theory about the origin of Hispania, for example, came from his contemporary Diego Hurtado de Mendoza. Hurtado de Mendoza believed that he had located the most reliable tradition regarding the true origin of Spain's name among the collection of Greek manuscripts that he had carefully amassed while serving the emperor Charles V as ambassador to Venice in the 1540s. In their Latin works Pliny and Varro narrated the story of the famous Greek captain Dionysus, also known as Bacchus, who traveled to Spain to advance his conquests. Upon his return to Greece, he granted authority over the recently subdued areas in Iberia to one of his captains, a commander named Pan. The ancient Greek article (*-is*), attached to the commander's name to denote dominion over his colony, transformed the word to Ispanos, or "that which is of Pan." Morales enthusiastically embraced Hurtado de Mendoza's theory.[58] Although Morales's concerns about the lack of early documentation prevented him from definitively identifying Greek as Spain's most ancient language, his selection of this etymology did allow him to highlight the Greek settlement of Iberia as the formative moment in the peninsula's primitive history. Morales's and Valdés's preference for the Greek etymology, however, was not destined to be the final word on the subject. For many, Ocampo's controversial genealogies showed that the postdiluvian and most ancient dwellers of Spain could be traced directly to Noah's progeny. The problem then became determining what language the biblical patriarch's descendants had spoken, and whether they had maintained it intact after reaching Iberian shores.

The Basque hypothesis, though popular, had to compete with the increasingly appealing notion that Hebrew had been the first language of the peninsula. In his work on the city of Córdoba, the painter Pablo de Céspedes (1538/1548–1608) explicitly challenged the Basque hypothesis. Noah's descendants, he thought, must have spoken Hebrew, the first language of humankind. They had settled many urban centers throughout Spain, most notably Córdoba.[59] Céspedes offered both etymological and archeological evidence to link his birthplace of Córdoba to ancient Jews.

Through an elaborate succession of etymologies, Céspedes claimed that Córdoba was properly interpreted not as a Latin or Greek name but rather as a Hebrew derivative. He reduced the word to two "Hebrew" syllables, *car* and *daba*. Together, he argued, these meant "plain of strength, of fertility, of richness, of beauty, of fecundity, and virtue." With the expressive power that characterized the holy language, the Hebrew word conveyed the city's main properties—its fertility and temperate climate. These features allowed the city to nurture countless virtuous men and women and to encourage the blossoming of philosophy, medicine, and the other arts.[60] The toponym was most likely reused by Greek and Latin geographers, he suggested, who routinely adopted the preexisting names of the provinces where they settled, changing the words slightly to pronounce them more comfortably. It was no accident, according to Céspedes, that classical authors like Strabo and Virgil described the city using the same succession of adjectives implied in the Hebrew etymology: at some point they must have learned the meaning of the native appellation from the locals. Céspedes imagined Roman authors following the same procedures as contemporary crown officials who conducted the royal censuses of the *relaciones topográficas* starting in the 1570s, querying the local inhabitants about the names of their towns and their traditions.[61]

Céspedes's contemporary, the biblical scholar Benito Arias Montano, also believed that the multiplicity of "Hebrew" toponyms scattered throughout the Iberian Peninsula could be explained only by the arrival of Jewish settlers in ancient times. This included the name of Spain itself. He proposed that Hispania or España was not a corrupt Greek word but, rather, originated in the Aramaic Spamia, a place-name found in the Aramaic paraphrase (Targum) of the biblical book of Obadiah. Following a line of argument found in Sephardic Jewish commentaries, Arias Montano used this reference to Spamia, the supposed destination of one of the several Israelite diasporas, to claim that Jewish settlers had known and inhabited many provinces of Spain in antiquity. He buttressed his claim by identifying numerous Hebraic toponyms scattered throughout the Iberian Peninsula.[62] In his *Commentaria in duodecim prophetas* (1571), however, Arias Montano opted for a chronology that diverged from that of Céspedes. He emphasized that the ancient Hebrews who settled some of Spain's most distinguished cities came with the Babylonian king Nebuchadnezzar II after the destruction of the First Temple in

Jerusalem. As Dominique Reyre has argued, Arias Montano's etymological exercises, lifted from medieval Jewish commentaries, aimed to show that the Jews who settled these primitive communities had arrived in the Iberian Peninsula at least five centuries before the birth and death of Christ. Hence, they and their descendants were exempt from deicide.[63]

Arias Montano found etymological hints about the New World as well as the Old in the Hebrew Bible, which he thought foreshadowed the discovery of the Americas. He and his contemporaries scoured the Bible in search of American toponyms and identified certain places in the text as their American equivalents. Luis de León (1527–1591), among others, recognized the Yucatán Peninsula as the Joktan of Genesis (the second of the sons of Eber) and the regions that comprised Peru with the lands of Parvaim, where King Solomon's gold was found.[64]

The Basque lawyer Andrés Poza used a method similar to that of Céspedes and Arias Montano but reached different conclusions.[65] He believed that the appellation Spain predated the arrival of Greek settlers in the peninsula and signified something else altogether. Poza pointed to the phonetic similarities between España and the Basque word *esbana*, meaning "land of good lips and tongue." Insisting that the Basques must have been the peninsula's most ancient inhabitants, Poza defended the appropriateness of his etymology by noting that the "Spaniards naturally possessed great eloquence."[66]

Poza composed his treatise *Of the Ancient Language, Settlements, and Provinces of Spain* (1587, *De la antigua lengua, poblaciones, y comarcas de las Españas*) to prove (like Garibay) that Basque, not Hebrew, was the most ancient language of the peninsula and had been spoken throughout its entire territory. He argued that the Basque territories were first settled by Tubal, who brought the Basque language with him to Spain. The main evidence was once again etymological. Poza contended that the suffix *briga* in the names of some of Spain's oldest cities was a Basque word. Cantabria, for instance, was a composite of *brigo* and *canto*. *Abrigo* means "shelter and company," from which the Castilians took the verb *abrigar* (to shelter). From this meaning, "it can be deduced that the ancients called *brigas* the settlements that were fenced or sheltered, because these provide shelter or *abrigo* from those who live in the roughness of the cliffs, edges, and mountains of these regions." In his *Chronicle*, Ocampo confirmed that *briga* was part of the ancient language of Cantabria, theorizing that the Roman emperor Vespasian had established a city in Spain on the edges of the Bay of Biscay named Flavio briga. This was "a combination of his name, with the speech of the region, in which towns were referred as to as *brigas*."[67]

Basque's expressive capabilities marked it out as one of the seventy-two languages that emerged immediately after the confusion at the Tower of Babel, since these ancient tongues possessed special powers. "An excellent language," Poza reasoned, is "one in which the names themselves teach their own cause, [or] the definition and nature of the thing named." He invoked Plato's *Cratylus* to argue his case. Languages could only be described as "elegant, substantial, and philosophical" if their names were able to transmit the qualities of the objects that they signified.

By contrast, languages that were not part of the seventy-two but were "rather *mestizos* [mixed] and imperfect," possessed names "without any mystery," and thus were conventional. Poza's use of the word *mestizo* is telling, for he also brought up images from the Americas to support his linguistic genealogy.[68] Many ancient Basque and Hebrew toponyms survived across the peninsula even though speakers of these languages no longer resided in their former lands or had abandoned their original idioms in favor of others. The reasons for this were identical, Poza hypothesized, to what was occurring to the Spaniards who established themselves in the Indies. The Spanish conquerors, even though they were mainly speakers of Castilian, still referred to the provinces of the Americas "with their first names in the Indian languages." Mexico, Peru, Chile, and Cuzco were all Indigenous placenames. Poza believed that these names were so resilient to the passing of time, and even to the demise of the regions' Indigenous inhabitants, because of their antiquity and their language of origin: the original, shared tongue of humankind. Although Poza did not identify the specific settlers who brought this language to the Americas, his chronology implies that the West Indies must have been settled by the survivors of the Flood.[69]

The impetus to proclaim the nobility of Basque extended to the provinces of New Spain, or modern-day Mexico, which had boasted a thriving Basque community since the earliest days of the conquest.[70] Baltasar de Echave Orio the Elder (1548–?), a distinguished painter of Mexico City, enjoined his fellow Spaniards not to disregard their most ancient language.[71] The Basque tongue, in his *Discurso de la lengua cántabra-bascongada* (1607), took the form of an old lady who wore rags and complained "that although she was the first language in Spain and common to the entire peninsula, her native speakers have forgotten her," for in their fascination with the foreign they had accepted other languages.[72]

Echave Orio found support and even amplification among his Basque peers in the Americas. The Basque lawyer Hernando de Ojeda composed a short prologue to introduce his friend's work. In his opening remarks, Ojeda evaluated the notion that the spread of empire is always accompanied by the complete linguistic assimilation of those defeated.

> Even though it is true that, as it has always been said, conquerors and their
> languages consume the speeches of vanquished peoples: this does not apply to
> the names of provinces, mountains, rivers, and springs, even if they are a bit
> altered, as we experience in the infinite provinces of these Indies, which still
> conserve with little variation its ancient names, because even though we
> renamed many of them in the Spanish fashion: these have been forgotten or
> have fallen out of use altogether and the old names of the Indians prevail,
> even after all of them have died-out in some parts. This is verified in the
> Island of Cuba, which the Spaniards named in the beginning Fernandina, in
> Habana, Bayamo, Jamayca, Yucatan, Chapultepec, Campeche, Mexico,
> Mechuacan, Tezcuco, Tlaxcala and Cholula, which are all Indian words.[73]

Using the Americas as a case study, Ojeda aimed to prove that Basque had been the first language of Spain, spoken before the arrival of Greek, Roman, Gothic, and Arab conquerors in the Iberian Peninsula's remote past. This was confirmed by the survival of what he believed to be numerous Basque toponyms in all of Spain's regions. The American examples helped to explain the simultaneous abundance of Basque place-names and yet the very small number of remaining speakers.

Ojeda claimed that the American examples were fundamental for Echave's argument. The many social, religious, and demographic transformations brought about by Spanish rule, which contemporary scholars could witness with their own eyes, coupled with an understanding of the workings of ancient Indigenous societies like those in Mexico and Peru, could act "as examples and live portraits of what it was once like in the Old World."[74]

The Americas, however, not only provided examples to understand the past but actually helped to define the fate of Basque. Scholars not only saw the experience and languages of the Americas as parallel to those of Europe but also used the former to imagine and explain the past development of the latter. In the book's final chapter, the personified Basque language warned its speakers of something that she "had learned through [her] great antiquity and long experience"—namely, that the Basque people's honor was "entirely dependent on" their language's survival. The Americas and their promises of gold were leading many young people into perpetual exile, enticing them to abandon their ancestral estates. When these youths reached the Americas, their descendants most likely would forget the Basque language or become disenchanted with its "style and ancient plainness."

Population decline and migration, according to Echave, threatened to bring this ancient and noble language to its end. In the same way that the inhabitants of Hispaniola had ceased to speak Taíno by the end of the sixteenth century, Echave feared that Basque, too, would fade away.[75]

Arabic and the History of Spain

Poza's establishment of Basque as the first general language of Spain, as opposed to Greek, Latin, or Hebrew, stressed an autochthonous genealogy, one connected to the events at the Tower of Babel, permanent since its inception and immune to the changes wrought by the conquering polities that settled Spain, most notably the Muslim overlords who ruled parts the peninsula for almost eight hundred years. The sudden ruptures brought about at Babel were enough to explain all linguistic divergences. In contrast to Poza, Céspedes and Montano sought to stress the continuities with the biblical past and in this manner vindicate Spain's Jewish heritage.

The presence of Arabic in Spain, like that of Hebrew, offered numerous conceptual problems and possibilities. Its influence on modern Castilian seemed ubiquitous. Some scholars, like Morales, struggled to include Arabic and Islamic sources alongside the more conventional Greek and Roman sources familiar to most Renaissance antiquarians. Morales was keenly aware of how profoundly speakers of Arabic had transformed Spain's linguistic landscape. Other scholars struggled to defend Arabic-speaking Christian communities, their religious commitments, and their customs from increasingly restrictive legislation. Indeed, by 1567, moriscos were already forbidden to write or speak Arabic, ordered to learn Castilian within three years, and denied the use of Arabic surnames, clothing, baths, and any other distinctive markers.[76]

The Valencian chronicler Rafael Martí de Viciana (1502–1582) condemned the use of Arabic in Spain and lamented the debt speakers of Castilian owed to what he considered to be an infidel tongue. In his 1574 *Book of Praises of the Hebrew, Greek, Latin, Spanish, and Valencian languages* (*Libro de las alabanças de las lenguas hebrea, griega, latina, castellana, y valenciana*), Viciana's principal goal was to prove the nobility of Valencian.[77] He believed that Latin was the mother of both Valencian and Castilian. because of their dealings in war and peace with the "Hagarenes" (that is, Muslims), Castilians had regrettably introduced many Arabic words into their language.[78] For Viciana, Castilian neglect "ha[d] allowed for the loss of their own and natural words, adopting strange ones," especially from this enemy tongue. This was reprehensible because there were many wise men in Castile who could have turned to the Latin language, or to Greek or Hebrew, to enrich their vocabulary.[79]

According to Viciana, Valencian had not engaged in this irresponsible borrowing even though "in the kingdom of Valencia two thirds [of the people] were Hagarenes who spoke Arabic and to this day one-third are converted [Muslims] who still speak Arabic." Viciana asked his reader to consider that, unlike Castilian, "never has the Valencian language taken . . . any Arabic word"; "on the contrary, the Arabic language being such an enemy of Christianity, it abhors it." Those residents of Valencia who had converted to Christianity from Islam still refused to abandon Arabic, even fifty years after their baptisms. Whenever the authorities pressured these converts to speak Valencian, Viciana recalled, they responded by questioning the authorities' intentions: "Why do you want us to abandon the Arabic language? Is it because it is evil? And if it is evil, why do the Castilians speak it mixed in with their language? Leave us our language and we will leave it little by little."[80] Implicit in this rebuttal was the notion that Castilian was capable of seamlessly incorporating Arabic words into its lexicon because the Arabic language could be separated from the religion of Islam. The Granadan nobleman Francisco Núñez Muley had made a similar argument to Philip II in 1567, in response to legislation that sought to prohibit the use of Arabic. Among other arguments, Núñez Muley evoked the existence of the Arabic-speaking Christians of the East.[81]

By contrast, Viciana believed that the separation between language and religion was impossible. To make matters worse, Castilian's borrowing from Arabic was so extensive, Viciana noted, that even all of the main rivers of the region, such as the Guadiana and the Guadalquivir, retained their Arabic names even after Christians recovered the territories. In contrast to Castilian, Valencian preserved its proximity to Latin and borrowed words as needed from its mother tongue. The scholar included tables demonstrating that Valencian words were closer to Latin in form and meaning than Castilian ones. He called for purifying Castilian of its Arabic components. Rather than the common usage of what he considered to be ignorant people, Castilian writers and scholars should direct the growth of their language and select from Latin, Greek, or even Hebrew any additional words that they required. This proposition differed significantly from Valdés's understanding of both linguistic borrowing and the setting of a language's standard on the basis of collective usage. Valdés believed that spoken languages always fluctuated and its standards had to emanate from usage. Viciana, on the contrary, believed that scholars even create neologisms and regulate the functioning of languages.

Similarly, Ambrosio de Morales's surveys of Castile's countryside led him to recognize that the omnipresence of Arabic toponyms made the reconstruction of Spain's Roman history challenging.[82] This was especially true of the names of towns, rivers, and mountains, which had been corrupted or completely effaced by the Arab conquerors. Morales cited the surviving histories of Saint Isidore and Saint Ildefonso, the records of religious councils, and the writings of the Archbishop Don Rodrigo, all authors who were "grave and trustworthy" and who had lived after the Romans and before the arrival of Muslims. His goal was to demonstrate that neither the Visigoths nor any other post-Roman conquering nation had significantly transformed Iberian toponyms until the arrival of the Arabic speakers. Only in the Muslim period did the most significant mutations occur and cities and entire regions acquire entirely new denominations.[83]

Like many other historians and antiquaries of his generation, Morales was concerned about the rupture that the Islamic conquest had caused. He tried to compensate for this discontinuity by compiling a study, *The Antiquities of the Cities of Spain* (1575, *Las antigüedades de las ciudades de España*). Combined with his desire to elucidate the reasoning behind the study of antiquities, his anxiety about the loss represented by the Muslim conquest of Iberia compelled Morales to formulate thirteen methodological considerations for the student of Spanish stones, statues, coins, and inscriptions to bear in mind when performing research.[84] These he explained at the outset of his treatise. A companion volume to the author's continuation of Ocampo's *Chronicle*, the *Antiquities* sought to address the technical aspects of the study of Roman antiquities that Morales had omitted from the *General Chronicle* so as to not bore or distract his eager readers from the narrative of rulers, wars, and religious events that formed the core of this official work.[85]

The "General Discourse on the Antiquities of Spain" began with a disclaimer about the kinds of knowledge that antiquarian research could yield. The antiquities of Spain were "buried in the darkness of old age and oblivion," given the numerous invasions that the Iberian Peninsula had suffered, and "when reason reaches the [form] of a good and possible conjecture, not more is possible nor should be expected." Morales would assemble evidence from multiple sources to produce probable conjectures about Spanish cities and towns that had disappeared, or became new places altogether, in the wake of the Islamic conquests. The evidence offered by linguistic research, he cautioned, offered plausible but not definitive proof. Although often a step toward greater certainty about the foundation of a settlement, etymology could not serve on its own as proof of origins.[86]

In the search for Roman antiquities, the first circumstance to consider was the survival of "signs or remains of antiquity from the time of the Romans," which could range from ruined buildings to statues, coins, medals, and sculptures. What these things looked like could not be taught with words but rather through guided observations.[87] Ptolemy offered a second source from which to reconstruct knowledge of an ancient town. The ancient geographer had, however, recorded only major cities and land formations, and sometimes his measurements were incorrect either because of corrupt translations and manuscript transcriptions or his own miscalculations. Despite these errors, Ptolemy did provide the names of many towns and could serve as a preliminary guide. Morales also included as sources the *Antonine Itinerary*, other geographers like Strabo, ancient historians and writers like Pliny, and the records of the religious councils held in Spain.

In addition to these sources, Morales proposed examining the forms of the settlements and the rivers that passed through them, and the martyrdom, lives, and legends of saints. He also recommended a process similar to the royal survey, the *relaciones topográficas*: relying on "the authority of some people to whom credit should be given and on the common opinion of vulgar people, which sometimes is right."[88] He thus codified the uses of oral testimony and the questioning of trustworthy informants to make "good and possible conjecture[s]" about vanished cities and peoples. Morales's example shows that reliance on local experts and oral knowledge was a feature of antiquarian research in Spain as much as in the Americas.

Morales also addressed the use of language specifically. The former names of cities, and the ones that they currently possessed, could also serve to establish correspondences between ancient and modern towns. Morales could reconstruct an ancient name by detecting phonetic similarities with a modern name, for example, noting that the ancient city of Larissa, a small expanse of land with a single house and no surviving antiquities, corresponded to the Carixa of modern times. He attributed the phonetic change to the passage of time. Finally, he could study Arabic etymology and compare its meaning with the actual formation or landscape in question, as in the case of the Guadalquivir (big river valley), which was an apt descriptor for the waterway and the surrounding territories that the ancients had known as the Betis.

The "Moors," Morales explained, changed many of the names of Spain. He elaborated extensively on some of their etymologies. Almagro, for instance, "a principal town and head of the fields of Calatrava" was named after the Arabic for "acid water, which is true about almost all of [the water] of that place." They named the town Alcántara in Extremadura after the word for "bridge," as there was a wonderful bridge in that region. Many smaller locales between Alcalá de Henares and Guadalajara likewise possessed Arabic names "that very much agree with the places or other properties of the land." Guadalajara meant "river of stones," and the Henares River, when it crosses that area, was full of rubble and stones. The dwellers of the region were famous for their goat's milk butter, especially those of Irepar, which in Arabic meant "milk or fat of goat." The town of Buje, located deep in a valley between hills, was named after the word for hills. Likewise, Benalque meant "house of wine," and it was a place where much wine was produced. Morales learned some of these Arabic etymologies from his friend and correspondent Francisco de Medina Mendoza (1516–1577), whom he described as "a principal man of Guadalajara."[89] Medina had become blind, yet, Morales claimed, "all he lost in sight, he gained in his wonderful memory."[90]

Morales's main concern was to identify Roman settlements that had become concealed by Arabic toponyms. His method openly acknowledged that Arabic had to be reckoned with and incorporated into the study of Spain's, and in particular Castile's, history. To him, Arabic served mainly as a useful tool to reach back toward earlier, more desirable origins, but on occasion he admitted the deliberateness and appropriateness of the Arabic place-names.

Some cities, like Morales's beloved Córdoba, even possessed an extensive Muslim patrimony that amplified their Roman legacy.[91] When Morales described the fertility and beauty of Córdoba, like Céspedes he emphasized its temperate weather. Córdoba possessed magnificent monuments such as the "very religious monastery of the discalced Franciscans named Arrizafa, which in Arabic means royal orchard; and this is a place of enjoyment and freshness that very well honors its name."[92] He claimed that it was no wonder that when the Arabic captain Muza, a reliable witness on account of his social stature, left the city after the Christians conquered it, the Muslims could do nothing more than lament the terrible loss of the city.[93] Córdoba's fertility and its propitious landscape provided a suitable environment for people of intelligence and talent to flourish, regardless of their religious commitments. In the city numerous martyrs had immolated themselves in defense of their Christian faith after the Islamic conquest of 711, but the landscape had positive effects even on the infidels, enabling the famous philosopher "Averroes, and with him Abezoar, Rasis, Abenragel, and many others" to achieve great things. Morales conceded that "even though they were Moors, they were born in Córdoba and the fact that they were infidels does not detract from their greatness and their high accomplishment in natural goodness."[94] Before 1609, when, as Mercedes García-Arenal and Fernando Rodríguez Mediano have argued, the expulsion of the moriscos made the use of Arabic among Spanish historians a less loaded issue, Morales framed his discussion of Córdoba by praising the city. By relying on the tradition of *laudes hispaniae*, Morales highlighted the powerful effect of Spain's natural goodness even on infidels and their language, thus tempering possible objections over the use of Arabic toponyms in his reconstruction of Spain's past.

In other sections of the *Antiquities*, Morales pointed to what could be possible remnants of the first language spoken in the peninsula. Linguistic traces of the primordial tongue remained in this most ancient of place-names, which Morales believed either to begin with the prefix *ili-* or to conclude, as Poza had argued, with the suffix *-briga*. The city of Iliturgi, for instance, was often cited by ancient authors. Besides in Pliny and Ptolemy, some of these place-names were mentioned in the works of Livy and even Polybius. Morales came to the conclusion that "Ili in the ancient language of our Spaniards meant basically, place, village, or city, or words to that effect as we see many places that in their names of those times have this word in the beginning." This could be seen, for instance, in the names Iliturgi, Ilipa, Ilipula, and Iliberi. Morales concluded that some claimed "that Briga in our ancient language meant this same thing." However, Morales conjectured that perhaps the prefix *ili-* belonged only to the ancient language of Andalusia, since he was able to trace these lexical elements only there.[95]

While Morales understood the presence of Arabic names throughout Spain as one of the lingering consequences of a medieval political rupture, the Arabic lexicographer and Franciscan Diego de Guadix (?–1615) conceived of these words as vestiges of a much more ancient history. For Guadix, Arabic exceeded any other language in the world in antiquity "because it is the Hebrew language, though corrupted, and the Hebrew language was the one spoken by Adam, Noah, and Abraham."[96] More than a thousand years before the birth of Muḥammad, the world was already replete with Arabic verbs and nouns. The threats assailing Arabic in the sixteenth century compelled Guadix to finally clarify the language's genealogy and prove its nobility as the most immediate ancestor of Hebrew.

In his emphasis on the antiquity of Arabic, Guadix had something in common with the forgers of the Lead Books of Granada, spurious documents attributed to first-century Christian martyrs in an effort to place the arrival of Arabic in Iberia centuries before the birth of Muḥammad, the rise of Islam, and the arrival of

Muslim conquerors from northern Africa. The ultimate goal of the forgeries, as García-Arenal and Rodríguez Mediano have shown, was to demonstrate the antiquity of moriscos and conversos and to dissociate the cultural characteristics of these communities from religion.[97]

Rather than focus, like the forgers, on the putatively ancient arrival of distinguished Hebrew- and Arabic-speaking Christians in Iberia, Guadix emphasized the similarities between Hebrew and Arabic, outlining a theory of linguistic corruption which posited a greater proximity between the two languages. Greek and Latin, in this monogenetic understanding of linguistic origins, emerged much later in history of humankind.

This argument was not novel; it had already been made in an anonymous grammar published in Louvain in 1559. The *Grammar of the Vulgar Language of Spain* (*Gramática de la lengua vulgar de España*) sought to explain the particularities of Castilian to students of this language. The author's hierarchy of languages of Spain employed two criteria: a particular language's proximity to Hebrew and the availability of works in it. After Basque, the author contended, the second language of Spain was Arabic, "which is truly Hebrew." This language possessed its rank "not only because of its ancient and noble descent" but because "many Spaniards had composed in it many useful books in all the liberal arts."[98] Castilian, by contrast, came fourth, after Basque, Arabic, and Catalan.[99] In sum, the author, like Guadix, assigned Arabic more prestige than Castilian on the basis of its identity with Hebrew.

While controversies about the Lead Books of Sacromonte of Granada long raged in scholarly circles in Spain, Guadix was appointed interpreter of the Arabic language tribunal of the Inquisition in the city of Granada and its kingdom in 1582. Singled out for his alleged linguistic expertise, he traveled to Rome in the 1590s to serve as an interpreter and translator for the papacy.[100] Around this time, he composed his *Compilation of Some Arabic Names that the Arabs Gave to Some Cities and Many Other Things* (1593, *Recopilación de algunos nombres arábigos*). In the work's inaugural pages, Guadix presented a number of disclaimers in anticipation of criticisms of his etymological artistry. According to Guadix, because Arabic resembled Hebrew, the first language of humankind, more closely than Greek, Latin, or Basque, it must predate them. Consequently, Arabic must have been spoken in Spain, Italy, France, and the rest of Europe prior to the birth of Muḥammad, the emergence of Islam, and the Islamic conquest of Visigothic Hispania in the eighth century. If a word appeared both in Arabic and in Castilian, Italian, or any other language, it was incorrect, given Arabic's greater age, to claim that Arabic borrowed it from any of these derivative languages. Even though their primary identity might be difficult to detect "because their letters, syllables and accents are so altered and moved," all these words were in fact originally Arabic locutions.[101]

Using this linguistic chronology, Guadix rejected the Greek and Basque etymologies of the name of Spain. He argued that scholars like Morales and Poza should have known that Arabic was, in fact, the most primitive tongue spoken in ancient Iberia and thus the source of the country's name. Hispania was a composite word made up of *ex*, or thing, and *bania*, or building, together signifying "thing that has been built." Ex-bania, with the passing of time, became the corrupted word España. As this example showed, time brought about the decomposition and corruption of words. This process was aggravated by the fact that those who did not know words' true meanings (or even how to articulate Arabic sounds) progressively transformed them. Reconstructing the original uncorrupted forms required the etymologist to give them "a thousand turns . . . or to guess, in order to divine their meaning[s] and integrity in Arabic."[102] Even though the process might seem arbitrary, it was supported by the genetic relationship that existed between Arabic, Hebrew, and their descendant languages. The verb *abrigar*, for instance, which Poza had believed to be a Basque suffix of Spain's most ancient language, was clearly a composite of two Arabic units, *berr* and *gar*. In Arabic, *berr* meant "field or desert," and *gar* meant "cave or lair." The word *berrgar* therefore signified "field cave," and "to this algarabía," Castilian speakers prefixed the sound "a," turning the word into *abrigar*.[103]

Another important proof of Arabic's antiquity and its independence from Islam was the fact that Arabic toponyms persisted even in places where Iberia's Muslim inhabitants had never set foot. When the Spanish conquerors arrived in the Caribbean, they found many words that, as Guadix could easily show, possessed Arabic ancestry. The plant *caçabí*, for instance, which the Indigenous people of Hispaniola ate, originated in the Arabic *caçab*, which signified "reed." Likewise on the basis of phonetic similarity, Guadix contended that the word *cacique*, "lord of the town," emanated from the Arabic *caciq*, which meant "religious [man].'"[104] The name was appropriate because as the chronicles reported, "the principal lord of the town, while ruling over the republic, also had to teach religion and good customs." Even the names of regions had Arabic origins: like *abrigar*, Peru also came from the Arabic *berr*. In this case, however, the word had been transformed differently, the (u) in the end corresponded to the "the third person [su]ffix," so together the word signified "his field or desert."[105] The origins of toponyms like Guatemala and

Mexico, and of names of objects like the boats called *canoas*, could similarly be located in Arabic, proving incontrovertibly that in these distant lands where no Muslims had ever been, and which Christians had only recently reached, Arabic had once been spoken. It was the original language of the Americas just as it was of Iberia.

Challenges to the Study of Toponyms and the Etymological Method

In 1592, the Jesuit historian Juan de Mariana took issue with the idea that toponyms contain immutable traces of their founders. Mariana believed that based on the name alone it was incorrect to assert that, for instance, the Portuguese region of Setubal was first established by the biblical king Tubal. While Mariana did concede that Tubal had settled Spain, it was not possible to determine where he had established himself, what language he had spoken, and what rulers had succeeded him.[106] Many authors, both Spanish and foreign had made grave errors by following etymologies and the writings of Berosus (one of Annius's authors). "What else is it," Mariana asked, "but nonsense and error, to reduce the origins of Spain to Latin derivation and in this way tarnish its venerable antiquity with lies and nonsensical dreams as these [scholars] do?"[107] By the early seventeenth century, as Mariana's reflections demonstrate, doubts had emerged regarding the reliability of the study of etymologies and their uses in historical writing. Detractors condemned the practice, arguing that the correspondence between toponyms and the intention of their earliest creators was impossible to prove.

In his vast compilatory work *Republics of the World* (1575, 1595, *Repúblicas del Mundo*), a rare history of the Americas to be published after 1556, the Augustinian friar Jerónimo Román Zamora had already condemned etymological derivations. Román criticized writers that connected Indigenous American communities to ancient founders based on linguistic evidence alone. Román pointed out how onomatopoeia was a flawed method for establishing lineages. He compared many examples of etymological proofs to reach his conclusion. A certain jurist, Román complained, for instance, claimed that Indigenous Americans descended from ancient Hebrew speakers because of linguistic similitude. This jurist offered as evidence the name of a former queen of Hispaniola, Anacaona, who was known as a skillful performer of *areitos*. "Ana," according to this author, meant "in the Hebrew language graceful, or merciful, or [she] who sings, or answers." Román believed that this was flimsy evidence. In his reading of the early histories of the

Americas, he found words that resembled Latin dictions in the Yucatán and words that sounded "Tuscan, French, and Spanish" among other nations. He determined that it "could not be said that [Indigenous Americans] descended from Europeans," even if to some it seemed likely. Róman's conclusion was unambiguous: "Nothing can make history less truthful than conjecturing, if [the conjecture] does not have first a truth upon which to establish itself."[108]

The grammarian Bernardo de Aldrete also believed the etymological method was a "risky business" because place-names, like languages, were subject to relentless and unpredictable change.[109] In *The Origins and Beginning of the Castilian Language* (1606, *Del origen y principio de la lengua castellana*), Aldrete criticized scholars who relied exclusively on etymological derivations to prove either the primitive nature of the Basque language or the identification of certain geographical spaces in Spain and the Americas with biblical toponyms.[110]

For Aldrete, phonetic similarities were insufficient to establish historical concordance. The name Peru, which had inspired the Arabic etymologies of Guadix, for instance, not only did not come from Arabic but had also been misinterpreted by those who traced it back to Hebrew. According to Aldrete, Peru was not a corrupted form of Parvaim, and there was no evidence to show that the "Gold of Ophir was brought to King Solomon from Pirú." José de Acosta (1539–1600) and Garcilaso de la Vega (1539–1616), two authorities on the languages and histories of the Americas, also denied the Old World etymology of Peru. Rather, they asserted that when Spaniards reached South America, they asked an Indigenous man what land they were in and "without understanding, he answered *Beru, Pelu*." The inhabitants of those territories referred to their kingdom as the Tuantinsúiu, which signified the four parts of their reign, and had never used the word Peru to describe their domains.

Francisco Cervantes de Salazar (d. 1575), the first official chronicler of Mexico City, made a similar claim about the toponym Joktan. When the Spanish first reached the coast of Mesoamerica and the Yucatán Peninsula, they met "certain men who, when asked the name of the large town nearby, responded by saying 'Tectetlan' which meant, according to him, 'I do not understand you.' The Spanish, thinking that the town was called this way, corrupted the word and have referred to the region as Yucatán down to the present."[111] Names, then, could be assigned haphazardly; they could result from the errors of name-makers and were not always vestiges of the past.

Aldrete also argued that languages borrowed words unscrupulously. The dense etymological histories of the previous decades had shown that Castilian itself contained lexical units from Gothic, Arabic, Greek, and even the languages of the Americas. Toponyms, then, might not necessarily represent a direct path toward origins, since a locale could be named with borrowed or translated words, presenting numerous and unpredictable intermediary factors that muddied the relationship between the intention of the namers and the ancient or original meaning of a word. For Aldrete, etymological arguments were therefore a dangerous topic to deal with "because [one] walked only on . . . uncertain proofs, depending on words so inclined to changing."[112]

Aldrete argued that the reasons that compel people to name places are diverse and that words change in response to a multiplicity of factors, which are often unaccountable or unpredictable. Furthermore, as Kathryn Woolard has shown, Aldrete's investigations of the causes of linguistic change and their correlation with political, climactic, and social conditions led him to a drastic conclusion. Origins did not determine the future of communities. They did not have an eternal explanatory value. Under the proper circumstances, Woolard writes, Aldrete found that "communities give up their deeply held languages and customs, acquire new ones, and form new social bonds, and loyalties, to the point of becoming indistinguishable from former enemies."[113] Ancient Iberians adopted Latin just as the inhabitants of the Americas were abandoning their languages in favor of Castilian.

While Aldrete sought to understand the specific principles that governed linguistic change, he challenged the idea that linguistic remains could provide a reliable trail to origins, for "words alone are bound to be forever, being the lightest of things, lighter than the wind, the most subject to change." The study of ancient names and their meanings could not serve as proof of origins unless their presence and usage was attested to in multiple "truthful histories," and even then they were to be approached with much caution. He who takes names as the sole proof of origins, Aldrete concluded, "fools himself truly, to seek in the most unstable and meager thing, perpetuity and firmness."[114]

Conclusions

Etymological histories proliferated in the religious and social controversies of the sixteenth century, but the method was used with contrasting objectives. Some scholars endeavored to integrate Jewish and Muslim converts, and their languages, into a general history of Spain. Others found it necessary to explain away the significance of the many centuries of Islamic rule for Spanish identity and to justify the many vestiges of Arabic in the Castilian language, since speaking this language was now prohibited and a marker of religious alterity. To promote their vision of Spain's ancient past, these writers developed innovative methods of understanding and theorizing linguistic change. Concepts like "linguistic corruption," the idea that languages mutate over time, and that Hebrew was the first language of humankind were more or less omnipresent in these histories. However, the specific ways through which languages changed in response to historical transformations and how societies should regulate their languages varied from author to author.[115] Etymologists drew from different textual traditions, and even ideas about climate and its relationship to the human body, to make their arguments. Increasingly after the 1550s, Spanish authors also relied on the knowledge that circulated in the early printed histories of the Americas. As their histories accumulated into an interrelated body of knowledge, it became obvious that a single word, such as Hispania, Peru, or *briga*, could possess multiple and equally plausible trajectories.

In the long run, the sheer mass of linguistic polemic, the perspective offered by the evolving linguistic landscape of the Americas, and the many contradictory accounts that linked a word to particular histories all destabilized the idea that linguistic vestiges could transmit the original qualities of their earliest speakers. Challenges to the etymological method forced scholars of subsequent generations to employ other strategies to contend with Spain's linguistic diversity and to highlight and explain the legacy of its Arabic and Hebraic traditions.[116]

By the eighteenth century, the etymological method had fallen into disrepute. The Royal Academy of Language's *Diccionario de Autoridades* (1726–1739) dismissed the practice as almost always conducive to mistakes.[117] The academy added that the study of etymologies was difficult and, what was more, that it was "hopeless," because "the most advanced progress of this study is to gain a useless erudition of knowing the root of one word, which is generally achieved through a method that resembles rather an apparent divination."[118]

As Bibliander had already noted in the sixteenth century, however, etymologies, and the promises they seemed to offer, were hard to abandon. In his *Origins of the Spanish Language* (1737, *Orígenes de la lengua española*), Gregorio Mayans y Siscar sought to reinvigorate their study by establishing a method with a canon of specific rules and principles that would allow the etymologist to correctly find the origin of a word.[119] Mayans believed that knowledge of a word's true origin was essential because primitive names corresponded to the objects that they signified. This was why Plato and even Aristotle, in his *Metaphysics*, had applied themselves to the study of etymologies. "If we knew the meanings of primitive names,"

Mayans explained, "there would not be anything that presented itself to our gaze that we would not recognize later."[120] This power of primitive names was especially true of Hebrew, the first language of humankind, and its cognates. Mayans combined the theory put forward by Joseph Justus Scaliger (1540–1609), that there were a limited number of *linguae matrices*, not connected to one another and from which the others derived, with Juan de Valdés's ideas about categories of words that languages borrow in response to the particular interactions between societies.[121] A sensible etymological method demanded that the etymologist trace patterns of sound changes in a particular tongue and then find them attested in books or glossaries. Only then could the student of etymologies arrive at an understanding of the structure of words and thus determine their primary letters and servile consonants, which were more likely to mutate.

Rigorous etymologies also required knowledge of history, as Valdés had proposed in his study of language, to comprehend the contacts and political events that would have facilitated linguistic mutations. In the case of Castilian, it forced Mayans to establish hierarchies that identified in descending order of influence the discrete languages, like Greek, Gothic, or Arabic, from which Castilian had borrowed many words and made them its own.[122] Yet, for all of Mayans's faith in etymology, he still asserted that it was a separate endeavor from the search for historical truth. In 1737 Mayans claimed that the "etymological art is not part of history." The discipline of history demanded a study of "the truth of past events and to learn how to accommodate them . . . to public teaching."[123] In so doing, he refuted the notion that searching for the origins of a language was the worthiest historical enterprise, or that it could lead, as many had argued in the preceding centuries, to incontrovertible proofs about the origin of a nation.

CHAPTER THREE

Language and the Ancient History of the Americas

The judge Alonso de Zorita began his history of the viceroyalty of New Spain by presenting the etymology of that territory's Indigenous name, Anahuac. According to Zorita (c. 1512–c. 1585), *anauac* meant "large land enclosed or surrounded by water." The word, he explained, was composed of "*atl*, which means water, and *nauac*, which means inside or around." Mexicans, Zorita added, refer to the whole world as Çemanauac, from *çem*, a "copulative word," and *anauac*. This compound word "means everything that is under the sky, without making any division, in accordance with the true meaning of the word *çem*." The name was appropriate because the whole world was, in fact, surrounded by water.[1]

In his choice to open with Anahuac, as in many others, Zorita followed the direct example of Fray Toribio de Benavente (c. 1482–1569), better known as Motolinía, one of the first Franciscan missionaries to arrive in Mexico in 1524, shortly after Hernán Cortés conquered Tenochtitlan. Zorita, however, provided additional reasons to justify the extensive etymological attention that he dedicated to Indigenous words. He explained that the Indigenous inhabitants of Mexico had great skill in naming things "in accordance with the quality and property" of each thing signified. This was a noteworthy ability that Zorita related to Adam's naming of all animals in Genesis.[2] To name objects in accordance with their properties was a skill that Plato had praised in his *Cratylus* as a sign of "divine wit" and that ancient authorities like Cicero, Pythagoras, Priscian, and Aulus Gellius, among many others, had presented as essential to the knowledge of all things. Because of this tradition, Zorita conferred great explanatory power on the etymology of Anahuac.

Zorita's introduction conveyed the importance of local nomenclature in shaping all aspects of his historical reconstructions. He focused on understanding pre-Hispanic institutions of governance and tributary systems by glossing local terminology. The ability of names to convey the functions performed by different members of Mexica society compelled Zorita to comment, more than once, on the characteristics of Nahuatl (which the Spaniards often referred to as the Mexican language), sometimes even in his administrative writings, where linguistic discussions appear out of place from a modern perspective. For instance, in an official response to a royal query about the tribute that the Mexican rulers had extracted from their subjects in pre-Hispanic times, Zorita thought it relevant to include another aside: "And it is appropriate to note something about the names," he stated, "and it is that the titles and trades, and the names of towns and mountains ranges or hills, they impose them in conformity with the quality or fertility or sterility of what was abundant in each part." The names of the regions of Michoacan and Tehuantepec, for example, meant, according to Zorita, "land of many fish," and "snake mountains," respectively.[3] The etymologies offered clues to the settlements' natural characteristics and the resources they contained.

Zorita was among the Spanish scholars, missionaries, and crown officials who devoted themselves to studying the human and natural history of Indigenous America in the middle of the sixteenth century. Their works, completed after 1556, circulated mainly in manuscript. Zorita finished the *Account of New Spain* (*Relación de la Nueva España*) in 1585. Tight restrictions on the publication of books about the Americas limited the dissemination of his writings, with a few notable exceptions, to networks of administrators, members of the missionary orders, and scholars who were either connected to the Council of Indies or under the direct patronage of Philip II.[4] Their mostly unpublished works belong to a variety of genres, ranging from official memoranda to responses to royal questionnaires, to historical accounts (*relaciones*), to works concerned with religious conversion, and to the collection of Indigenous antiquities in various types of compendia. In spite of these limitations, Zorita and his colleagues saw themselves as contributing to broader debates about some of the most central linguistic and historical questions of their time. Scholars like Zorita and the secular priest Miguel Cabello Valboa (c. 1535–c. 1606/1608), among others, also explicitly addressed the polemical theories of their peninsular counterparts to explain Spain's linguistic diversity and to identify the language of its most ancient inhabitants.

This chapter traces debates in the American viceroyalties about language and its relationship to origins and history writing. It shows that scholars concerned with the Americas also turned to the archives of language. In their efforts to mine linguistic information, these scholars applied the same methods, such as the etymology, the questioning of local authorities, and the use of conjecture in reconstructing Spain's past. Etymology was a common tool, its prestige assured both by the long tradition of applying it to ancient European and Semitic languages and by its flexibility. In theory, any language could be subject to its analytical methods. The collection of etymological knowledge even became official royal policy with the distribution of the questionnaires known subsequently as the *relaciones geográficas* or *topográficas*.[5] The questionnaires were distributed between the late 1560s and the 1580s in the Iberian kingdoms (Castile) and the American viceroyalties.

The scholars who pursued research in the Americas, in contrast to those who wrote about Spain, had to work solely with nonalphabetic forms of storing and transmitting historical information. In Spain, scholars relied on both oral and written sources, including written sources that they presumed had once been oral. As this chapter demonstrates, for the writers who studied American materials, oral knowledge could be translated into written evidence. This was true even of scholars who defended the written word as one of the most effective forms of protecting information from the passing of time.

To make sense of oral traditions in the Americas, scholars drew on an ample set of approaches, including antiquarianism, the collation of legal testimony, the use of historical conjecture, and the compilation of sayings or proverbs. In so doing, they engaged in forms of experimentation with oral traditions similar to those taking place in Spain and other parts of Europe to study vernacular languages and their histories. The scholars' attempts to parse American oral traditions, however, would lead to new conclusions about the existence of formal systems of oral historical transmission that relied on specialized groups of knowledge keepers.

These linguistic studies differed from the histories of the Americas published in the first half of the sixteenth century, studied in chapter 1, which relied on word assemblages that had been textually or orally transmitted and were generally detached from systematic attempts to understand the workings of various languages. In contrast, the accounts completed after the middle decades of the sixteenth century could make use of grammars, lexica, and various types of Spanish and Indigenous linguistic experts. Missionaries of the Mendicant, and later of the Jesuit, orders worked to "reduce" what they identified as some of the most widely spoken languages of each region to rules that could be taught to their peers in their efforts to advance conversion agendas.[6]

Linguistic evidence embedded in the names of things held a special place in these historical reconstructions. Like their Iberian counterparts, scholars concerned with the Americas used linguistic evidence to make arguments about the origins and future of various speech communities. Through arguments about language, they debated the political and religious integration of formerly non-Christian subjects into a Spanish monarchy of global pretensions. Unlike the controversies surrounding the status of speakers of Arabic in the peninsular kingdoms, the debate about the Americas was characterized by universal agreement that the Spanish king should impose Christianity on his Indigenous American subjects and provide them, at least in theory, with his protection from excessive tributary obligations and the abuses of local interest groups. Vassals of the crown, the Indigenous inhabitants of the America were rendered by law into something akin to legal minors, subject to the king's paternalistic protection.[7]

Intertwined commitments to convert and to protect lay at the foundation of Spanish claims of possession over the Americas.[8] In practice, however, throughout the second half of the sixteenth century, a perennially bankrupt Philip II pursued economic reforms to maximize American revenues by substantially increasing Indigenous tributary obligations, both in kind and in labor. These reforms exacerbated the demographic collapse that was already underway.[9] In the works of authors like Zorita and missionaries like Bernardino de Sahagún (c. 1499–1590), language became a live repository of constantly fluctuating historical evidence that demonstrated to Spanish authorities how the conquest and the colonial project had, irretrievably and almost always negatively, transformed Indigenous societies.

A Polyglot Word

The astounding linguistic diversity of the Americas became a commonplace in histories of the viceroyalties, which also abounded in the paratexts (prologues, dedicatory epistles) of the missionaries who wrote lexicons and grammars of American languages. Various contrasting views emerged by the 1550s about the relationship between polyglossia, the power of pre-Hispanic rulers, and the use the Spanish king could make of languages as a resource, both symbolic and material, to govern, convert, and extract resources from these new territorial possessions.

To harness the information embedded in language for historical writing, scholars studied the origins of toponyms, the names of Indigenous deities, and all social, political, and natural terms that might illuminate Indigenous customs. Francisco Cervantes de Salazar's unpublished *Chronicle of New Spain* (*Crónica de la Nueva*

España) is replete with etymologies that provide his intended European audience with images of events, landscapes, and people.

For Cervantes de Salazar, the words themselves contained important traces of ancient Mexican peoples, their ways of life, and origin. For instance, he presented two distinct accounts of Mexico City's Indigenous name, Tenochtitlan: "There are some that say this renowned city in this New World took its name from its first founder, who was Tenuch, second son of Yztacmixcoatl, whose children and descendants then populated this land of Anauac, which is called now and will always be called New Spain. Also, others say that it was called Tenuchtitlan on account of the *tuna* [cactus] of *grana* or *cochinilla* that is born in other kinds of *tunas*. *Nuchtli* is the color of *grana*, so bright that the Spaniards call it crimson."[10] Tenochtitlan was not the name of the whole city but only part of it. The Indigenous residents of this city referred to it as Mexico-Tenochtitlan. This composite title, Cervantes de Salazar asserted, was kept in subsequent Spanish royal provisions. Mexico meant "the same as water or spring, for the many water springs that exist around the mainland." Mexico had other possible meanings: "The first founders, called themselves mexiti, and even those who originate in that neighborhood or people are called mexica." Cervantes de Salazar explained that the "mexiti" took their name from their main deity, "Mexitli, who is the same as Huiçilopuchtli." Cervantes de Salazar believed that these interpretations were worth relating even if he could not discern which of the traditions was the most accurate.[11]

A trained Latin scholar and an admirer of Hernán Cortés, Cervantes de Salazar arrived in Mexico in 1551. He eventually won the post of Latin professor at the newly created University of Mexico City (1553) and also was appointed the city's first official chronicler.[12] His linguistic research went beyond etymological evidence. Before departing for Mexico, he was committed to devising ways to exalt the Castilian language. He was interested in the works of Juan Luis Vives. Cervantes de Salazar is perhaps the Valencian humanist's first biographer, and he translated a few of his works into Castilian. One of Vives's well-known interests was the education of children, the relationship between Latin and vernacular languages, and the role that these languages were to play in literate culture and in everyday life.[13]

Given the attention that he devoted to Vives's writings and his interest in the relationship between Latin and Castilian in the Iberian Peninsula, it was natural for Cervantes de Salazar to comment about the linguistic diversity of New Spain: "It seems, as experience teaches us and the Divine Scriptures manifest, that even to these distant parts the confusion of tongues spilled because of the sin of pride.

For the number of languages that there are in New Spain, cannot be counted." In a town called "Tacuba," only a "league away from Mexico," Cervantes de Salazar asserted, six different languages were spoken. The first was Mexican (Nahuatl) "although corrupted, because it is spoken on the mountains." In this assertion about linguistic corruption away from urban centers, Cervantes de Salazar conveyed the Spanish conviction that languages reached their most elegant form in cities, which were the cradle of civility. In addition, the people of Tacuba spoke the "otomí, the guata, the mazava, the chuchume and the Chichimeca" languages. In a small geographical extension, many languages coexisted.[14]

The difficulties that this multilingualism might have created were averted by Nahuatl's wide currency: "In all of New Spain and outside of it, the Mexican language is so universal, that in all parts of it there are Indians who speak it, like Latin in the kingdoms of Europe and Africa, and the Mexican language is so esteemed as French is in Flanders and in Germany. In this way, the lords and principal men learn it to ask and answer the questions of the Indians from distant lands." Nahuatl's spread evidenced the power of the last pre-Hispanic ruler, Moctezuma II, whom Cervantes de Salazar described as the emperor of a vast polyglot kingdom that comprised many nations. Cervantes de Salazar related that "it was [owing to his] greatness and an argument for [his] great majesty that when an embassy appeared before Montezuma or before a prince not as great as him, those who brought the [embassy] would speak in their own language." An interpreter who understood this language would then convey the message to another and so forth, "and in this manner it would pass through six or seven interpreters, until it reached Montezuma in the Mexican language and answering in that same language, the answer was also transmitted to the embassy through the same interpreters."[15] This chain of interpreters bespoke the power that this empire had achieved since it demonstrated all the language groups over which Moctezuma had ruled.

Even though he relied on Indigenous sources to write the first sections of the *Chronicle*, Cervantes de Salazar was critical of Indigenous scholars' ability to maintain and transmit historical information. He complained that while paintings existed to store these histories, the images could be read by only very few. As a result, "only the priests understood something," and in "the passing of the years and the months there was no certitude." The "Indians were so barbarous," he claimed, "that they lacked the principal order, which is writing and knowledge of the liberal arts, which are what lead and guide men to understand the truth of things." He also compared the images to Egyptian letters. Cervantes de Salazar called both systems of inscription "paintings that those who are absent use to convey their concepts." He did not specify whether one form of inscription was superior to the other but emphasized that only few could access the information that these paintings contained.[16] In doing so, he was drawing from a tradition that treated symbolic and hieroglyphic writing as primitive.[17] Nonetheless, Cervantes de Salazar accepted the information that these etymologies contained.

The Franciscan friar Alonso de Molina (d. 1585) addressed the challenges of polyglossia in his work but encouraged the Spanish king to cultivate Indigenous languages not only for religious conversion but also for governance. He presented his ideas in the prologue of his *Vocabulary in the Castilian and Mexican Language* (1571, *Vocabulario en lengua castellana y mexicana*), the first dictionary ever printed in the Americas. A Franciscan friar from Extremadura, Molina arrived in New Spain as a child in 1522, where he learned Nahuatl from interacting with Indigenous children. Like other missionaries, Molina defended the idea that learning languages was a way to surmount the evils engendered by the Tower of Babel. In doing so, missionaries embraced an interpretation of the miracle of Pentecost (when Jesus's apostles miraculously gained the gift of languages) in which grammar and linguistic competency were the best tools to remedy the enduring historical sins brought about by human pride in Babylon. In accordance with the apostolic ideal, the prologues of these two works emphasized the importance of preaching in tongues. Molina described how faith was achieved through listening. Preachers, therefore, must speak the languages of their audiences, "because otherwise [as Saint Paul himself says] he who speaks will be seen as a barbarian." Moreover, "to declare the mysteries of our faith, it is not enough to know the language . . . but to understand the property of their words and their ways of speaking."[18]

For Molina, learning Indigenous languages mattered not only for conversion efforts. It was also fundamental for the Spanish Crown's governance: to govern justly and maintain good order, administrators must understand their subjects. He provided an immediate example. "The anguish of our Spain was not small when the undefeated Caesar [Charles V] began to reign, since he could not understand his [people] due to their different languages. And on the contrary, there was great relief and joy when he could understand and speak our language without interpreters. Because often, even though the water is clean and clear, the conduits through which it passes make it turbulent."[19] In contrast to Cervantes de Salazar, Molina portrayed interpreters negatively. This passage also demonstrates how, in the 1550s, Molina believed that the colonial elite had to understand the language of the vanquished for both governance and religious indoctrination to succeed. Yet Spanish administrators did not follow this path and resorted instead to the labor of interpreters to impose justice. This was the case even in New Spain, where documents in Indigenous languages were produced in significant numbers and admitted in court. However, Molina's prologue reminds us that in the 1550s the adoption of Castilian as the language of administration in Indigenous communities was still not universally accepted by Hispanophone intellectuals.[20]

Even the earliest writings about the Incas discussed the political and administrative cohesion brought about by what scholars of early modernity referred to problematically as "general languages"—that is, languages they believed to be widespread in a region. However, as many chroniclers were careful to point out, the Spanish believed that Quechua, the language of the Incas, coexisted with many other languages in various hierarchies. The early grammarian of Quechua, Domingo de Santo Tomás, one of Pedro Cieza de León's main informants, also praised Quechua's spread over vast mountainous domains. Santo Tomás justified the composition of his lexicon by highlighting Quechua's "abundance of words." He also emphasized how adeptly Quechua's words conveyed what they signified. Their etymologies forcefully brought to the mind features of the objects that they denoted. According to Santo Tomás, Quechua naturally possessed "artifice," meaning that the language signified with complexity, and Quechua speakers had "good ways of speaking." In these respects, the language resembled both Latin and Castilian. Santo Tomás even argued that these qualities were prophetic, foretelling that "the Spaniards were to possess [Quechua]." Thus, he conceptualized Quechua as a possession, like a territory, that could become a resource for the Spanish king.[21]

The Etymologies of Philip II: Collecting Place-Names and Languages

Spanish administrators sought to record and catalog the languages of the Spanish Empire through a variety of projects that Philip II sponsored during his reign between 1556 and 1598. Using oral interviews and etymologies to access local knowledge became an officially codified method in the questionnaires known in Spanish America as the *relaciones geográficas* and in the peninsular realms as the *relaciones topográficas*. They continued to be undertaken in Spanish America until the nineteenth century. These censuses broadly aimed to gather information about the realms' populations, administrative divisions, botanical wealth, and languages. Their immediate uses depended on the administrators collecting the information.

The exact origin of the questionnaire has been difficult to establish. It may date back to 1555, when Juan Páez de Castro wrote his *Method*. The questions and the instruction for their distribution are congruent with the program he outlined for the execution of a complete history of the Spanish realms.[22] The lawyer and president of the Council of the Indies Juan de Ovando (d. 1575) put this program into practice when he initiated multiple information-gathering projects that aimed to catalog the laws, languages, geographies, and natural knowledge of the Americas.[23] In 1573, the Council of the Indies sent a questionnaire with 135 queries to the American realms. The following year, the first peninsular questionnaire was sent to Ovando's native city of Coria, in Cacéres, Extremadura.[24] Several questionnaires of varying lengths were tested in the Americas and then in Spain, some containing more than one hundred questions. In 1577, the cosmographer of the Indies Juan López de Velasco (1530–1598) issued a version with fifty queries.[25] The instructions specified that the questions were to be sent to *corregidores* (district governors), local officials, priests, or influential people, and reliable witnesses, to procure the information and then diligently return the responses to the council.[26]

Philip II proclaimed his intentions in the instructions for the questionnaires of 1575, asserting the Crown's need to possess more accurate information about the resources and histories of the places under its jurisdiction.[27] "Having understood that until now a particular description of the towns in these kingdoms has not been made," the king declared, "which is suitable to their authority and greatness, we have agreed to commission this said description and a history of the particularities of said towns."[28] To acquire the necessary information, the first question asked the compiler to "declare and say the name of the town . . . what it is called at present and why it is named in this manner. If it has possessed any other name; and, why it was named in this way if it is known."[29] Five of the fifty questions in the 1577 questionnaire inquire directly about the names of cities or geographical formations, the histories of these names, the changes that they have undergone, and their language of origin.

In other words, the questionnaire investigated etymological matters.[30] Question 1, for instance, stated that in American "towns of Spaniards [*pueblos de españoles*], the name of the province should be established, as well as its meaning in the Indian language and why it is called this way."[31] Question 9 sought more specific details, including the name and surname each city or town has or has had and why it was so named (if known), who named it, and who founded it, under whose order it was populated, the year of its foundation, how many first populated it, and how many it has at present.[32] Question 13 asked for the name of the language spoken by the Indigenous inhabitants of the town.[33] Question 14 elucidated the etymological information by also asking the witnesses about ancient tributary systems, the old Indigenous deities, and "their good and bad customs." The additional insights provided by this question could help to contextualize the meaning of the name.[34]

The *relaciones* offer glimpses into local historical traditions as recorded in language. They also show that Spanish administrators generally preserved the Indigenous names of places; renaming was not part of legal possession of a territory.[35] Even when the name of a patron saint was attached to the ancient toponym, the Indigenous name remained in use. Administrators and scholars alike considered knowledge of the original designation and its meaning useful. This was also common in Spain where once-Muslim cities retained their former appellations, often in Castilianized Arabic, after passing into the hands of Christian monarchs.[36]

The interrogation carried out in the town of Culhuacan, an important pre-Hispanic city and subsequently the location of a prominent mission, exemplifies this principle. In 1580, after summoning the most elderly of the Indigenous men for questioning, the priest Juan Nuñez, who spoke Nahuatl fluently, and the lieutenant of Diego de Paz, also a competent interpreter, proceeded to explain the meaning of the locale's name. Culhuacan, the Indigenous informants explained, "in our language, is 'point of curved mountain.'"[37] The nearby town of Iztapalapan was named in honor of one of its main resources: limestone. The priest, "Doctor Loya, who had administered for very long the sacraments of this said town of Iztapalapan," learned directly from the "elderly and principal Indians of this town" that Iztapalapan meant in *romance* "town situated in a place of paving stone and water." These stones "they call in their language Iztapaltetl."[38] Other towns, their elders would relate, were named after games that their ancestors had invented. The mines of Tlachco were referred to in this way because they were discovered close to the town of Indians who bore the same designation. "In ancient times the Indians of said town," the elders reported, "would play a game with a ball, thick as [a] bowling ball, that in their tongue they call Tlachtli, and in the same Mexican language corrupted, it is called Tlachco." These particular informants offered further insight to their interrogators. "In this land," they proclaimed, "it is very common to take the name of the towns from something significant in the land that is raised, sold, or used."[39] The name thus provided important clues as to the geographical or natural historical features of an area.[40]

Questionnaires in the viceroyalty of Peru similarly convey the etymologies of place-names through the testimonies of Indigenous authorities.[41] In the province of Xauxa, in the Peruvian central highlands, the questionnaire was completed in 1582 through the mediation of Don Felipe Guacra Páucar, a "ladino Indian who had been in Spain, and who is the brother of the principal *cacique* of this *repartimiento* [distribution] of *Hurin Guanca.*" The name Xauxa was imposed upon the province after the Spanish conquest because the Spaniards settled in the town and *tambo* (pre-Hispanic inn) named Hatun Xauxa, and the Spaniards thus named the province Xauxa. Under Inca rule, however, the informants related, "this valley was called *Guanca Guamaní,* because the Inca Capac Yupanqui arrived at the beginning of this valley, and in an open field found a long stone which was the size of a man, and the Indians generally call these stones *guaca . . . rumis.*" So the Inca named the valley Guancas, and the Guamaní means "Valley" or "Province."[42] The Spaniards, then, might have misunderstood the name of the town and applied an Indigenous toponym incorrectly to the whole area on the basis of the name of what they referred to as a *tambo.*

What happened to the questionnaires once they reached Spain is difficult to determine. Following the culture of scientific secrecy espoused by Philip II, the responses were to be consulted only by the royal chronicler cosmographer. While López de Velasco did not use these questionnaires to elaborate a new synthetic cosmographical account, he did preserve and inventory them. María Portuondo has convincingly argued that López de Velasco most likely conceived of the replies in themselves as the "cosmographical corpus, standing as individual and unmediated testaments of the reality of the New World as articulated by first-hand observers." Instead of synthesizing them in his own works, he filed them and recorded them in a careful inventory, readily available for consultation by the members of the Council of Indies.[43]

The testimonies of unmediated firsthand observers, although regulated by the demands of a standard questionnaire, allowed for the incorporation of local knowledge, names, and categories of authority. Moreover, collecting the names of places was a crucial part of visualizing the territories under the rule of the Spanish king. It points to the sixteenth-century preference for collecting and assembling local information under broad meta-categories that allowed for composite compendia. These assemblages, or catalogs, were, in themselves, evidence of royal power.

The Spanish Crown undertook the project of the *relaciones* in the Iberian Peninsula after initial versions of the questionnaires had been tested in the realms of the Americas. From surviving records in the Escorial, where the reports were deposited, it appears that more than seven hundred jurisdictions were surveyed.[44] The first Iberian question was identical to that of the American census. It allowed the eldest and most respected men of a locale to narrate their historical traditions as preserved in the etymology of their town's name. Eighty-year-old Alonso Ro-dríguez was among the elders of the town of Alamo whom the priest Diego Suárez summoned to respond to the two royal commissioners in 1576. According to Ro-dríguez's and his fellow witnesses' memories, the town "had always been called this way and that the foundation and choosing of the name was taken from a big *alamo* [poplar tree] that stood in this place before the establishment."[45] If the dwell-ers of Alamo owed their name to a venerable tree, the residents of the town of Aravaca (Plowcow) claimed that theirs originated in their founder's confrontation with an insubordinate animal. They heard from the elders that the man who founded Aravaca named it this way "because plowing with a cow that did not want to, he could not plow and would say 'plow cow!' [*ara vaca!*] and they have never heard that it was called by anything else."[46] The town of Casarrubios (Blondes' house) was named after its first dwellers, two blond brothers who lived in the set-tlement's first house.[47] Other towns were named after patron saints or to commem-orate episodes of religious martyrdom. The elder men of the town of Santorcaz declared that their people believe that the name of their town originated in their filiation with Saint Torcaz, who came from Rome to preach and convert the gen-tiles of the village centuries ago.[48]

In other localities, some of the witnesses interviewed asserted that the founda-tion of their cities could be traced back to the Romans, as their names proved. In Extremadura, for instance, Juan de Ovando's region of origin, the people of Co-ria informed the king that their city was once called in Latin Cauria. The passing of time and the "diverse inhabitants of Spain," caused the word to degenerate into Curia, meaning "council of judges." The witness explained that he had read in the cathedral's ancient records that the city had once even obtained privileges from Emperor Don Alfonso XI. The Roman antiquity of Coria could also be seen in its numerous surviving stones and their Latin inscriptions, too numerous to describe in the answers to the survey.[49]

The *relaciones* provide a window into how the most distinguished or elderly people of the towns of Castile and the viceroyalties of the Americas remembered their town's history through the etymology of its name. The *relaciones* also capture how historical knowledge circulated between intellectual elites and popular tra-ditions. In Villamanta (near Madrid) in 1576, members of the town's clergy gathered in the presence of the scribe. The witnesses claimed that even though they did not know exactly why the town was named in this way, on the authority of the antiquarian Ambrosio de Morales and the mathematician Pedro de Esquivel (d. c. 1570), they suspected that perhaps it had been founded in Roman times. Philip

II had charged Esquivel with making a complete map of the Spanish kingdoms in 1566. The informants of Villamanta recalled Esquivel's efforts: "Maestro Esquivel, the chronicler of Charles V . . . came to this place to look at the antiquities that there were here and the disposition of the land, and fallen buildings, and stones and the signs on them." Esquivel, "unfolding [his] astrolabe[,] measured it and found the north and declared that this settlement was the real Mantua Carpetana named by the ancient cosmographers and historians and that villa of Madrid had stolen this name claiming to be Mantua the Carpetana and the same claimed the maestro Ambrosio de Morales who succeeded him in this history." According to the witnesses, Morales also came to study the locale's ruins and inscriptions and determined that the farmers who once settled in Mantua probably corrupted its name to Villamanta.[50] The surveys, as this example demonstrates, did not just collect local knowledge; they shaped it. Through the questionnaires, scholarly interpretations were absorbed locally and became part of local self-understanding.[51]

Historical and cosmographical works also relied on the study of place-names. Like other sixteenth-century cosmographers, López de Velasco was interested in toponyms. His contemporary Abraham Ortelius (1527–1598) would dedicate an entire work, the *Synonymia geographica* of 1578, to historical and current place-names, as part of his effort to create historical maps that would aid readers of ancient accounts.[52] López de Velasco's interest in etymologies was perhaps augmented by his direct involvement in the project to produce a complete edition of the works of Isidore of Seville, which prominently included the *Etymologies*.

Spanish authorities cared about names of towns because they understood towns to be the units that made up their kingdoms. As Richard Kagan has shown, the Spanish Empire was an "empire of towns."[53] The histories of these communities contained their customs, the foundation of customary law.[54] The Spanish Crown commanded the further distribution of questionnaires in the American viceroyalties several times during the later decades of the sixteenth century. By 1604, however, the etymological questions that had appeared in the 1577 list were no longer part of the census. This might be due to the fact that López de Velasco was no longer involved in their formulation. The transformations in the questionnaire, however, also coincided with the many intellectual critiques that Spanish scholars articulated about the historical information that etymologies could provide.[55]

Although questions about the origins of names were no longer included in the formal census after 1604, etymologies continued to be useful to scholars who sought to investigate the histories of places. As late as 1639, when the Jesuit Bernabé

Cobo (1580–1657) completed his history of Lima, in order to discover what had compelled ancient peoples to establish themselves in the place that had become the viceregal capital, Cobo employed two methods widely used in the early histories of the Americas: the study of etymologies and the questioning of Indigenous experts. Cobo claimed at first that he did not wish to "trace etymologies" and "obscure beginnings," alluding to the conjectures that Spanish authors elaborated to establish the founders of cities.[56] Even so, he dedicated an entire chapter to the name of Lima.

Looking back on more than one hundred years of Spanish settlement in the Americas, Cobo explained that the Spaniards always established their cities on preexisting Indigenous settlements because they understood that the locals had settled in the best possible lands.[57] Lima was not a purely Indigenous name, he continued: the Indigenous inhabitants of the coast called the settlement Limac, while the mountain dwellers referred to it as Bimac. Cobo believed the reason for the divergence was that when the "Spaniards entered this land only a few years had passed since the Inca Kings of Cuzco had conquered these maritime provinces, introducing their language to their inhabitants . . . and they did not speak [Quechua] as elegantly as those from the mountains."[58]

He also remarked that some pronounced the name as Rimac and others interchangeably with an *L*, Limac. The name might have come from the name of the Rimac River, which is a "participle and means to speak." According to Cobo, this was a good name for the river because it could be heard all over the city, though he added that it was unclear whether the settlement or the river had been named in this manner first. Despite the long disquisition and his praise of the appropriateness of the name, Cobo concluded by having his etymologically skeptical cake and eating it, too. He claimed that in the "antiquities of Indians little clarity can be found and least of all in things that they did not pay much attention to, like the etymology of words."[59] Thus Cobo signaled his distrust of Indigenous etymologies while still relying on the method to reconstruct Lima's origins.

Parsing Oral Traditions: The Experienced Truth of Sayings

Toponyms, and their etymologies, did not suffice to study the early historical traditions of a society. To complement that information—or sometimes to correctly interpret a toponym—the laws, customs, and religious beliefs of a locale also had to be reconstructed. In the absence of written sources, sayings, known in Castilian as *refranes*, could reveal information about ancient times.

The collection of proverbs was a common pursuit among sixteenth-century European scholars, starting with Erasmus of Rotterdam's *Adages* (1500–1536), the earliest and most influential collection of proverbs.[60] Erasmus pursued the collection of proverbs continuously from 1500 until his death in 1536. His book eventually included more than four thousand glossed Greek and Latin proverbs.[61] In the introduction of the *Adages*, Erasmus defined proverbs, described the kind of wisdom that they contained, and explained why collecting them was a worthwhile endeavor. Proverbs, he argued on the authority of Aristotle, were part of the "science of philosophy" because they were "simply the vestiges of that earliest philosophy which was destroyed by the calamities of human history."[62] In this aspect, proverbs resembled antiquities, but they were in fact more resilient than paper or stone. The reason for proverbs' resilience was "that what vanishes from written sources, what could not be preserved by inscriptions, colossal statues, and marble tablets, is preserved intact in a proverb."[63] Proverbs encoded the collective wisdom of societies because of the nature of transmission itself: they were "approved by the consensus, the unanimous vote as it were, of so many epochs, and so many people."[64]

Following the example of the *Adages*, Spanish scholars set out to collect proverbs in vernacular Iberian languages, justifying their undertakings along the lines that Erasmus had laid out. They also sought to highlight the virtues of various vernacular languages. Esteban de Garibay, for instance, assembled and translated into Castilian a collection of Basque proverbs to demonstrate the latter language's richness in moral dictums.[65] In their shared interest in proverbs and the kind of information they could reveal, these scholars resembled Juan de Valdés, whose *Dialogue* had argued that proverbs were one of the main sources upon which to base the standard of spoken languages that were in constant flux.[66]

A disciple of Elio Antonio de Nebrija, Hernán Núñez served as professor of Greek, Hebrew, and rhetoric at the University of Salamanca. He began collecting proverbs as early as 1508.[67] Even though Núñez intended to gloss the sayings to show when they derived from the Greco-Roman tradition, he did not attempt to demonstrate as Erasmus had that ancient sayings could express Christian beliefs and values. Rather, his efforts were directed toward showcasing the richness of Castilian. To do so, he relied on printed compendia of sayings like the *Trezientos adagios y fábulas de Fernando de Arce* (Salamanca, 1533), and the *Trezientos proverbios de Pedro Luis Sanz* (Valencia, 1535), among others. For proverbs in other languages, Núñez drew on books in French and Italian.[68] He complemented these materials with other sayings that he had heard himself.[69]

Núñez's *Sayings or Proverbs in Romance* (*Refranes o proverbios en romance*) was published posthumously in 1555 without many of the glosses or introductory materials that its compiler and editor had intended. Nevertheless, the collection included more than eight thousand proverbs, which were organized alphabetically and not thematically like those of Erasmus. These proverbs originated from Castilian, Portuguese, French, Galician, Catalan, Greek, and Latin.[70] Many were related to individual moral improvement, but others reflected the everyday experiences of lords, doctors, priests, farmers, and other common people, sometimes even including crass or anticlerical statements that subsequent readers expurgated. For instance, a proverb advised the reader to keep away from clerics: "Neither priest, not friar, nor Jew will ever be a friend to you." Other sayings presented misogyny as prudence: "From the sea the salt, from women much evil."[71] Figures 3.1 and 3.2 show how a reader crossed out a saying that explicitly mocked the marriages of couples who enjoyed premarital encounters. Núñez believed that the sayings demonstrated the many qualities of the Romance languages like Catalan, French, and Portuguese. They also captured numerous words and metaphors that had never been recorded in writing before.[72]

Núñez's student and successor as professor of Greek at Salamanca, León de Castro (1505–1585), composed the work's prologue and claimed in its dedication that "paroemiology," the study of sayings, was a worthy pursuit because it offered wisdom on any topic.[73] But many of the *refranes* were the words not of wise men but of common people in the vulgar tongue. To confront any possible objection, Castro offered an erudite rebuttal, noting that Erasmus had already extensively explained the value of collecting Latin and Greek proverbs, and how they preserved a common tradition that could be reconciled with Christianity. The publication was a great success: Núñez's *Sayings* appeared in five editions over the course of the sixteenth and seventeenth centuries.

This collection of proverbs could do more than prove the richness and variety of Castilian and other Romance languages. In the eyes of one of Núñez's students, it proved the pre-Hellenic and Roman antiquity of Spanish wisdom and culture. In 1568, the Sevillian scholar Juan de Mal Lara (d. 1571), a former student of Núñez and Castro, reprinted many of the sayings in his teacher's collection, added new entries, and included longs glosses to the proverbs. Mal Lara entitled his work *On Vulgar Philosophy* (*De la philosophia vulgar*), dedicated it to Philip II, and introduced it with a detailed method for collecting and interpreting sayings.[74] He revealed that Núñez had gone so far as to pay for interesting or novel sayings.[75]

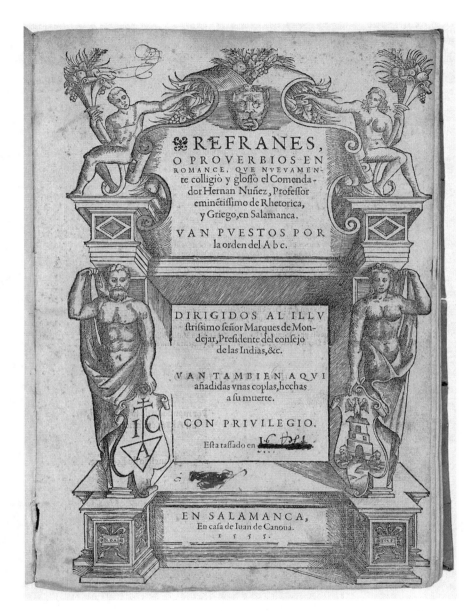

Figure 3.1. Title page of a 1555 edition of Núñez's *Refranes*. Hernán Núñez, *Refranes o proverbios en romance* [. . .] (Salamanca: Juan de Canova, 1555), EXOV 3166.675.21. Department of Rare Books and Special Collections, courtesy of the Princeton University Library.

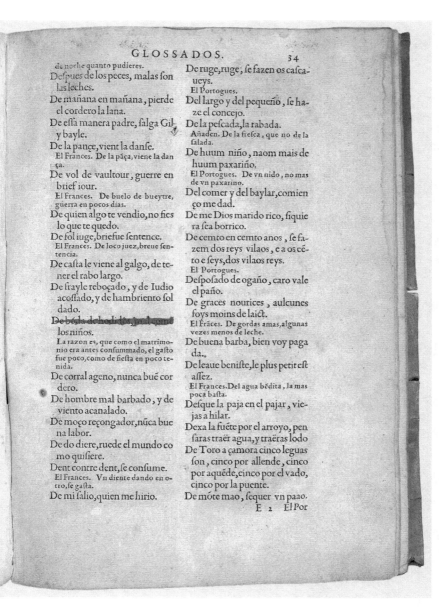

de noche quanto pudieres.

Despues de los peces, malas son
las leches.

De mañana en mañana, pierde
el cordero la lana.

De essa manera padre, salga Gil
y bayle.

De la pançe, vient la danse.
El Frances. De la pãça, viene la dan
ça.

De vol de vaultour, guerre en
brief iour.
El Frances. De buelo de bueytre,
guerra en pocos dias.

De quien algo te vendio, no fies
lo que te quedo.

De fol iuge, briefue sentence.
El Frances. De loco juez, breue sen-
tencia.

De casta le viene al galgo, de tener el rabo largo.

De frayle reboçado, y de Iudio
acossado, y de hambriento sol
dado.

De boda de hodidos, mal come
los niños.
La razon es, que como el matrimo-
nio era antes consummado, el gasto
fue poco, como de fiesta en poco te-
nida.

De corral ageno, nunca bué cor
dero.

De hombre mal barbado, y de
viento acanalado.

De moço reçongador, nũca bue
na labor.

De do diere, ruede el mundo co
mo quisiere.

Dent contre dent, se consume.
El Frances. Vn diente dando en o-
tro, se gasta.

De mi salio, quien me hirio.

De ruge, ruge, se fazen os casca-
ueys.
El Portogues.

Del largo y del pequeño, se ha-
ze el concejo.

De la pescada, la rabada.
Añaden. De la fiesca, que no de la
salada.

De huum niño, naom mais de
huum paxariño.
El Portogues. De vn nido, no mas
de vn paxarino.

Del comer y del baylar, comien
ço me dad.

De me Dios marido rico, siquie
ra sea borrico.

De cemto en cemto anos, se fa-
zem dos reys vilaos, e a os cé-
to e seys, dos vilaos reys.
El Portogues.

Desposado de ogaño, caro vale
el paño.

De graces nourices, aulcunes
foys moins de laict.
El Fráces. De gordas amas, algunas
vezes menos de leche.

De buena barba, bien voy paga
da.

De leaue beniste, le plus petit est
assez.
El Frances. Del agua bédita, la mas
poca basta.

Desque la paja en el pajar, viejas a hilar.

Dexa la fuéte por el arroyo, pen
saras traér agua, y traéras lodo

De Toro a çamora cinco leguas
son, cinco por allende, cinco
por aquéde, cinco por el vado,
cinco por la puente.

De móte mao, sequer vn paao.

E 2　　El Por

Figure 3.2. On page 34r. of Núñez's *Refranes*, a reader crossed out the proverb "De boda de hodidos, mal comen los niños." Department of Rare Books and Special Collections, courtesy of the Princeton University Library.

Witnessing Núñez's fervor convinced Mal Lara of the worthiness of the enterprise. His preambles echoed some of Castro's arguments for the importance of collecting "vulgar" sayings but expanded on the methods of acquiring them and investigating their meanings. Most importantly, he offered an alternative genealogy of Spaniards' philosophical knowledge, tracing the origins of Spanish "vulgar" philosophy all the way back to Adam and to the establishment of Tubal in the Iberian Peninsula. His goal was to demonstrate that the inhabitants of Spain possessed an ancient philosophical patrimony that predated the Greeks and the Romans. When Tubal arrived in Spain, he "set upon it all civility [*policia*] of good customs, and holy laws." Tubal had passed on to his companions the "doctrine and arts" he had inherited from his ancestors.[76] All of this essentially Adamic knowledge of nature and the arts was preserved in these ages, according to Mal Lara "not so much in books written" by these early people but "bequeathed to their great memories" and transmitted from generation to generation through the spoken word.[77] Thus, Mal Lara believed, this oral knowledge was the most ancient that could be collected and parsed.

Although *refranes* contained collective and common knowledge, glossing sayings was not easy. Mal Lara made clear that one of the main difficulties lay in properly understanding the words in each of these expressions. To overcome this potential difficulty, he proposed a connection between etymology and paroemiology. Sayings exhibited the multiple and metaphorical ways a language signified.[78] Sayings preserved forms of speech that often relied on metaphors or analogies to create meaning. In addition to this, they recorded the diverse forms that words took over time. Sayings constituted an archive of linguistic history.

While Iberian scholars hunted for *refranes* to include in their compendia, the Franciscan missionary Andrés Olmos (c. 1485–1571) developed a theory of how different kinds of sayings could preserve the fundamental linguistic, cultural, and philosophical patrimony of a society for posterity. Olmos arrived in Mexico in 1528, four years after the first Franciscan friars had settled in Mesoamerica.[79] He came in the service of the first bishop of Mexico City and eventually the leading benefactor of the College of Santa Cruz de Tlatelolco, the Basque Franciscan Juan de Zumárraga (d. 1548).[80] The college was built on the site on the pre-Hispanic school known as the *calmecac* and opened in 1533. Before emigrating to New Spain, Olmos and Zumárraga had served together in the Inquisition tribunal at Biscay. Zumárraga came to be known both for his initial campaigns to burn Indigenous codices and for his defense of Indigenous communities from increasing tributary demands and abuses.[81] His disciples tended to follow the latter example rather than the former. Olmos remained in the Americas until his death in 1571.

His location in Mexico gave Olmos a unique perspective on Indigenous culture. In his more than thirty years of missionary labors in Mexico, Olmos learned languages as different as Nahuatl, Totonac, Huastec, and Tepehuan and composed the first treatise on Nahuatl, the unpublished *Art of the Mexican Language* (c. 1547, *Arte de la lengua Mexicana*).[82] But his efforts went beyond this. In the College of Tlatelolco, he translated many of his sermons into Nahuatl—the most commonly spoken language of the former Triple Alliance of Mexico, Tenochtitlan, Texcoco and Tlacopan—including treatises on witchcraft, which reflected his Inquisitorial experiences in Spain. At the college he taught Latin to his students, all members of the Indigenous nobility. Olmos also worked on the transliteration of spoken Nahuatl into the Latin alphabet. The students at the college worked alongside Olmos as informants, grammarians, translators, scribes, and painters.[83]

Olmos's writings would be some of the most important sources for learning the "Mexican" language. They exemplify the goals and methods of missionaries in middle decades of the sixteenth century. The scholar of Nahuatl literature Ángel María Garibay credits Olmos with first developing the method that the Franciscan Bernardino de Sahagún, the Dominican Diego Durán (c. 1537–c. 1588), and later missionaries would use to collect data for their ethnographic and historical works. The missionary and royal authorities commissioned Olmos to write a book about the antiquities of the Indigenous people of New Spain, especially of the city-states of Mexico, Tenochtitlan, Texcoco, and Tlaxcala, "so that the bad could be refuted, and if there was something good so that it could be noted, as the things of many other gentiles are noted and preserved."[84] To do so, Olmos visited the pre-Hispanic centers of power in New Spain, examining the Indigenous codices that survived the conquest and initial burnings of religious authorities. He interrogated the eldest and most distinguished members of their communities as to the meanings of the illustrations in these manuscripts.

While most of Olmos's works have been lost, passages of his writings survive in the works of his contemporaries, and they include some of his work on Indigenous sayings. Between 1546 and 1554, he is believed to have composed a summary of his work on Indigenous antiquities, perhaps in response to a request by Fray Bartolomé de las Casas (c. 1484–1565) and the Council of Indies. Alonso de Zorita relied on Olmos's summary to write parts of his *Account of New Spain*.[85] Part of this effort was a grammar (*arte*) that adapted features from Nebrija's *Introductiones latinae* (1481) and *Gramática de la lengua castellana* (1492). It included a section about how the forms of speech used by "elders" in "ancient speeches," which aimed to explain "metaphors and ways of speaking, because words mean once thing and their sentences another."[86] Olmos derived many of the materials for this section from his collection of *huehuetlatolli*, which Victoria Ríos Castaño has defined as "exhortations, prayers, and salutations in which Nahua traditional religious, moral, and social concepts were conveyed in a beautiful and persuasive style."[87] The inclusion of a section on "manners of speaking" would eventually become a conventional practice in the dictionaries of Indigenous American languages.[88] The *huehuetlatolli*, or "words of the elder(s)," from *huehue* (old) and *tlatolli* (speech or language), were intergenerational speeches that included prayers to the gods, professional advice, and courtly statements.[89] As such, they were formally taught and transmitted, and differed substantially from proverbs, which were popularly used and informally conveyed.[90] The missionaries, however, were also interested in collecting proverbs. They would eventually group *huehuetlatolli* and proverbs together, as forms of rhetoric; however, they differentiated between their mode of transmission and their features.

Writing a generation later, Bernardino de Sahagún generally followed Olmos's example, with the exception that he requested his Indigenous pupils write their own history in the alphabetical transliteration of Nahuatl that the missionaries had developed and taught them.[91] Sahagún's elderly informants in Tlatelolco and Tepepulco would remain anonymous, as they largely were in the formal questionnaires of the *relaciones geográficas*. Sahagún's students and collaborators, however, make named appearances throughout his writings.[92] Perhaps relying in part on Olmos's collection of *huehuetlatolli* as well as on his own investigations, Sahagún and his Indigenous collaborators produced, over the course of decades, the *General History of the Things of New Spain* (c. 1575/1577, *Historia general de las cosas de Nueva España*). The work, initially composed in Nahuatl and later translated into Castilian in the middle of the 1570s, was preserved in various manuscript versions, the best known being the illustrated manuscript now known as the Florentine Codex. It remained unpublished until the nineteenth century, although manuscript copies of it circulated in Spain.[93]

Sahagún's translation into Castilian of the *General History* was part of a larger effort to collect and compile accurate local information about the Americas.[94] His broader goal was to advance his fellow missionaries' understanding of Nahuatl metaphors and ways of speaking. The ultimate objective, as he explained in the introduction of the *General History*, was to extirpate idolatry more effectively but also to preserve what was good about Indigenous culture, just as Olmos's initial orders had conveyed. Sahagún probably found inspiration in the works of Isidore of Seville, Pliny, and the medieval encyclopedist Bartholomeus Anglicus, since these assemblages, in their various ways, used words as the basic categories around which to collect and organize historical information. Inventories attest to these works' presence in Tlatelolco's library.[95]

In the prologue of the *General History*, Sahagún stated that he wished his history to serve as a "Calepino." He wanted it, that is, to mirror Ambrosio Calepino's (d. 1510) compilations for students of Latin. Calepino, however, took lexical items, and their meanings and metaphors, from canonical Latin authors.[96] Sahagún could not emulate this method because there were no similar books in Mexican. However, through his interrogation of the principal *tlamatinime*, or scholars, trained in the *calmecac*, Sahagún was able to set down in writing a group of texts that he believed served the same purpose in Nahua culture as the writings of Virgil and Cicero, and others that functioned as adages.[97]

Sahagún turned to proverbs and other forms of oral tradition to ascertain etymologies and the metaphorical meaning of words in the same way that contemporary scholars studied the Castilian vernacular, a language with a still incipient and debated literary canon. Book 6 of the *General History*, or the *Book of Rhetoric* (*Libro de la retorica*), was dedicated to the moral philosophy and ancient social order of the Mexican people.[98] Before the arrival of the Spaniards, the Indigenous Mexicans had devoted themselves to cultivating their language and crafting speeches and other compositions that Sahagún struggled to classify with Latin or Castilian terms.[99] The *huehuetlatolli*, demonstrated that elaborate rhetorical traditions existed in pre-Hispanic New Spain. These expressions were full "of beautiful metaphors and ways of speaking." They were made up of polysemic words that the Castilian language could not properly render. In the first forty chapters of book 6, Sahagún presented the *huehuetlatolli*, as Jeanette Favrot Peterson has argued, as "civic speeches that are epideictic, setting societal standards through techniques of praise or blame and thus conditioning honorable behavior."[100] The last three chapters included, translated into Castilian, eighty-three *tlatolli*, or sayings; forty-six *çaçanilli*, or conundrums; and ninety-one *machiotlatolli*, or metaphors.[101]

Sahagún, who may have studied at the University of Salamanca during Núñez's tenure, employed the model of the Erasmian book of sayings and ideas of classical rhetoric, both in his conceptualization of book 6 and in his actual translation of the sayings into Castilian. He often sought *romance*, or Castilian, equivalents for Mexican phrases. This was especially important in the Castilian edition, which was aimed at a European readership in the Council of Indies. Moreover, Sahagún would have also been familiar with the books of the Hebrew Bible that contained sayings like the Psalms, the Proverbs, and the Song of Songs.[102] To express, for instance, that one had experienced terrible luck, the Mexicans would declare the Castilian equivalent of "It cannot be any worse, it could not be any blacker than the wings of the crow." The literal translation of this proverb was closer to "misery is complete." Sahagún explained that this saying would be used if one made a bad investment, for instance, or lost an entire shipment at sea.[103] To criticize a bad messenger who returned without a response, the phrase "messenger of the crow" could be used (rendered more literally as "servants are sent"). This adage originated in a legend about the god Quetzalcoatl, king of Tula. When he noticed two women washing themselves in the stream where he normally bathed, he sent up to three messengers to report on their identity. Too busy spying on the women, none of the messengers bothered to return.[104] The adages that Sahagún collected and rendered into Castilian included moral advice, details about the natural world, and the geography of Mexico. As such, they were a rich trove of information that contained linguistic evidence but also, in Erasmus's words, the approved truth of epochs and peoples of those regions. Both the missionaries and scholars in the Iberian Peninsula endeavored, at the same time, to parse oral traditions and to describe the transmission of certain kinds of information, linguistic, natural, medical, and historical, that these compositions encapsulated. The missionaries like Sahagún built on the idea that sayings contained the approved wisdom of people to describe more elaborate speeches that they believed to be faithfully transmitted among generations of students in specialized institutions.

Language and the Problem of Origins: Alonso de Zorita and Miguel Cabello Valboa

Many of the scholars who wrote about the history of the Americas contended with the identity of its first inhabitants. This mattered to several controversies: first and foremost, the legitimacy of Spanish possession of the "New World." So too, as the historian Giuliano Gliozzi demonstrated, the existence of the Americas, their nature, peoples, and languages, had to be reconciled with the Bible and its account of human history, and the knowledge contained in ancient works. Moreover, Spanish rulers needed to justify the exploitation of the Indigenous inhabitants of the new viceroyalties. And origins were, in the early modern European tradition, an important source to make arguments about the status of communities.[105] Francisco López de Gómara and Agustín de Zárate had argued that the Americas were the Atlantis that Plato described in the *Timaeus*; thus, they were known to the ancients.[106] By contrast, Gonzalo Fernández de Oviedo proposed that the Americas had been settled in antiquity by the descendants of Hesper, the twelfth Spanish king of Annius of Viterbo's fictional genealogies. In Oviedo's view, the Americas were the Hesperides, islands that had been populated by ancient Spaniards. Hence, Columbus had merely rediscovered the Americas, not found them anew. For Oviedo, this ancient connection gave Spain an ancestral claim over these regions.[107]

Chroniclers like Florián de Ocampo and Esteban de Garibay adopted from the forged genealogies of Annius the notion that Tubal, the grandson of Noah, had been the first to settle Spain. This claim became a source of debate in several historical works about the Americas in the second half of the sixteenth century, especially when it came to reconstructing the first ages of the world. Alonso de Zorita and Miguel Cabello Valboa, among others, wrote their unpublished works about the viceroyalties of Peru and New Spain in dialogue with Spanish histories of these early ages. To reconstruct the earliest history of the Americas, they had to rely on the same tools that their Spanish counterparts used to study the unrecorded past of Iberia, including etymologies, speculation about the causes of linguistic change, the study of antiquities, and conjecture. On the textual front, they scoured the Bible and Annius's spurious authors.

At the same time, the authority of these Spanish scholars depended on their use of Indigenous sources, which they believed to be reliable. The abundance of Indigenous words in their writings, together with their accompanying etymologies, formed an essential part of the scholarly apparatus that underlay these histories and gave them philological legitimacy. For the periods to which Indigenous historical accounts bore witness, the scholars attempted to access these sources, whether oral, pictorial, or material, and to understand their contents with the aid of local experts. Spanish chroniclers admitted the difficulties they had faced because of the lack of alphabetic writing in the Americas, often in order to highlight their achievements.[108] Their translation of Indigenous sources into their own historical narratives invariably transformed Indigenous historicities to fit with European assumptions, chronological and otherwise, about history writing. In this respect, Zorita and Cabello Valboa resembled the early lexicographers of Quechua who, Bruce Mannheim writes, "often provided important and subtle insights into the nature of" Indigenous languages and histories but "also distorted"

Indigenous language "terms and cultural patterns by forcing them into Western frameworks."[109]

Zorita's writings were the product of a distinguished legal career that took him to different parts of the Americas as a *letrado*, or "man of letters," in the service of the Spanish Crown. A native of Córdoba, he studied at the University of Salamanca between 1537 and 1540. Zorita arrived in Hispaniola in 1548 to serve as a judge (*oidor*) in the high court of Santo Domingo. Between 1550 and 1556, he worked as an itinerant judge first in the province of New Granada (modern-day Colombia) and then in what would become the Audiencia of Guatemala (a jurisdiction that stretched from southern Mexico to the north of present-day Panama). Between 1556 and 1566, he became a judge in the high court of New Spain in Mexico City, where he also served as a member of the recently founded Royal and Pontifical University of Mexico.[110] Zorita corresponded with Fray Bartolomé de las Casas and was sympathetic to his advocacy.[111] One of his tasks as *oidor* in the Audiencia of Mexico was the implementation of the New Laws of 1542, which were meant to regulate compulsory Indigenous labor and limit the amount of tribute that local authorities could extract from Indigenous communities. Their application, however, was controversial, as Spanish settlers and *encomenderos* (those who held *encomiendas*, or grants of Indigenous vassals) resented the Crown's attempt to rein in their power. In Peru, the implementation of the New Laws brought about armed confrontations and even the assassination of the first viceroy. Zorita returned to Spain in 1566, spending the last years of his life in Granada in financial penury, working to complete the *Account of New Spain*, which he did in the year he died, 1585.[112]

Zorita's two main works, the *Brief and Summary Relation of the Lords of New Spain* (c. 1570) and the *Account of New Spain*, sought to reconstruct pre-Hispanic customary law and tributary mechanisms by defining local nomenclature. In the *Account*, Zorita described modes of historical commemoration that existed in New Spain and how missionaries like Motolinía and Olmos had learned local histories from Indigenous scholars. He explained that these missionaries obtained their information from the memory of people "who knew how to relate the passing of times and the lineages of the lords" and from "their books and paintings." For instance, he learned from Olmos that the first inhabitants of Anahuac came from a "cave or town called Chicomostol, which means seven caves and is [located] towards the land of Gelisco."[113] By providing these sources and carefully studying the origin place-names, Zorita proposed a corrective to the bibliographic panorama offered by Jerónimo Román Zamora in his *Republics of the World*.[114]

Román argued that it was doubtful that anyone had consulted as many works about the ancient Mexicans as he had.[115] Zorita, however, questioned this, as Román was clearly unaware of the writings of Olmos and Motolinía, two of the central pillars upon which Zorita's *Account of New Spain* was based.[116]

Indeed, Zorita engaged numerous published and manuscript writings about the Americas. By employing the works of the Franciscan missionaries, Zorita also could draw on Indigenous sources and expert interpreters. For his discussion of Indigenous origins, for instance, he relied not only on Motolinía and Olmos but also on the testimony of the Indigenous *cacique*, or ruler, of Xaltocan, Pablo Nazareo.[117] Zorita's sources indicate how manuscript materials circulated informally among the administrative and missionary elites of New Spain. In his "Catalogue of Works Consulted," which preceded the historical narrative, he informed the reader that while working on the *Account* in Granada, he had to return Motolinía's manuscript to the Franciscan missionary Jerónimo Mendieta (1525–1604). He did so by entrusting the manuscript to a fellow Franciscan who was returning to New Spain.[118]

Zorita believed that the Indigenous inhabitants of the Americas descended from the ancient Israelites. This claim was not unique in New Spain in the mid-sixteenth century: the Dominican missionary Diego Durán had likewise made it his *History of New Spain* (completed c. 1579).[119] The chronicler Mendieta had also advanced a similar proposition.[120] For Zorita the clearest indication of the connection to ancient Jews was the evidence embedded in the Mexican language, both in the force of its words and in its capacity to signify. This language, he believed, had an abundance of words suited to conveying the properties of things. In this it resembled Hebrew, the first and richest language of humankind. When describing Mexico City, for instance, Zorita noted the presence of many Indigenous herbalists who had "roots and medicinal herbs with which they naturally and swiftly cure." The Indigenous doctors named plants in accordance with their effects, calling an herb, for instance, "medicine of the spleen" in their own language. All the names of herbs, then, conveyed their medical utility, which had been learned through experience. This resembled how the Mexicans named places: toponyms described the most important resources and geographical features of an area. Words' powers of signification could easily be unraveled by the mind.[121]

Zorita's argument for the link between ancient Israelites and modern Mexicans was not merely linguistic. Throughout the *Brief and Summary Relation* and the *Account*, he speculated about the origins of the Mexicans and even of all Indigenous Americans. Although most of his evidence about origins was linguistic, he also occasionally compared the people of New Spain, their ceremonies and customs, to what he found in the Hebrew Bible and in other religious writers like Eusebius. He found similarities in land tenure patterns and in public celebrations. Although his etymologies appear fanciful to present-day readers, they rested on a tradition, shared by most of his contemporaries, who understood Hebrew to be the first language of humankind. When explaining how the Mexicans managed land tenure and inheritance, for example, he noted that they never passed their lands on to anyone outside their own "calpulli or neighborhood." This he believed resembled the customs of the ancient Israelites, for whom it was illegal to "grant the lands or possessions from one tribe to another." Zorita concluded this passage by reaffirming that what "they call in New Spain calpullec, is the same that among the Israelites they call tribes."[122]

Zorita further claimed that he had learned from other unnamed sources that the language of Michoacan contained many Hebrew words, and "that this language, like almost all others [Indigenous languages] is similar in pronunciation to Hebrew." From other sources, which he also did not identify, he had heard that "those who have been in the province of Peru affirm the same thing, as those who have been in the other parts of the Indies." But how did the ancient Israelites wind up in the Americas? Zorita considered three possibilities, all of them allegedly taken from Olmos, who had learned from the elders of "Tlezcuco" that to get to their current lands, their ancestors had crossed the sea. According to the first hypothesis, they might have come from Babylon after the fall of the Tower of Babel. The second possibility: they came to the Americas when "the children of Israel entered into the promised land." According to the third hypothesis, perhaps they had come later "from [the ancient biblical city] Sichem." Zorita favored this latter interpretation because he believed, but did not explain why, the name of Sichem was closest to that of Mexico, no matter that the letters were "corrupted." He defended his perplexing conjecture by offering the example of "Cuhuanauac," for instance, "which the Spaniards call Cuernavaca." Many other names were corrupted beyond recognition. Languages, after all, mutated on account of unions with other people, wars, and military conquests. A combination of these factors must have transformed Hebrew so profoundly in the Americas as to make the current languages found there so different from it.[123]

In their dances, celebrations, and *areitos*, the Indigenous inhabitants of Mexico also resembled ancient Jews. Drawing on Eusebius's *Ecclesiastical History*, the Hebrew Bible, and the etymologies of the names of the dances celebrated in Mexico, Zorita established a connection between the songs and dances of the Nahuas and those of ancient Hebrews. He explained that although the Spaniards referred to the dances as *areitos*, this was "a word from the islands," and the dances of the islands were very coarse compared to those of New Spain. The people of New Spain danced with "great beauty and grace."[124] In the language of Anahuac, the dances had two names: "One is maçeualiztli and the other is netotiliztli. This latter one properly means dance of joy, of which they take pleasure from and when they dance the second and main name is maçeualiztli, which means penitence and merit or confession, because in their hearts they called upon their gods and offered them those dances and movements that they did."[125] This explanation of dances also highlighted a quality that Zorita believed extant in many Indigenous institutions and customs, that is, that they were in accordance with natural law. Also, in this form of sung praise, they exhibited a naturally existing capacity for religious devotion. Zorita explained that in many places in the sacred scriptures people danced to praise the divinity. For instance, Miriam, the prophet Aaron's sister, took a timbrel in her hand and danced and sang praises to God with the other women when the Egyptian enslavement ended and the Egyptians drowned in the Red Sea.[126] David was also received with various songs and dances by the most distinguished women of Israel's cities, including call-and-response lyrics. Likewise, in Anahuac, the "masters of songs of the Indians in their dances would sing a verse," and the "multitude would answer like the Israelites." Zorita noted that this was another "conjecture to believe that they descend from them." At the end of the passage, he admitted that call-and-response singing was also prominent in Old Castile.[127]

Contemporary scholars have debated the accuracy of Zorita's reconstructions of precolonial Mexica society and whether he idealized the Indigenous nobility, their modes of tributary extraction, and their fiscal demands. Some of Zorita's explanations may have served Indigenous nobles by establishing their claims to land and dependents against the Spanish-style governance based on townships.[128] In the 1550s, Spanish governance was spreading in New Spain in an effort to centralize land and labor under the authority of the Crown. Zorita's arguments may have also appealed to certain missionaries and Spanish administrators who preferred to deal with fewer intermediaries in their efforts toward evangelization and governance because they believed the most efficient form of Christianization was to first convert elites; then through their example commoners would follow suit.[129]

In the *Brief and Summary Relation*, Zorita did admit that generalizations about the Americas were difficult. Although his definitions came directly from testimonies of "old and principal Indians, of whom it is plausible to believe that they would tell the truth," it remained impossible to establish a "general rule" about anything because of "the great differences that there are in everything" in each province. In some towns, two or three languages were spoken. The Indigenous inhabitants of the same region would also offer differing accounts of the same topic. These discrepancies, Zorita believed, could often be attributed to mistakes by the interpreters, but they also resulted from the lack of "letters and writing," and the fact that "all of their antiquities were in paintings, and of these many have been lost and destroyed, and memory is fragile, and the eldest who knew these [histories]" have died.[130] His was not a rejection of oral traditions and the forms of inscription that existed in New Spain, which Zorita believed to be trustworthy, but of the difficulties of accessing this material.

Zorita's studies of pre-Hispanic origins did not happen in a political vacuum; on the contrary, his scholarship had a clear politics. His linguistic research supported his arguments about the governance of New Spain and endorsed a particular relationship of the Spanish Crown with its Indigenous subjects. Zorita completed his *Brief and Summary Relation* in response to a royal questionnaire that the Council of Indies had issued almost twenty years earlier, in 1553.[131] With his response, he aimed to intervene in debates about the future governance of New Spain. The *Relation* attempted to persuade Philip II that the most harmonious form of government in the Americas had to be based on pre-Hispanic systems of rule and on the maintenance of the social hierarchies to which locals were already accustomed.[132] In other words, to govern New Spain and to ensure the sustainable collection of tribute, local fiscal customs, which had worked so well for pre-Hispanic rulers, had to be studied, properly understood, and then preserved, albeit in the service of the Spanish monarchy. For Zorita, grafting the monarchy's laws and expectations onto local customs constituted good governance. There are similarities between Zorita's proposals and Molina's call in his *Grammar*'s prologue for secular administrators to learn local languages to better exercise justice.

In the *Brief and Summary Relation*, Zorita justified his favorable view of pre-Hispanic institutions by claiming that with the exception of their religious ceremonies, the governments of Mesoamerica were in "accordance with natural law, which resembles divine law, and even civil law and canonical law, even if they did not know it."[133] Las Casas defended a similar stance in his unpublished *Apologetic History* of the Indies.[134] As further proof of the order that had existed in New Spain before the conquest, Zorita transcribed in the summary the ancient "manners of speaking" that he had learned from Olmos about the rigorous and moral ways in which Indigenous children were educated. This understanding of pre-Hispanic *policía* (civility) collided, however, with official interpretations. The Ordinances of the Council of Indies of 1571, for instance, expressed how necessary it was for Spanish administrators to convince the Indigenous populations that they were now under the protection of the Spanish king "for their wellbeing," because the Crown sought "to remove them from the tyranny and servitude in which they lived before."[135]

In the *Brief and Summary Relation*, Zorita asserted once more that the Indigenous people of Mexico needed to be governed in accordance with their local customs. Under Spanish rule, the law and its practice were discordant. Zorita observed that laws proliferated; there were judges in excess, countless administrators, a "million lieutenants and a million more sheriffs, but this is not what the Indians need nor what will remedy their misery."[136] What was necessary was the restoration of Indigenous *fueros* (customary laws and privileges).[137] To those who insisted that the Indigenous people of the Americas were "barbarians," Zorita offered a long etymological refutation. He found a deep contradiction in Spanish writings that described the Indigenous people's institutions and all they had achieved, sometimes even with admiration, but then concluded by calling them "barbarians." This assignation, Zorita believed, betrayed a misunderstanding of the concept of barbarism and the term's etymology. Even in ancient times, the term "barbarian" did not denote an absence of *policía* or reason. Ancient and current scholars used the word "barbarian" to mean people who spoke a language different from their own. Zorita observed that Spaniards in his day called the "Indians barbarians because of their great simplicity" and "because they lacked malice." Historically and at present, according to Zorita, "barbarian" designated not a lack of civility but rather unfaithfulness and foreign speech. Through his etymological discussion of "barbarian," Zorita challenged Spaniards who claimed that Indigenous Americans lacked civil institutions.[138]

The problem of origins led the secular priest Miguel Cabello Valboa to very different conclusions. Cabello Valboa finished writing his *Antarctic Miscellany* (*Miscelánea antártica*) in 1586, only a year after Zorita completed his *Account*. Born in Archidona, Málaga, Cabello Valboa was part of a Lima literary circle known as the Antarctic Academy. He descended from Vasco Núñez de Balboa (d. 1519), the first European to cross the isthmus of Panama and to explore the Pacific Ocean from the American coast. Cabello Valboa travelled to the Americas in 1566, where he remained for forty years. He was ordained as a priest in Quito in 1571 and then moved to Peru. He participated in the Third Council of Lima of 1582 and might have met Acosta at this time.[139]

In the *Antarctic Miscellany*, Cabello Valboa aimed to refute the origin theories of writers like Zorita, offering his own interpretation of the "origins and beginnings" of the inhabitants of the Americas.[140] He admitted numerous times that the sources to uncover this history were difficult. However, through a combination of conjecture, readings in ancient and biblical sources, and consultation of quipus and their keepers, Cabello Valboa believed that he had established "with as much certainty as human diligence" could accomplish the date when the Inca dynasty was founded. His chronology agreed with the "quipos and Indian reckoning that have remained from that time until ours." He also included the years in which each Inca ruler lived and the most noteworthy events of their reigns.

Given the range of materials that Cabello Valboa consulted, he decided to name his work the *Miscellany*.[141] He argued that the Americas had been first populated by the descendants of Ophir, the son of Joktan, fourth grandson of Sem, Noah's first son.[142] He identified similarities between the people of the East and West Indies. And he synchronized the history of the Inca kings and contemporaneous events in Spain until the arrival of the Spaniards and the death of Atahuallpa Inca.

Cabello Valboa had concluded that Indigenous Americans descended from Ophir from his own readings, but his hypothesis gained strength from the writings of Benito Arias Montano in the *Apparatus* of the Antwerp Polyglot Bible (published 1568–1573). Using Montano's research, he concluded that the descendants of Ophir passed from the East Indies to the Americas.[143] The West Indies were known to ancient Israelites and appeared in the Hebrew Bible as the land of Ophir, which was rich in gold, silver, and precious stones. This land was mentioned by the author of the Paralipomenon, who wrote of the ships of Solomon that sailed east. Cabello Valboa cited Arias Montano in Castilian: "These lands were called Pirú, because in Hebrew they are written in dual pronunciation, which is Peruain or Paruain, which came from Ophir or Opir." Scholars had doubted this etymology because the appellation Peru was not known when the Incas ruled but rather originated as a result of Spanish misunderstanding. Moreover, in New Spain this name was completely unknown. Cabello Valboa thought this insufficient to disprove Montano's hypothesis. If one asked the inhabitants of Toledo, for instance, whom Cabello Valboa regarded as "less barbarous than the people of our Indies," who the Carpetani (pre-Roman inhabitants of the Iberian Peninsula) had been, they would be unable to say unless they were versed in ancient scriptures, because "that is the nature of time." However, Cabello Valboa claimed that there were, in fact, geographical formations with the name of Peru in the Indies. Close to the city of Ancerma, in Popayán, he offered, there was a big river called Berú or Perú.[144] Subsequent navigations from the East Indies brought new people to the Americas, including the Nayres, from whom the Indigenous people of the lands of Chile originated, although this claim Cabello Valboa described as a conjecture.

Cabello Valboa rejected the Jewish origins of the Americans that Zorita and others had defended, arguing instead for the extreme linguistic diversity of the Americas (which did not support a linguistic derivation from Hebrew). His explanations relied on both natural and political causes. He expressed skepticism about all Hebrew etymologies of New World words: "There are so many different languages that they speak and treat in our Indies" that "there were not enough letters in the arithmetic to number them since there are so many." This was a feature of all provinces. Down the Magdalena River, in the territories of New Granada, so many languages were spoken that even in single towns more than three could be heard. Some of them were gendered, so that men and women in the same settlements spoke different languages.

Even in Quechua, the most widespread language of South America, Cabello Valboa could not find any words similar to Hebrew. Moreover, Quechua words that were phonetically analogous to lexical items in Spanish or French had profoundly different meanings. For instance, speakers of Quechua referred to the chest as *casco*, the word that the Castilians use for a helmet.[145] Phonetic similarity was not enough to establish etymological derivation. The most universal word that Cabello Valboa could find that had the same "sound and meaning" in the languages of Europe and the Indies was the word *marcha*, or *marca* without the *h*. This word existed in both Quechua and Aymara, the most widely spoken languages of the viceroyalty of Peru, and meant "province or payment of land." It meant the same thing in Italian, Flemish, German, and Hungarian, although in Castilian it had acquired a slightly different form as *comarca*.[146]

To explain American polyglossia, Cabello Valboa proposed that once the descendants of Ophir had reached the Western Indies and spread throughout its vast lands, they became entangled in conflict with one another. Political division would then generate linguistically autonomous communities. Each family sought with "all of their strength and study, to not name usual things with the same words and terms" that their enemies used. In this manner, Cabello Valboa continued, they would invent words and verbs to teach to their children. They were able to do this because "they did not have great copiousness in words, only those that they could use to understand each other and their domestic affairs." The passing of time would bring people together and two of these "primitive languages" would merge, thus generating a third, "composite one."[147] Cabello Valboa found evidence of this in the town of Olmos, in the plains of Peru, "where the Indigenous inhabitants still seek to create new words so that others will not understand them."[148]

To explain how languages intermixed, Cabello Valboa offered the additional example of *mitimas*, "those nations that Incas would remove from the provinces in which they were from and made them live among other nations." The *mitimas* (*mitmaq*) were communities uprooted from their ancestral homes and moved to different locations to establish productive exchanges between highlands and lowlands, and thus manage the mountain environment.[149] Cabello Valboa believed that in their new places of settlement, they would mix their words with the names of the places where they had been transplanted, often corrupting their language and making a different language altogether from the one they had originally spoken. It was precisely this situation that the Inca kings sought to remedy, according to Cabello Valboa, "when they commanded the use of one general language that did not admit any others." However, Cabello Valboa lamented that since the Incas no longer ruled, "millions of corruptions were introduced into the language and in some places of Peru, it is almost completely altered and made into a different language."[150] He did not elaborate extensively on why this was the case but suggested that these corruptions were a characteristic of postconquest life.

A variety of conflicting theories existed to explain the origins and antiquity of the people of the Americas. As the examples of Zorita and Cabello Valboa reveal, scholars endeavored to use linguistic evidence eclectically to defend different interpretations of a common textual tradition. Their investigations led them to formulate, as scholars also did in the Iberian Peninsula, models to account for the ways languages change in response to natural and political stimuli.

Language and the Spanish Conquest

The historic understanding of language in the Americas shared by the subjects of this chapter was limited but influential. Reflections on linguistic change tended to adopt the metaphor of corruption. European scholars used this term since the fifteenth century to describe the transformation of Latin.[151] In humanist fashion, Spanish scholars tended to link linguistic efflorescence with political stability and linguistic corruption with political and moral disorder. Thus, the decline of the Roman Empire had corrupted Latin. Moreover, for sixteenth- and seventeenth-century scholars, although languages might provide evidence of historical change, they did not do so in accordance with consistent patterns.[152] Transformations could happen in ways specific to each context and language. Sahagún believed, for instance, that the rich adages of Nahuatl speakers reflected the appropriateness of the laws and institutions that governed the pre-Hispanic inhabitants of Mesoamerica. In book 10, dedicated to the "vices and virtues" of the Indigenous inhabitants of Mexico, he described how the Mexican emperors had crafted a form of governance fitted to the needs, temperament, climate, and geography of the peoples of these regions. In this, he expressed a vision like that laid out by contemporary European scholars. For example, in his *Method* Jean Bodin claimed that political systems had to be adjusted to the temperaments of their subjects, and that these dispositions were in great part determined by the climate of a region and its latitude and position on the globe. According to Sahagún, the pre-Hispanic forms of temporal governance were much better suited to the temperament of Mesoamerican peoples than the systems currently enforced by the Spaniards.

Ancient speeches and their eloquent pronouncements were evidence of the fitness of pre-Hispanic governance to the peoples of Mesoamerica. The strict manner of raising children and ruling over society in "past times" was appropriate for the moral and natural philosophy of New Spain because the climate of that land, according to Sahagún, and the "constellations that rule over it greatly help human nature become full of vices and idle and given to the sensual vices." Moral philosophy and experience, Sahagún continued, taught Indigenous people that to live "morally and virtuously the rigor of austerity was necessary," as was continuous occupation in activities that that were useful to the "republic."[153] In this assessment, Sahagún drew from the same cosmological paradigm that Oviedo had used to explain the customs and nature of the people of the Americas. Despite the profound differences between these two writers, they agreed that the Indigenous Americans were ruled by a climate and stars that made them prone to vices and that required a strict form of temporal governance.

Pre-Hispanic rulers, Sahagún believed, had developed the ideal form of governance for their subjects. While he claimed that it had been necessary for the Spaniards to tear down all temples and to abolish all the "customs of the republic" that were mixed with idolatry, this indiscriminate destruction of the old ways of governance had engendered people of "very bad inclinations."[154]

The purported decline in eloquence among Nahuatl speakers was also diagnosed by the *Huehuehtlahtolli* (1600), the only pre-Hispanic text besides the grammars (*artes*) to be published during the colonial period. The Franciscan *fray* Juan Bautista de Viseo, a disciple of Olmos, prepared it for print. In the prologue, he declared that its speeches had been collected by Olmos and used by Fray

Bartolomé de las Casas to demonstrate the natural *policía*, or civility, of the Indigenous inhabitants of Mexico. Bautista observed the changes that the most el-
evated forms of speech had undergone in the previous decades: "Considering
[Christian reader]," he averred, "the poor upbringing, scant respect and consideration, barbarity in speech, and little orderliness of young Indians of our time,
and by contrast the good upbringing, urbanity, respect, courtesy, and elegance in
speaking of the elderly Indians, finding some of the speeches that the ancient
Indians used to give their sons and daughters, it seemed to me important to
bring them to light."[155] Bautista explained that traditional speeches, which sometimes contained "funny sayings," were pronounced with "as much art and elegance as in any other language."[156] Scholars and distinguished men would get together in "royal houses." There, "all of the consuls and noble people and officials
who held public offices and the High Priests and elder ministers and officials of
the temple," he wrote, "would all together confer about whom was to be elected
as Lord or King and who had all the necessary parts to rule."[157] If taught to
children as they once had been, the "sayings of the elders" could help to advance
society in the "good treatment and preservation of the common people and the
poor."[158] Bautista's goal was to elevate traditional Nahuatl language and by doing
so to exalt the moral capabilities and "natural eloquence" of Indigenous Mexicans. He concluded that from these texts it would be understood "that the ancient Indians had much order in governance and achieved much in morality."[159]

Like Sahagún's before him, Bautista's goal was to demonstrate the capabilities
of the Indigenous Mexicans to become Christians as well as exalt the virtues of the
Franciscan order. However, he also echoed a now-commonplace lament over the
loss of eloquence. Both missionaries and secular clergy like Cabello Valboa were
overwhelmingly aware of the rapid decline afflicting American societies and the
urgency of reform. These lamentations extended beyond the clergy. The lawyer
Antonio León Pinelo wrote in 1629 that the Quechua language of Cuzco "was very
courtly and elegant, and it still is today, even though it is very corrupt because it
has been Hispanicized, and it has been lost due to the lack of care of the governors
of the provinces."[160]

Conclusion

Etymologies played a number of roles in efforts to write the natural and human
history of the Americas. The Spanish crown collected toponyms as part of its projects to visualize and catalog its global domains. In historical works and in govern-

ment censuses, scholars in the American viceroyalties and in the Iberian kingdoms relied on similar, interconnected methods to consider spoken languages and to harness the information they contained. Etymologies in scholarship about the American viceroyalties provided evidence of scholars' authoritative knowledge and could sometimes be treated as vestiges of origins, for example, establishing a connection between Indigenous languages and Hebrew. While most of these histories of the Americas written between 1550 and 1600 remained unpublished, their authors nevertheless contributed to central questions about language, origins, and the secular and spiritual future of the Spanish Empire.

These writers' historical understanding of languages inflected how they interpreted the speech communities of the Americas. With their sixteenth-century tools, they understood linguistic change as a live archive of social transformation. Their discussions usually took the form of lamentation for the loss of ornate speech, as in the prologue of Bautista's 1600 edition of the *Huehuehtlahtolli*. An understanding of linguistic change over time would influence grammarians like Bernardo de Aldrete and even a Basque apologist like Andrés Poza, who sought to analyze the changes afflicting their own speech communities.

In 1606 Aldrete noted that just as the ancient Spaniards learned Latin from the Romans, many Indigenous people from Cuba to the Philippines now knew Castilian. Yet this was happening through the social forces unleashed by the conquest and not because of the deliberate and disciplined teaching of the language. In this, Aldrete believed, the Spaniards had much to learn from the Inca king "Guaiancapa" and from the Romans. However, linguistic diversity persisted because of the "devotion that [the Indigenous people] have for their language." No one could force them to speak another tongue.[161]

In New Spain Indigenous people knew how to speak Castilian but often chose not to do so. Before the arrival of Cortés, Aldrete claimed that the Mexicans, "surrounded by many nations who had different languages, they did not want to use them, admitting only the Mexican."[162] For all the similarities, Aldrete perceived a great difference between ancient Spaniards and Indigenous Americans: the use of letters, which he equated with sciences and political living. In doing so he drew from a similar tradition than Cervantes de Salazar, and like his predecessor, he saw no contradiction in denying the people of the Americas the "study of sciences and *policía* that accompanied them" while also discussing the imperial accomplishments of the Incas. More detailed knowledge of languages and history, as in the case of Aldrete, did not always result in more nuanced interpretations.[163]

CHAPTER FOUR

Language and the Secrets of Nature

Located on the eastern shore of a lake of the same name, the city-state of Texcoco was part of the Triple Alliance, along with Tlacopan and Tenochtitlan, that the Spaniards defeated in 1521. In a treatise composed during the 1570s entitled *Antiquities of New Spain*, the Toledan doctor Francisco Hernández (1515–1587) described the vestiges of the palaces of its former kings. One of the palaces of the fifteenth-century ruler Nezahualcoyotl (1402–1472) was now "inside the city and next to the plaza where the Indians hold their weekly markets."[1] The building, Hernández claimed, "was admirable for the spaciousness of its rooms, for the number, of its patios and architraves, as the ruins and vestiges of the ancient buildings indicate, for the firmness of the work, for the size of its columns and beams, for the consistency, splendor, and longevity of the limestone and *tezontli* stone surfaces."[2] He proceeded to describe other, perhaps more ancient, royal dwellings in the area, including one with "more than two thousand stone steps through which to ascend to each floor." Some of these magnificent staircases, he clarified, seemed to be carved from a single rock, revealing the vast labor force Texcoco's rulers had once commanded. Aqueducts, gardens, and many other rooms comprised part of the complex.[3] A patio contained an elderly tree that remained verdant "after seven hundred years."[4] And there were other noteworthy remains: the people of Texcoco had preserved with "great, almost religious respect" Nezahualcoyotl's "statue, his shield, flags, trumpets, flutes, weapons and other ornaments that he used in war and in public dances." Hernández "had them painted," to "put, as much as I can, before the eyes of ours, the things of the past and for those who have not been able to see, [these] distant people, so that they can know

[them] as much as possible." As he dictated his observations of ancient Indigenous building practices to a scribe, he claimed to be standing inside the remains of the palace of King Nezahualpilli (1464–1515), Nezahualcoyotl's son (figs. 4.1 and 4.2).[5]

Among the various animals painted on a wall of another Texcoco residence was a strange bird. Hernández described the creature as possessing blue and white wings with some yellowish feathers, a red chest and belly, and a white, featherless, and rugged head that seemed almost scarlet colored. Under its beak, the bird had something odd: a circle formed by many white and reddish spots. From questioning the locals Hernández learned that this bird was called a *cuechtototl*. Although he admitted to never having seen it with his own eyes, the testimony of local authorities, the painting, and the bird's name seemed, in combination, authoritative enough that he included the creature in his compilation of the animals and plants of New Spain. The locals had assured Hernández that the bird could be spotted close to the ocean and that it sustained itself by scavenging, although he could not learn anything else about its "nature or customs."[6]

In this case, as in his description of numerous other animals and plants, and of Indigenous antiquities, Hernández relied simultaneously on multiple approaches to study the natural and human history of Mexico. They included firsthand observations of Indigenous material culture, the testimony of local authorities, the information encoded in Indigenous names, and his own experiences. For Hernández as for other sixteenth-century natural historians, experience was defined broadly. While it included direct, empirical observations of plants and animals—both in situ and as prepared specimens, animal dissections, and even experiments with various plants—it also encompassed the collection of manuscripts and ancient books, the practice of textual emendation, and the oral knowledge of largely illiterate common people. Gianna Pomata and Nancy Siraisi have shown how "closely intertwined . . . the practices of natural history, medicine, philology and antiquarianism [were]," and a scholar like Hernández could use them interchangeably, or even simultaneously, in his efforts to describe the nature of a new continent.[7]

Trained in medicine at the University of Alcalá de Henares and later at the Royal Hospital of the Monastery of Guadalupe (1556–1560), Hernández was appointed *protomédico* of the Indies in 1570. The *protomédico* was the highest authority in charge of regulating medical practice in a region.[8] Hernández's main assignment, as King Philip II specified in a royal decree that same year, was to travel to the Indies and "gather information about herbs, trees and medicinal plants." The decree commanded Hernández to consult Indigenous physicians as

Figure 4.1. Portrait of Nezahualcoyotl (c. 1580–1584). *Codex Ixtlilxochitl*, Ms Mexicain, 65–71, 106r. Courtesy of gallica.bnf.fr / Bibliothèque nationale de France.

Figure 4.2. Portrait of Nezahualpilli (c. 1580–1584). *Codex Ixtlilxochitl*, Ms Mexicain, 65–71, 108r. Courtesy of gallica.bnf.fr / Bibliothèque nationale de France.

to how these plants could be harnessed for medical use.[9] The botanical search would take Hernández far and wide into the viceroyalty of New Spain, often along the path of hospitals and monasteries that Franciscan, Augustinian, and Dominican missionaries had erected in the earlier days of the conquest.[10]

Hernández's journey forced him to confront Indigenous medicine, language, and culture at a time of momentous transformation. In the middle decades of the sixteenth century, the Spanish introduced the *repartimiento* (allotment of Indigenous laborers) labor system to phase out the hereditary *encomiendas* (grant of Indigenous tributary labor in exchange for protection and Christian instruction) of the first generation of conquerors and created new Hispanic-style institutions of town councils, or *cabildos*, led by an Indigenous *tlatoani* (a former ruler of the *altepetl*, or a pre-Hispanic city-state) and nobles to administer the labor draft. The monastery complexes of the missionary orders flourished in this period, while Indigenous languages increasingly came to be written alphabetically.[11]

At the same time, wave after wave of epidemics decimated the Indigenous population, bringing the mortality rate to over 50 percent in epidemic years.[12] Hernández's many writings, in particular his Castilian translation and commentaries to Pliny's *Natural History*—a work he began before his American sojourn in 1571 and that he continued to develop after his return to Spain in 1577 until his death almost a decade later—brim with personal anecdotes from these years. Their reflections on method foreshadow the contents of his most important work, the *History of the Plants of New Spain (Historia plantarum Novae Hispaniae)*.[13]

This chapter argues that Hernández's investigations into the nature of the Americas relied on the collection of authoritative local knowledge. Hernández began by researching Indigenous nomenclature. In his view, Indigenous doctors possessed "centuries of experience" with the region's flora and fauna; their explanations were the obvious starting point for his investigations.[14] In adopting this approach, Hernández explicitly deployed a version of natural history inherited from Pliny. To truly understand the natural history of a place, its human history had to be taken into consideration, including how local experts had come to learn about the natural world. Their knowledge was often embedded in the most noteworthy, but also the most mundane, features of each society, such as its customs, dress, laws, and even its religious history. Information about the properties of animals and plants was often encoded in the names of things. Name-making practices, or nomenclature, therefore, could be studied through the etymological method to reveal the properties of plants and animals and to mine the expertise of those who had named them.

Hernández's translation and commentaries on Pliny's *Natural History* exhibit the philological awareness of a humanist of his time. In both his Plinian commentaries and his other writings—which include the book on Mexican antiquities, a number of philosophical treatises on Stoicism and Aristotelian maxims, a translation of the works of Dionysius the Areopagite from the Greek, and a manual of Christian doctrine—he often combined etymological analysis of local names with reports and experimental observations to produce descriptions of plants and animals and their medicinal uses, to speculate on the origins of Indigenous antiquities, and to make geographical conjectures.

This chapter sketches the linguistic beliefs that underpinned Hernández's etymologies. It contextualizes his ideas within broader debates about the links between language, natural history, and knowledge gathering. Thinking about "Old and New World" etymologies in the context of Pliny's compendium allowed Hernández to incorporate significant aspects of Indigenous medical knowledge into an expanded natural history of universal ambition. To do so, he relied equally on humanist writings about the natural world and on missionaries' understanding of the languages of Mexico.

Unsurprisingly, Hernández's priorities, prejudices, misunderstandings, and mistranslations shaped the form and content of his works, as did the interests of his patron, Philip II. Aware of the king's imperial ambitions, and of the Spanish Crown's mission to impose Christianity on the Indigenous people of the Americas, Hernandez also attempted to assert his own authority and to defend the worthiness of his undertaking. He often subordinated the knowledge of Indigenous doctors to his own causal explanations of disease and dismissed the non-Christian religious meanings of certain plants and animals, and the worldview they betrayed, as mere superstition or idolatry. This was not unlike the way contemporary natural historians dealt with the natural knowledge of European peasants or those whom they considered rustic.

In translating and commenting on the Pliny's *Natural History*, Hernández found a classical model to emulate with an ambition and scope that matched his own. The commentaries often express a tremendous eagerness for new knowledge animated by Hernández's awareness of Spain's imperial position and its seemingly relentless capacity to expand further into unknown lands. Hernández's emphasis on local knowledge and his insistence on the importance of the Indigenous names of plants and animals shaped his natural-historical compendium. Confronted with the diversity of the Americas, he transformed classical models, like that of Pliny, to accommodate a deluge of new information under broad categories for describing the natural world. The commentaries' composite qualities were premised on the collection of the Indigenous names of plants and animals. Moreover significantly, in the course of the expedition, Hernández developed an understanding of local scholarly traditions and methods for passing down information. He believed that Mexican botanical knowledge was ancient and could be studied using descriptive and experimental techniques analogous to those employed by contemporary European naturalists in the study of pharmacopoeia, despite it not being inscribed alphabetically. Hernández would serve as the mediator, selecting the important information, and organizing and transforming it to compose a compendium that would bring better knowledge of the Americas to all—Spanish patron and European scholars alike.[15]

Collecting and Cataloguing the World: Hernández, Pliny's *Natural History*, and the Spanish Empire

By the time Hernández began working on his translations of Pliny in 1567, Philip II claimed possession of vast kingdoms.[16] Under the auspices of the Spanish Crown, two hundred ships and more than forty thousand people had sailed back and forth between the Iberian Peninsula and the Americas between 1492 and the middle of the sixteenth century.[17] It was impossible for a contemporary observer like Hernández, reflecting on the *mirabilia* (wonders) of the ancient world that Pliny reported, not to wonder at the world broadening before his very eyes. In a comment to the opening section of book 7 of the *Natural History*, he urged his readers to consider how anyone could have "ever thought that there was a world as big as this" and that "thousands of leagues away" from Europe, there were "new peoples, new rites and religions," and even "human settlement south of the equator against the opinion of so many and such grave authors?" The Americas contained a great diversity of "sky and ground," of "admirable" and even "monstrous plants and animals," and Spaniards were able to "walk on it, so boldly and skillfully trusting fickle winds and fragile vessels."[18]

Like all of Hernández's works, his commentaries on Pliny were not published until centuries after his death. The only known manuscripts survive in the National Library of Spain and include translations and commentaries on books 1–25 of the *Natural History*. It is unknown whether Hernández completed translations for all thirty-seven books, since the remaining sections, if they ever existed, do not survive. The translations and commentaries that we have are bound in ten manuscript volumes. Two versions, or drafts, of several of Pliny's books exist in this collection, allowing for comparisons between earlier and later renditions of various sections of the *Natural History*. Hernández's commentaries generally follow his translation of each of Pliny's chapters under the subheading "the interpreter." In combination with the main translation, these sections provide a rich trove of information not only about Hernández's practices of reading, writing, and his interpretation of a variety of subjects but also about his editorial choices, the instances in which he learned new information about a particular topic, corrected previous assessments, or simply opted for a new phrasing. For instance, in a comment on book 7, on the human animal, in which Pliny describes humans as the only animals that cover themselves using borrowed trappings, Hernández recounts the different kinds of animals that possess feathers, hair, spines, and fur. After porcupines and hedgehogs, which cover themselves in spines, and goats, which possess hair, Hernández mentions an additional type of creature, those that have both "hair and spines like the hoitztlacuatzin [spiny opossum of Mexico]."[19] The inclusion of the "hoitztlacuatzin" shows how Hernández expanded the *Natural History* on the basis of new knowledge.[20]

With a few noteworthy exceptions, modern scholars have tended to scour the commentaries on Pliny in search of details to illuminate the doctor's otherwise fragmentary biography. This chapter instead employs the commentaries' multiple drafts to reconstruct Hernández's processes of translation and interpretation. The commentaries illuminate concrete links between Hernández and contemporary Spanish and American scholars, such as the editor of the Antwerp Polyglot, Benito Arias Montano, revealing their shared antiquarian, cartographic, natural-historical, and linguistic interests. Although specific sections of the manuscripts are difficult to date, these writings span at least twenty years. They richly document how Hernández read ancient and modern authors, how he engaged in textual criticism, some of the manuscripts he consulted, and even significant events in his life—like seeing a porcupine for the first time at the royal court—as well as the scholars with whom he collaborated. They offer a glimpse into the working methods of the men charged with directing the various knowledge-gathering projects that Philip II commissioned. The commentaries also demonstrate that Pliny, despite his errors and his lack of a global view, remained a powerful model of inquiry in sixteenth-century Spain, as well as a useful resource for integrating natural and human knowledge about Europe, Asia, Africa, and the Americas.

In Hernández's view modern Spain resembled imperial Rome, which in Pliny's day had ruled the greater part of the known world.[21] The commentaries explicitly establish Spain's standing as the head of a new empire of even greater immensity and extraordinary natural variety and Hernández's standing as the royally sanctioned doctor in charge of recording and cataloging its many wonders—a new Pliny. Considering Rome's political achievements, Hernández explained what Pliny meant by proclaiming it the "Conqueror of the World" in book 3 of the *Natural History*.[22] Pliny had discussed an early Roman omen recorded by Livy and Dionysius of Halicarnassus: when the ground was dug to erect a temple of Jupiter on the hill that would be later known as the Capitol, a human head (*caput*) was found. The soothsayers took it as a portent: Rome would become the head of the world. Beyond Rome and its domains, according to Pliny, lay "the rest of the universe," miniscule when compared to the vast Roman territories.[23] Hernández invoked Castile's American domains as proof of Spain's role as the successor to the Romans, conquerors of the "rest of the universe" and inheritors of their imperial legacy.[24]

Pliny's approach to studying the world in its totality and celebrating its strangeness resonated with Hernández's motives and outlook.[25] Increasing knowledge of the world beyond Europe engaged him even before he left for New Spain. News of previously unknown animals and plants reached the Iberian Peninsula via the early reports of travelers and the writings of Christopher Columbus, Pietro Martire d'Anghiera, and Gonzalo Fernández de Oviedo.[26] While working in the Hospital de Santa Cruz in Toledo, and after 1567 as a court doctor in the city of Madrid, Hernández listened to and read scores of accounts, some first person and some at second hand, of the fauna and flora of the Americas. Referring to Pliny's entry on the varieties of dogs, for instance, Hernández discussed the auxiliary role that canines played in the conquest of the Americas.[27] From the renowned mathematician and alchemist Bernardo Pérez de Vargas, he also cited the unusual behavior of canines in the north of New Spain. When Spanish soldiers reached Cybola, he learned, they found a breed of docile dogs that carried loads of corn, or *tlaolli*, on their backs.[28]

Creatures with even stranger features also seemed to exist farther south. When a soldier returning from the viceroyalty of Peru informed Hernández of the fearsome birds of the province of Homagua, which "are so big that they snatch Indians with their claws, and they eat them and tear them apart in the air," Hernández speculated that they might be Pliny's gryphs or vultures, and that although the account might seem marvelous to a cautious reader, it was worth mentioning because "[news of] monstrosities arrive each day from those lands." In Seville he had also seen what seemed to be the claw of an animal that, if not a gryph, certainly was a bird of equal proportions and ferocity.[29] Hernández inserted this witness report to elaborate on Pliny's description of the Arimaspi. He used it to suggest to those who had challenged the existence of the gryph that perhaps the bird really dwelled in the Andes, and not in the unknown Riphean Mountains. In doing so, he tried to identify an American animal with its still unattested literary equivalent. But perhaps more significantly, Hernández took advantage of the tale to claim that despite Pliny's mistakes—the Arimaspi were not in fact a one-eyed people—the startling marvels and mysteries of the Americas left Pliny "very well credited." Hernández had heard about, and would eventually see with his own eyes, the gigantic snakes of Tepoztlan, or the lizard with the round body called *tapayaxin*, known to the Indians as a "friend of man," because it liked to be held and if mistreated cried tears of blood."[30] Echoing Pliny almost word for word, he proclaimed "that the nature of things seems at each step incredible, especially to those that think about them constantly."[31] Hernández assured his reader that new findings glorified, rather than discredited, the methods and learning of the ancients.[32] Besides the rhetorical function of such statements to enhance the urgency of his work, they also emphasized that as a representative of the Spanish Crown, Hernández saw himself occupying a privileged role. He was about to record, for the benefit of his king, the Americas' natural phenomena and their precious pharmacopeia. The doctor would also have the means to pursue, through his own observations and firsthand investigations, the resolution of numerous botanical, geographic, and ethnographic questions debated from antiquity until his day.

Pliny had been of interest to Spanish humanists since the early sixteenth century. Starting in 1513, at the University of Alcalá de Henares the grammarian Antonio de Nebrija and his successor taught Pliny's writings in addition to those of Augustine and Aristotle. Their main aim was philological: to gain new insight into the sacred scriptures.[33] Numerous letters also demonstrate the existence of an active group of scholars devoted to the study of Pliny's work around the royal court. This self-proclaimed *gens pliniana* included prominent Toledan doctors. They exchanged materials and dispatched frequent carriers between Madrid and Toledo. Their goal was to discuss the Plinian texts but also to collect, compare, and correct copies of the manuscripts extant in Spanish libraries.[34] Among the *gens pliniana* was Alvár Gómez de Castro, whose regular correspondent Juan Páez de Castro expressed his desire to produce a new edition of the *Natural History* that would improve on the many mistakes of Hernán Núñez's version. Hernández often cited Núñez, sometimes praising and sometimes correcting him. He attributed Núñez's mistakes to the defective Salamanca manuscript with which he had worked, rather than to his predecessor's inability to properly understand Pliny's writings.[35]

Various scholars before Hernández had composed medical works in Castilian, including Andrés Laguna with his 1555 translation of Dioscorides. Even so, Hernández perceived his own undertaking as an innovative project. He knew of no commentaries to the complete works of Pliny except those of the French humanist and ambassador Etienne de L'Aigue (d. 1538), whose work of 1530 had come to seem outdated and unreliable. Hernández considered Cristoforo Landino's Italian version of 1475 "not translation but confusion," and the more recent Italian edition of Ludovico Domenico a mere copy of Landino.[36] He does not seem to have known the French version of Antoine du Pinet (1562).

Italian humanists had studied the *Natural History* since the mid-fifteenth century. Among the most important commentators was the doctor Niccolò Leoniceno (1428–1524), who figures prominently in Hernández's work. A student of Guarino da Verona, Leoniceno put forward several criteria for commenting on Pliny, which included reading the text in conjunction with other ancient botanical authorities such as Dioscorides and Theophrastus. Hernández would adopt this strategy as well. For instance, in a comment on book 27 dedicated to the ailments of trees, Hernández noted that since Pliny's text was "depraved," he would not attempt to "castigate" the text. Instead, since Pliny was clearly citing Theophrastus, he would reproduce the thirteenth chapter of the fifth book of *On the Causes of Plants* so that the reader could "restitute and emend" the Plinian text on their own.[37]

Leoniceno also emphasized that the study of materia medica required not only reading ancient texts but also collating their contents with experimental knowledge.[38] He embodied what Brian Ogilvie believes to be the fundamental "epistemological attitude of humanism," the "concern with the particular."[39] An emphasis on particular details is present everywhere in Hernández's commentaries. Like Leoniceno, Hernández was preoccupied with ascertaining and describing the details of particular natural phenomena through the study of texts, both ancient and modern, mediated through his own experiences.

The commentaries go beyond clarifying difficult passages in Pliny's prose. Although a significant portion of Hernández's explanations deal with technical issues of translation, the majority of the remarks attributed to "the interpreter" debated the Plinian text either by expanding on a specific topic, by challenging an assertion, or by complementing a description. In the process, Hernández used an astounding range of ancient and contemporary sources, including the works of the reformed humanist Conrad Gessner on animals (1551–1587) and plants (1541), the Venetian scholar Ermolao Barbaro (1492) on plants, and the more specialized writings of the English naturalist William Turner (1538) and his Swedish contemporary Olaus Magnus (1555) on northern Europe. When his own experiences were relevant, Hernández also invoked the observations and experiments that he undertook in hospitals, botanical gardens, and increasingly, in the second half of the book, in the Americas.

The entry on the sea calf, following the contents of Pliny's book 11 on aquatic animals, is representative of Hernández's commentarial style and contains the kinds of sources on which he habitually drew. He begins by citing Aristotle's *De animalibus* (*On Animals*). The sea calf, or *vitulus marinus* of the Latins, was the "*phoca* of the Greeks," so named on account of its bellows. He then cited the work of Guillaume Rondelet, a respected French naturalist and professor of anatomy of the University of Montpellier. Published in 1554, the *Books of Saltwater Fish* (*Libri de piscibus marinis*) described around 250 marine animals. Besides the two kinds of sea cows that Rondelet illustrated in his treatise, the *vitulus marinus mediterranei* and the *vitulus maris oceani*, Hernández believed there to be an additional kind or "species" that dwelled in the Indies that the Haitians call the manatee. In the language of the people of that island, *manati* meant "breast or udder."[40] He deduced the affiliation among the three variants from similarities in the animals' forms, including the presence of the almost identical "stone that they grow on their heads," but especially because they all lactate.[41]

Yet modern naturalists ignored this, Hernández lamented, just as they failed to note the "reverse fish" that lived stuck on the belly of sharks. This species he himself had seen and drawn as he crossed the ocean en route to New Spain. He concluded by reprimanding the French naturalist Pierre Belon, whose illustration of the *vitulus marinus* in his *De aquatilibus* (On aquatic animals) of 1553 did not correspond, in Hernández's view, to the *bezerro* (sea calf) or *vitulus* of the ancients, a fact "clear to anyone who compares it with the ancient descriptions."[42]

Hernández not only offered additional information about the *vitulus marinus* but also introduced some of the characteristics of this American "species" that he would discuss further in his own work. While in Haiti, Hernández learned that the manatee's meat caused pain to those suffering from syphilis (morbus gallicus), and that the little stone, or bezoar, that grew on its head served as a very efficient diuretic (fig. 4.3).[43] Besides his corrections and contributions to many European debates on natural-historical topics, like fellow Renaissance naturalists Gessner and Ulisse Aldrovandi (1522–1605), Hernández composed his treatises mindful of

CAPVT VII.

De schaubhut.

VOcantur schaubhut siue summere, pisces quidam pileati, tantæ magnitudinis & roboris, vt naues sæpè ingentes subuertant. Quando enim mare placidum est, & nullæ tempestates existunt,tum tantâ sæpè vi ad nauem feruntur, vt omnia in eâ concutiantur. Itaque nautæ hoc animaduertentes, mox dolium vnum atque alterum in mare proiiciunt, vt cum ijs, relictâ naui, sese oblectent.

CAPVT VIII.

De lupis & canibus marinis.

IN plerisque Indiarum regionibus vber luporum marinorum natio, formidabilis vlulatu,edulis. Barbæ eorum seruiunt purgandis dentibus. Fertur etiam,dentes eorum dolorem sedare dentium,si calentes contingant nostros. Lapidibus vescuntur,à vulturi-

bus necantur. Insiliunt illi in oculos luporum,& excæcant,sic facilè dilaniant.Vultures illi ingentis molis sunt, magno quoque odio & periculo cum tiburonibus pugnant. Luporum pelles, etiam iam aptatæ ad cingula vel crumenas, pelagus sentiunt: eriguntur pili eorum, exaltato mari, comprimuntur,mari depresso. Cingula lumborum ægritudines medicantur. Profunde dormiunt, vehementer stertentes:ex ronchis enim deprehenduntur, & necantur.Franciscus Bertie de canibus marinis hæc testatur : Cùm Thomas Candisch in suo itinere insulam quamdam ingressus esset,offendit ibi maximam copiam canum marinorum,qui anteriori sui parte,quoad caput, collum & corpus medium, leonem oblongis & crispis capillis crinibusque per omnia repræsentant. Singulis mensibus catulos progenerant, quos lacte suo enutriunt. Hos canes interficere nullis armis potuimus, donec fustibus capita ipsorum percuteremus. quin etiam tanti roboris erant ; vt nostrûm tres vel quatuor vnum superare & interficere vix possemus. Carnem habent boni cibi , quæ vernecinam sapore refert.

MANATI PHOCÆ GENVS.

CAPVT IX.

De manati.

INgenuum monstrum manati est. Eius historiam ex Ouiedo vsurpo. Ingens est piscis marinus, tametsi in magnis huius insulæ (Hispaniolæ videlicet) & aliarum prouinciarum fluminibus assidue capiatur.Maiorisestlongitudinis & amplitudinis tiburone. Qui plenam adquisiuere magnitudinem, foedissimi sunt aspectus, nec multum absimiles illis vtribus,in quibus Medinç del Campo & vicinis locis vehitur mustum.Caput habet piscis iste bouis instar, vel etiam maius: oculos exiguos pro corporis magni-

tudine, & binos crassos pedes tamquam alas brachiorum loco,quibus natat, circa caput sitos. Corio integitur,non squamis , valdeq; mansuetum est animal , aduersa flumina subiens; & secundû ripas, quas apprehendere potest, herbas depalcens,nec aquam egrediens. In Americæ continenti solent sagittarij ex cymbâ siue monoxylo huiusmodi pisces & alios plerosque occidere,quia in summâ aquâ natant:hamata itaque iacula conjiciunt, in quorum summo tenuis, sed validus alligatus est funiculus. Ictus piscis fugit, funiculum autem laxat sagittarius, in cuius extremo ligni aut suberis fragmentum est annexum,ne prorsus mergi queat funiculus, tum etiâ vt appareat eius extremitas. Piscis verò

X 4

their intended audience, an audience that included first and foremost his patron but also an abstract community of natural historians whom he hoped to reach through publication. Although none of Hernández's works were printed in his lifetime, they bear many traces of his intended dialogue with an emerging community of self-aware practitioners of the new discipline of natural history, of which he saw himself an active member.[44]

Another link to the emerging community of natural historians was Hernández's propensity to identify with Aristotle. In a passage in book 8 on terrestrial animals, following Pliny's description of the gigantic snake known as the *megasthenes*, Hernández mentioned that he had heard of similar snakes in the Indies. He hoped to write about them "in the Natural History, that by command of the undefeated Philip II, Our Lord," he "will travel to those parts to write. This history will deal with all that Pliny writes in that [book] of his on the Old World."[45]

In fact, Hernández compared Philip II to Alexander the Great and himself to Aristotle, as the former had commanded the latter to describe all the animals in the universe. According to Pliny, Aristotle's expedition was ambitious and "orders were given to some thousands of persons throughout the whole of Asia and Greece . . . so that he might not fail to be informed about any creature born anywhere." The tale of Aristotle's studies, though exaggerated, was a source of validation for Hernández, as it was for other Renaissance naturalists. These scholars appropriated the story to appeal to their patrons' generosity. Perhaps most significantly, as Ogilvie has argued about natural historians in northern Europe, the tale, in its hyperbolized version, presented natural history as a collective enterprise.[46] In the case of Hernández, this identification seemed all the more appropriate because, as Philip II's instructions underscored, his expedition was to be based primarily on questioning Indigenous physicians. Furthermore, by exerting his title of *protomédico* of the Indies, the king placed at the disposition of the botanist, at least theoretically, an infrastructure of hospitals and medical practices across New Spain.

Observation and the Study of Names in the Plinian Commentaries: Investigating Local Knowledge

Hernández's attention to the names of aquatic animals and their meanings was characteristic of early modern natural history. Names often conveyed the major characteristics of a creature or a plant. The bat, Hernández and other natural historians asserted, took its Latin name *vespertilio*, from vesper or evening, from the time it goes out to feed. In Spanish and Catalan, its name reflected its mouse-like shape and its poor eyesight: like *mur* (mouse), and *ciego* (blind), hence Castilian *murciélago*, or "blind mouse," and Catalan *ratpenat*, "winged mouse."[47] When describing the whalelike animal that Pliny called *physeter*, known in Latin as *flator*, Italian as *capidolio*, and Castilian as *bufeo*, which all mean "bellows," Hernández clarified that these names all referred to the water that the creature expelled from a thick canal of greater diameter than in any other variety of whale.[48] Names also provided natural historians with insight into the attested behaviors of animals. In every entry of Gessner's books on animals and plants, a section listed all of a creature's known names.[49] Rondelet too, in his work on marine animals, collected all known names of fish. Studying the etymology of names had either descriptive or classificatory purposes. In addition to witness reports and personal observations, the study of the names of plants, animals, and minerals formed an essential part of Hernández's scholarly apparatus and almost always introduced the subject being described.

Etymologies constituted, for Hernández, a complementary body of evidence and a resource with various kinds of descriptive value.[50] As Isidore of Seville explained, an etymology gathered in the mind associated words and concepts and served as the beginning of a "memory chain."[51] For instance, the Castilian name of the bat, *murciélago*, brought to mind a mental image of a blind mouse. This image would then generate numerous other associated images. The clearer and richer the image brought to mind, the greater the force, or capacity of a word to convey many features or characteristics of the object that it signified.[52]

More than any other definition, Aristotle's understanding of the nature of speech shaped medieval and early modern philosophy of language. Spoken words signify by convention and correspond to a mental language, which has natural meaning; the mental language is common to all human beings. Written words, by analogy, are also conventional and correspond to spoken words. Earline Jennifer Ashworth explains that although scholars like Thomas Aquinas and Henry of Ghent discussed alternative views—mainly the Stoic proposition that names might have natural origins because the sounds of words could sometimes resemble the nature of the thing signified—they generally accepted Aristotle's schema.[53]

Nebrija adopted an intermediate position; for him, words could emerge naturally and later be modified by convention. The Valencian humanist Juan Luis Vives also defended a similarly intermediate position. "Since language is born out of reason," he wrote in *De anima* (1538), "[it is] natural to man and reason itself. . . . [T]here is no fixed language by nature, they are all artificial."[54] However, an ideal, ecumenical language, which for Vives was Latin, would have the ability to be soft, or easily pronounced, erudite, and rich. "The softness would consist of the musicality of the voices. . . . [T]he doctrine consists of the property of words that express things; the richness [to] the variety and abundance of words and expressions." Although artificial, if properly cultivated, languages had the potential to resemble the objects signified.[55]

The fact that languages were artificial (*ex arte*) did not mean that words were arbitrary. Scholars usually linked their study of words to an understanding of the history of languages. They explained the origins of words by relying on a linguistic hierarchy broadly laid out in the Bible. In book 9 of the *Etymologies*, Isidore synthesized what continued to be the dominant position throughout the sixteenth century: linguistic diversity arose after the Flood when humans, overwhelmed with pride, built the Tower of Babel. Before the tower there was one original language, infinitely prolific, soft, and expressive, with which Adam named all creatures. For Isidore this primordial tongue was Hebrew. It was the language that the patriarchs and the prophets used in their speech and in their sacred writings. After Babel, many nations of many languages emerged. Soon there were even more nations than languages because a single "language stock" engendered a variety of nations.[56]

Like most of his contemporaries, Hernández accepted this linguistic genealogy. Invoking Augustine's *City of God*, chapter 11, book 16—which, as the commentaries to Pliny demonstrate, he read alongside Vives's commentaries—Hernández claimed that Hebrew was the primordial tongue because, as Genesis 11 stated, before the confusion of Babel all peoples used the same language.[57] This language, the most excellent of all, would prevail among those who did not "conspire in the deviated edification of the tower," and one of the "just [ones] was Heber, from whom the Hebrew tongue takes its name."[58]

While generally remaining steadfast in asserting the primacy of Hebrew, early modern scholars developed new approaches to etymological method to uncover historical truths.[59] Marie-Luce Demonet has traced how scholars changed the traditional parameters of ancient and medieval etymologizing. In the wake of Roger Bacon (1213–1291), scholars abandoned the ancient system of Marcus Varro—in which words were analyzed alphabetically using their first letter—and instead adopted the concept of "root" from the study of Greek and, more influentially, of Hebrew and Arabic. As Demonet demonstrates, this was a fundamental shift because grammarians began to conceptualize a word's morphology in the Aristotelian terms of "substance" and "accidents."[60] When applied to vernacular languages, this principle changed how linguistic analysis could be performed.[61]

The question whether a language could naturally signify acquired new relevance in debates about word study. New works raised the question of the capacity of sounds to mimic the properties of things. Hernández referred his readers to Lodovico Ricchieri's (Caelius Rhodiginus) *Lectiones antiquae* (1516), where they could read about "the strength that uttered words possess to cause changes and effects on things."[62] However appealing, the naturalist understanding of words did not displace conventionalist approaches to the history of languages, "especially when dealing with post-Babelian tongues" that were distinct from Hebrew.[63]

Historians of language developed new and more sophisticated research techniques to establish etymologies in vernacular languages.[64] They were aided in their efforts by newly composed grammatical treatises, histories, and dictionaries. In his effort to prove that the pre-Roman language of Gaul had been Germanic, Gessner, for instance, turned to the names of local plants as recorded in ancient authorities. In his view, these names preserved the language of the region's earliest inhabitants and their knowledge.[65] Following this same practice, Hernández examined the chains of information implicit in Spanish, French, and German words, among others, in the service of natural or historical studies. He also attempted to extend this methodological innovation further by systematically applying it to the radically different languages of New Spain.

Another use of etymology in multiple languages was to identify variations in the designation of a single species across a wide geographic landscape.[66] This problem plagued early modern natural history. Elaborating on Gessner's entry on the tiger, Hernández noted that this animal's name designated great strength and speed and came from the Armenian word *tigris*, or "arrow." The river Tigris was also named in this manner owing to the strength of its waters. Hernández wished to record all the names of the tiger in other languages and their meanings "and to put at the end of" the Plinian commentaries "a table that contains in alphabetical order the names of plants, animals and minerals in all of the tongues, and in this manner satisfy all of the nations."[67] A table of this kind would allow natural historians to determine all the different words used to identify a single animal and to communicate more effectively about their subjects of study.

Sometimes apprehending the properties of an animal through its name allowed Hernández to establish connections between the creatures of Afro-Eurasia and those of the Indies. Rondelet's sea calves were similar to manatees, for instance. In Mexico, some animals bore tremendous resemblances to the porcupine. Hernández had seen the porcupine, the *echinus terrestris* in Latin or *hystrix* in Greek, in Spain, especially at the court of Philip II. Medicines made from the *hys-*

trix and also its different names appeared in many books. Hernández planned to include the names and medicinal virtues of the porcupine, as "we will do for other animals," in their proper place, most likely meaning his planned table of concordances.

Discussing the porcupine gave Hernández the opportunity to mention a strange animal he had read about in Gessner's book. In 1550, the creature toured northern Europe alongside men who made a living showcasing its monstrosity. Although Hernández had also read the account in Girolamo Cardano's *De subtilitate rerum*, he mentioned in a note that he would cite Gessner instead, since Cardano's *De subtilitate rerum* was censored in Spain.[68] Hernández believed that the *hoitztlacuatzin* of New Spain shared many qualities with the porcupine. Its name, according to Hernández, signifies "spiny opossum." Hernández further claimed that the Indigenous inhabitants of Mexico "save their spines and believe that nine of these, heated with a little bit of fire . . . then reduced to dust and administered with water or wine, break kidney stones and clean the urinary tract." Through his description and analogy with the European porcupine, Hernández tried to convey some of the extraordinary features of this animal "which is worthy of being seen," for "the strength of its quills is not any less admirable" than those of its European counterpart (figs. 4.4 and 4.5).[69]

Hernández recognized that names could mislead, especially if the name-makers were unqualified and incapable of capturing the properties of things, or if names had suffered irretrievable corruption with the passing of time. He cautioned his reader about this possible difficulty when discussing Pliny 8.34 on terrestrial animals, which dealt with the *tarandus*, the *lycaon*, and the *thoes*. The *tarandus* of Pliny's description populated the lands of Scythia. Gessner, a "wise and very hardworking man," suspected that perhaps this was the same animal that the Polish call the *thur*, "persuaded by the resemblance of the name *Daran*, which some moderns impose on it on account of its mouth, different than the rest of the ox, because of the size of its body and its rarity."[70] These reasons did not convince Hernández, however. It was better "to not judge such a thing on account of these [reasons], specially about beasts that he has never seen." The etymological proof had "little strength, as these [names] are found at each step in diverse tongues meaning very diverse things, and mostly that some poor fellow [*paniaguado*] could have been the imposer of this idea."[71] Although names could be valuable for obtaining and transmitting information, they were not in themselves sufficient to make conjectures about the identity of a particular animal or plant. Hernández insisted that their contents had to be corroborated by other types of evidence. In

Tunc feræ circumiacentes, arcano naturæ documento conſciæ inuitationis, ad epulum ocyùs aduolant, prædâ ex ſigno inuentâ ſariantur: poſtremus ipſe ocotochtlus deſcendit, vt innocuè paſcantur conuiuæ, nam ſi ille prior veſceretur, corruptis veſieno ſuo reliquijs, ceteræ perirent feræ, quotquot de-

inde guſtarent. Tantâ vrbanitate charitati fauit, tam comi prouidentiâ ſaluti alienæ. Scilicet nemo eſt qui prodeſſe néqueat, ſi velit. Non ita viribus & ogibus beneficentia, quàm voluntate & ſtudio conſtat. Plerimi opulenti inutiles ſunt, vtinam non etiam noxij; at nemo beneuolus non beneficus.

HOITZLAQVATZIN.

C.I.

CAPVT II.
De animalibus ſpinoſis.

ADmirandum genus bruti ſagittatoris nouus orbis alit. Animal ſpinoſum, mediocris canis exæquans magnitudinem. Ab Indis hoitztlaquatzin dicitur, ſiue ſpinoſus tlacuatzin. Nomen forma meruit. Simile eſt illi, quod tlacuatzin vocant, de quo ſuo loco ſermo erit. Veſtitur canis & acuminatis ſpinis, tres vncias productis pilis quoque nonnullis (ſi caput excipias) moſtibus inſertis, ſiue atrâ lanugine; & circa exortum candenti: venter tamen, crura, & brachia aculeis omnino carent. Spicula tenuia ſunt, candentia, &lutea: vbi deſinunt nigricant, mucrones peracuti. Cauda breuior eſt quàm tlacuatzin, craſſiue tamen maiori compenſatur. Inops aculeorum eſt à mediâ parte vſque ad extremitatem. Pedes ſimiles habet quaſi pecotli; ſed latiores (miſceo ſæpè verba & ſententias Franciſci Hernandi, nec meo cerebro & ſtilo loquor ſemper) roſtrum pænè caninum, niſi quòd reſimum. Dentes ſuprà infráq; bini, quemadmodum ruccæ. Iaculatur ſpinas ſuas in canes inſectantes, nullâ excutiendas induſtriâ, nullo conatu: ſenſim magis magiſque vi quadam arcanâ ſubeunt, & iactatione plus figuntur,

donec viſcera perfodiant. Alliciunt & ebibunt innatum humorem, donec transfixum & perforatum canem tabefaciant. Quærunt Indi aculeos illos dolori capitis placando. Admoti verò fronti & temporibus, ſponte adhærent, nec priùs delabuntur, quàm hirudinum more hauſerint noxium ſanguinem, quo ſuam farciunt cauitatem. Cauſâ languoris remotâ, ægrotans valet. Cuicumque rei cuſpis eius admiſceatur, etiam poſtquam ab animantis cute eſt diuulſa, occultâ virtute, etiam cùm à nemine vrgeatur, ſponte penetrat carnem. Feminis eius vſus perforandis auriculis ad inaurium ſuſpenſionem. Perforatio prorſus fit nemine impellente, quaſi exuat ſetas. Experimento compertum, vnius noctis interuallo craſſiſſimo corio admotam, ſponte cornu totum, perinde atque acus hominis impulſa manu tranſmittitur, penetraſſe. Eorumdem aculeorum puluis dragmæ vnius penderes ſumptus, dyſenterijs médetur, vrinam euocat. Gaudet hoc animal montoſis & calidis locis Xonoltz & Yzotzocolci. Fortaſſe ad Bæicam noſtram deportatus, non iniquum ſibi cælum toleraret. Veſcitur fructibus: cicurari poteſt. Aliqui putarunt hyſtricem aut mantychoræ ſpeciem, cùm multis differat, & in nonnullis admirabilior fortaſſe ſit, vt ex collatione

TLAQVATZIN.

CAPVT IV.

De animalibus manticatis,
siue Tlaquatzinis.

NOn minùs admiráda fera, quam Indi
vocant *tlaquatzin.* Antonius Herrera
taquatzin dixit: recentes Hispani Scriptores
corrupto nonnihil nomine *tlaquacum,* Car-
danus *chiurtam,* siue *chuciam,* Stadenius
Seruoy, Nomenclator *semi-vulpam,*seu *lope-*
copithecum: Raphe Hamor in descriptione
Virginiæ *apoßumen* dixit; alij *aucham,* alij
sasapim, alij *cerigonem* dixere, quia sic voca-
tur species vulpeculæ apud Brasiliam & Ma-
lucas,sed ea,de quâ exordium sermonis insti-
tuimus, etiam in Dariene & Floridâ reperi-
tur.Animal est parui canis formâ & magnitudine,binos dodrantes longum, rostro te-
nui, prolixo atque depili,exili capite,tenuis-
simis mollissimisq; auriculis, ac pænè trans-
lucentibus, pilo longo & candido ,sed circa
extrema fusco aut nigro. Caudâ tereti,duos
dodrantes longâ , & persimili colubrinæ;
fuscâ, sed postremò candidâ, quâ mordicus
&

Figure 4.5. Tlaquatzin in Nieremberg's *Historia naturae, maxime peregrinae* [. . .]
(1635). Courtesy of the John Carter Brown Library at Brown University.

the case of the *tarandus* and the *thuron*, an actual eyewitness account of the creature and whether its properties matched those that Pliny recorded was missing.

Studying names as Hernández advocated required asking local speakers the meaning of the names they used or searching for attestations in ancient works. This line of questioning could sometimes yield fruitful results, but it could also mislead. Regardless of the outcomes, it was a common starting point for natural history. "Interpreters of natural phenomena," as Katharine Park has emphasized, studied material that lacked causal explanations, and as such, they tended to rely on the recollection of "similar events in oral or written memory that might offer guidance as to what to expect." Ancient authorities like Pliny, Park writes, "emphasized that literacy was not necessarily a prerequisite for the development of conjectural knowledge; generations' worth of observation might be orally transmitted."[72] The oral knowledge of illiterate peasants, or those who did not read books or record their knowledge in natural historical compendia, constituted a valid source of knowledge for a sixteenth-century scholar like Hernández. This knowledge was particularly important in the "experiential sciences," like alchemy or botany, because scholars tended to study topics that were "contingent and therefore not amenable to deductive explanation."[73]

Sayings and folktales were part of this trove of oral information, as were the names of things. Hernández's contemporaries debated the nature of the knowledge contained in words and sayings and developed tools to study them. For instance, as we have seen in chapter 3, in 1568 the Sevillian paroemiologist Juan de Mal Lara defended the utility of glossing sayings. Arguing for the value of sayings for natural philosophy, Mal Lara claimed that there was not a day in the calendar that lacked a proverb instructing peasants (*labradores*) how to tend their land. These sayings reflected the wisdom of generations who observed the sun, moon, and stars and learned the agricultural arts. The wealth of sayings in Spain led Mal Lara to claim "that even before Greece had its philosophers," Spain already possessed an ancient tradition of proverbs.[74]

Hernández's correction of Gessner's overreliance on names suggests how scholars sought to corroborate what they learned via oral routes with direct observation, the collation of other scholars' observations, or the procurement of what they considered reliable testimony. Gianna Pomata has shown how, throughout the course of the sixteenth century, the category of observation underwent a significant transformation. Observation began to imply mainly "experientia propria," or "autopsia," that is, first-person "authored observation." Sharing observations became a central part of emerging disciplines like medicine, natural history, and astronomy. "The ideal of collective empiricism," as Pomata further elaborates, "cut across disciplinary boundaries."[75] For Hernández, it was crucial to connect his own observations to those of other practitioners of natural history, medicine, astronomy, and even human history. The insights that he routinely gained by questioning "rustics" had to be validated by his own knowledge or that of his colleagues. After all, scholars were guardians "of the heritage of the ancients." Ogilvie has stressed that this effort to verify was a fundamental characteristic of sixteenth-century natural history in northern Europe, where scholars aspired to practice "a science that went beyond mere individual experience."[76] As his many unpublished writings demonstrate, the same was true of Hernández.

The study of human history and of antiquities likewise could employ various kinds of information, including oral knowledge. Substantial sections of Hernández's commentaries on Pliny discuss the ancient history of Europe. The commentaries on book 3, which deal with Spain's history and geographical features, are particularly rich with observations on how to derive historical information from names. They also show how Hernández's experiences in the scholarly circles patronized by Philip II's court, where he resided between 1567 and 1570, shaped how he approached the study of cities, landscapes, and human practices and would later influence his study of antiquities and botanical practices in New Spain.[77]

As we might expect, Hernández often used etymologies in combination with other kinds of evidence to confirm or undermine a historical narrative or to establish the origins of peoples. In a comment on Pliny's *Natural History* 3.1, which dealt with Spain's geographic divisions, Hernández explained why some Iberian cities had names that were identical to those of towns in Palestine. The original inhabitants of these areas "came with the king Nebuchadnezzar, second of this name"; they were "Persians and Chaldeans, as Josephus and Strabo testify, and also, as some say Hebrews. And in this manner [they] gave to Spanish cities not a few names that endure until today, like Aceca and Escalona, Yepes, Maqueda and others similar to these, which were first names of Palestinian cities."[78]

Hernández most likely learned this explanation from the theories, discussed in chapter 2, that the biblical scholar Benito Arias Montano and his followers formulated, on the basis of biblical exegesis and etymological derivation, to explain the presence of Hebrew toponyms in Spain.[79] Much has been speculated about the relationship between Hernández and Arias Montano. A well-known Latin poem, "Epistle to Arias Montano," in which Hernández praised Montano and asked for his help in improving his standing at the royal court, has been the only known document to attest to their relationship. However, in the comments on book 3 of

Pliny's *Natural History*, which dealt with the ancient cities of Spain, Hernández describes having had an exchange with Montano, whom he lauds as "distinguished among men in letters and tongues," about the lack of correspondence between the locations of some surviving inscriptions and Ptolemy's measurements. Hernández records that the conversation took place before Montano left for Antwerp to work on the Polyglot. He reports that Montano told him that Vertóbriga had once been a Celtic city. Montano had seen "a stone where *respublica vertodeicensis* is written" located at the same longitude as Fregenal in Extremadura. Even if this was not the location given by Ptolemy, "not little credit should be given to inscriptions." This was especially true because many of the "latitudes and longitudes of that great mathematician were corrupt, not to say mistaken." Between the physical evidence of stones and Ptolemy, Montano trusted the former more.[80]

The royal chronicler Ambrosio de Morales also worked alongside Hernández to investigate Roman inscriptions in the Castilian countryside. In the city of Tarragona, whose Roman name was Tarraco, the two scholars attempted to piece together parts of an epigram that the passing of time, inclement weather, and waves of invaders had almost worn from the stone surface. Morales concluded the brief section that described this excursion by declaring his admiration for Hernández, to whom a long and happy friendship bound him.[81] Hernández's collaborations with Montano and Morales offer context into the antiquarian and textual practices that Hernández would then adapt to study the Americas' pharmacopeia.

Etymological analysis of names could sometimes yield ambiguous results. Conjectures were often difficult to make even when inscriptions and coins were available. When explaining the name Africa, Hernández presented a few possible etymologies. Perhaps the continent owed its name to Afer, of the line of Abraham. Its name might be derived, according to Leo Africanus, from Iphrico, the king of Arabia "whom erudite Moors claim arrived there frightening off the Assyrian king." Or perhaps from the Arabic verb "farco," which meant "to divide," "because it is divided from Europe . . . by the Mediterranean and Egypt and Asia by the Nile."[82] Hernández found none of these theories persuasive enough to select one above the others. Despite their inconclusiveness, the etymologies were suggestive. Each linguistic thread contained valuable information that might aid in understanding the geography and history of that landmass; for this reason, they were worthy of being related in the commentary.

In his account of the multiple ways geographical names were given and their mutability with the passing of time, Hernández exhibited an attitude to language akin to those of Aristotle and Vives. Names were assigned to things and thus were conventional. Some names, however, were more deliberate than others and therefore more accurately resembled what they designated. Some knowledge could therefore be gained from etymologies because they could be exegetically correct even when the name-givers were imprecise. That is, the names might capture the knowledge or mindset of the namers. Thus, aspects of the experience of the namers could be retrieved.[83] Names could also offer a way to understand the history of name-makers and not just of the objects that they signified. This broadened the historical possibilities of etymologies. Yet names could also be misleading, especially if the name-makers were unqualified and incapable of capturing the properties of things.

New Natures, New Languages

Francisco Hernández set out from Seville in 1570 having developed a relatively consistent method for studying plants, animals, and antiquities.[84] His son Juan Hernández accompanied him, as did the cosmographer Francisco Domínguez, whose purpose was to describe the territory and record geographical information.[85] The doctor's fleet followed the conventional maritime trajectory of Spanish ships bound for Castile's overseas possessions: they first stopped in the Canaries and continued from there to the island of Santo Domingo. In the Caribbean, Hernández began to compile information for his natural history. His expedition aimed to gather more accurate information regarding how American plants grew and how locals made use of them.

Nicolás Monardes's famous herbal, published in 1565, regrettably did not inquire into the conditions of origin of American pharmacopoeia. Like Hernández, Monardes was trained in liberal arts and medicine at Alcalá de Henares and Seville.[86] In the ports of Seville, Monardes keenly sought out both plant specimens and firsthand accounts of how the Indigenous peoples of New Spain used these herbs to treat illnesses.[87] This enabled him to publish a rich collection that followed the conventions of a Renaissance herbal but also included a few descriptions of how Spaniards had learned from the diverse inhabitants of the new territories to recognize, prepare, and apply New World medications.[88] At the time of his death, his *Medicinal History of the Things That Are Brought from Our Western Indies* (*Historia medicinal de las cosas que se traen de nuestras Indias Occidentales*) had appeared in seventeen Spanish editions, five Latin editions, three French editions, and three English editions. Monardes concluded his *Medicinal History* with a forceful admonition: his contemporaries were "worthy of great blame for nobody

[was] writing anything about all the herbs and plants and other medicinal plants from New Spain." This was all the worse because proper investigation of the "medicines . . . the Indians sell in their markets . . . would be a thing of great utility and benefit."[89]

This was the case of the root known as the *mechoacan*, which took its name from the region in central Mexico where it grew. According to Galenic pathology, to balance the humors of the body, human beings had to physically expel the excess humor that was responsible for causing different kinds of diseases.[90] Purgatives and sometimes bleeding were used to free the body from noxious extremes.[91] When *mechoacan* reached European shores, it gained popularity almost immediately on account of its purgative qualities. Spanish physicians, however, could not truly understand this medicine or, perhaps more importantly, grow it and commercialize it from their own gardens, because in many instances Spanish traders would obtain the plants only after Indigenous American suppliers had already dried them. Monardes lamented the circumstance: "The lack of attention is such— since everyone is only concerned with their earnings—that they do not know more about it, except that the Indians in Michoacán sell the roots dried and cleaned in the same manner that they arrive in Spain."[92]

Hernández's investigations were meant to remedy this by directly and systematically inquiring into the conditions in which American plants and animals flourished.[93] What Monardes indiscriminately knew as the *mechoacan*, Hernández would later realize were actually three types of roots that the Mexicans called *tlallantlaquacuitlapilli*, which were used in various ways that went beyond mere purgation. Even so, improving on Monardes was easier in theory than in practice. Interrogating Indigenous physicians would prove to be a great challenge for Hernández and a constant source of contention and disappointment to him, as the poem he dedicated to Arias Montano poignantly reveals.[94]

To fulfill this investigative purpose, the doctor organized his expedition in three main phases. During the first, which lasted almost three years, he traveled around the Viceroyalty of New Spain through a series of short trips around the city of México. During the second, he engaged in slightly longer journeys to the central regions, from Santa Fé to Toluca, and from Cuernavaca to Huaxtepec, where Moctezuma II, the Mexica ruler at the time of the arrival of Cortés, had once kept a lavish botanical garden. Finally, during the third phase, he undertook much longer routes to Oaxaca, Michoacán, and Pánuco.[95]

Besides the cosmographer Domínguez, who was only sporadically present, Hernández was always accompanied by Spanish and Indigenous doctors, as well as Indigenous interpreters, scribes, and painters (three of whom he identified in his will). Through interpreters he administered questionnaires, sought out Indigenous physicians, and gathered plant samples and drawings. When ill, he experimented with the medications on his own body, especially with the most prominent herbs and purgatives. He found refuge in monasteries and in the houses of welcoming *encomenderos* and profited from the information-gathering networks of religious institutions, most of which were Franciscan, though others were Augustinian and Dominican. He also consulted the libraries of prominent scholars such as the official chronicler of Mexico City, Francisco Cervantes de Salazar.[96]

By the time Hernández arrived in New Spain, interrogating local informants had already become the standard approach of those Franciscan missionaries, like Bernardino de Sahagún and the grammarian Andrés Olmos, who sought to research the antiquities of central Mexico.[97] Their quest for historical or natural knowledge almost invariably began by investigating the etymologies of the Indigenous names of objects, places, flora and fauna, rituals, and even of individuals.

In a 1573 letter to Philip II, Hernández vividly described the process of ascertaining the characteristics of a plant or animal. The account (*relación*) of Indigenous physicians was the crucial first step because, as Hernández explained to the king, they possessed "centuries of experience." Not a single entry was committed to writing before Hernández individually consulted more than "twenty Indian doctors" who would describe the properties of the plant or animal in question. Hernández could then collate these testimonies, apply his own knowledge of "the rules of medicine," and proceed to "see," "smell and taste" the various parts of a plant.[98] In other words, even if Hernández exaggerated the number of experts summoned, the botanical investigation began with the collection of authoritative local knowledge. Hernández developed procedures to test the reliability of oral testimonies regarding flora and fauna.[99] His experience-based work alongside Indigenous doctors, as Jaime Marroquín Arredondo has demonstrated, sometimes even led him to question the validity of certain Galenic medical principles.[100]

From the spring of 1574 until February 1577, Hernández resided mainly in Mexico City, organizing the materials he had gathered. He translated his works from Latin into Spanish and also oversaw their translation into Nahuatl. While in the capital, Hernández, who was the Indies' main medical authority, attempted to regulate Spanish medical and surgical practices, but he met much resistance from local doctors unwilling to accept oversight from Madrid. At this time, he also worked in the Hospital Real de Naturales, both testing the effects of Mexican plants and trying to manage the catastrophic outbreaks of the disease known as the *cocoliztli* that ravaged the Indigenous population of Mexico City in 1576. Hernández's handling of the crisis had a powerful impact on the local medical community. Eminent New World physicians such as Juan de Barrios and Agustín Farfán praised the Spanish doctor's efforts in their medical treatises.[101]

The first step to studying a foreign item and understanding its medicinal uses was to inquire about its name.[102] By 1574, almost midway through the expedition, Hernández knew enough Nahuatl that he could begin to appreciate the meanings contained in the Indigenous designations, as Jesús Bustamante García argues. The force of a word tugged on a memory chain that would bring to the mind the object's multiple characteristics and uses. The plant known as *qualancapatli*, or "medicine of the angry man," for instance, was named in Nahuatl after its ability to heal people who have fallen ill on account of a fit of rage or who have been the victims of some kind of injury. Hernández reported that its leaves were "to be administered stained or dissolved in water." The *qualancapatli* possessed cold and humid humors and no taste or smell.[103] The *miatlacotl*, or "water rod that yields arrows," received its name from its appearance, because it was an herb "of long roots as thick as the little finger, from where stems of the length of four palms are born with rounded leaves and a small fruit in the tip of the stems." It cured fevers.[104] The *micaxihuitl*, or "death medicine," contained clues to its function in its name. This plant was of "hot and dry temperament in the third degree almost. It cures epileptics and those with heart disease," in other words, those who were at risk of death. The *chichimecapatli*, or the "medicine of the Chichimeca," "relieves flatulence and colics." The famous doctors of Cholula prepared a popular remedy with this plant "with which almost all diseases are combated" (fig. 4.6).[105] This medicine was named, not for its effects but in honor of those who discovered its uses.

Animal names likewise encapsulated important characteristics. In the book dedicated to quadrupeds, Hernández introduced the *mapach*, or "animal that touches everything with his hands" (fig. 4.7). The creature received its appellation from the fact that "it is similar to the fox in its wit and customs, it washes its face with its hands as cats or martens . . . and puts food in its mouth with its hands like monkeys."[106] Likewise, in the book dedicated to reptiles, Hernández described the *ecacoatl*, or "wind snake": a harmless creature "with a long, thin, forked and blackened tongue that vibrates very fast, black eyes that [are] extraordinarily big in relation with the body, and yellow iris."[107] In the sections dedicated to aquatic animals, Hernández described the strange *axolotl*, or "game of water," a kind of "lake fish that is covered with a soft skin, with four legs like a lizard." It provided

Eyſtettenſi horto vbi(vt audio)plurima inter innumeras omnium plantarum differe-
tias eluntur.

Monardes Capſicum Americanum deſcribit cap. 54. vna cum viribus. eiuſq́; ſa-
porem,quem cibis condimenti loco conciliat,orientali Piperi anteponit.

Ouetus Hiſtoriæ naturalis Indicæ lib.7 cap.7. vocat Aſci, & præfert etiam Pipe-
ri orientali. præſertim ſi piſcibus , & carnibus iungatur.

De CHICHIMECAPATLI, ſeu Chichimicæ gentis medicina.   Cap. IV.

HERBVLA *Chichimecapatli*
nuncupata , ab alijs verò *Tama-*
*capatlis,*quoniam primo guſtu radix dul-
cis , & temperata videatur , dicta , folia,
fert rara,longa,& tenuia, caulem verò te-
nuem , et cubitalem, radicem nuci Iugla-
di,forma & magnitudine ſimilem, lacte
manantem valde tenaci, & glutinoſo,in-
tus candidam extra nigram.Quæ regionibus prouenit temperatis , pauloue frigi-
dioribus, qualis eſt Mexicana , ac *Tetzo-*
*cana,*aſperis & môtoſis locis. Radix viret,
arenſue (quamquam ſiccato lacte ſit in-
nocentior) ex liquore quopiam deuora-
ſcrupuli vnius cum ſicca eſt pondere, hu-
mores omnes , per ſuperna & per inferna
euacuat , ſed maiori ex parte per inferna.
Calenti ordine quarto conſtat natura,vi-
ribusq́; vehementibus. Quamobrem non
cuiuis temerè exhibenda eſt, ſed eis tantù,
qui viribus valent , nec acuta vexantur
ægritudine aut his,quibus leuiora alia remedia tentata ſunt. flatum etiam dicitur diſcutere,& cholicis doloribus opportunè auxiliari. Ex huius radicis vncijs
quatuor,& drachma vna *Cocoztic,*parat Chululēſis medicus,nobile illud,acto-
ta noua Hiſpania celebre pharmacum,quo cunctis pene morbis obſiſtitur, ru-
ſis permixtisq́;,addito ſaccharo,& ad ſolè, donec humor abeat, expoſitis: deuo-
randum verò drachmæ vnius mēſura, quod ita eſſe iuridico patuit teſtimonio.
Ricinis forte annumeranda fuerit hæc planta.folia enim Cannabis habet,& ſemen
triangulare videtur. guſtuq́, vt & in Ricini ſpeciebus initio ſubdulci videtur,dein-
de vero guſtur vrit,vnde vix inter Eſulas reponi poteſt.omnes enim ſtatim acrimo-
niam ſuam produnt.Et licet Ricinus apud nos non lacteſcat,apud Americanos tamen
lacte manare poſſe, non eſt incredibile, cum regio illa quàm noſtra ſit longe humidior
calidiorq́; . Sic enim & Nerium apud ipſos lacte abundat , quod in noſtro orbe nun-
quam eſt viſum.

                                                                                     Vſus

---

Figure 4.6. Chichimecapatli in Hernández's *Nova plantarum animalium et
mineralium Mexicanorum historia* (1651). Francisco, *Nova plantarum
animalium et mineralium Mexicanorum historia* (Rome: Vitale Mascardi,
1651), 138. Courtesy of the John Carter Brown Library at Brown University.

belluam primùm aftu fatigauit, deinde cum ſummo ſpectantiùm horrore terræ affixit. Hoc genus canum creſcit ad formam ampliſſimam, terrificis latratibus, vltra leones rugitibus inſonans.

CAPVT XL.
Miranda de canibus.

MEntio canum aliquam celebritatem ingenij meretur. In libro venationis Alfonſi Regis Hiſpaniæ ſcribitur, grauidam canem, vocatam Buſteram, dum inſequeretur ceruum, correptam partu peperiſſe vnum catulum, quem ore ſumens percurrebat aliquod ſpatium: poſteà adlatrabat feram, interim pariens alium; quo cum alio catulo ſeruato, recurrebat mater, & relatrabat. Hoc pacto reliquos effudit. Occiſo demù ceruo, ſtatim accurrit vbi catulos collegerat. Mirabiliora narrabo. Quemdam noſtri Ordinis Procuratorem canis comitabatur. Accidit vt cùm Concham peruenillet, canis pareret quinque catulos. Diſceſſit Procurator repetiturus Domum probationis Villaregenſem, reliquens canem Conchæ commiſſam noſtris. At vnâ nocte tranſtulit canis omnes catulos ex Conchâ in Villaregum. Diſtant

nouem leucas; itaque pluſquam octoginta leucas vnâ nocte percurrit. Alia canis puerpera etiam dimiſſa Valliſoleto, vt nuntium fauſtum latæ ſententiæ pro Naualcarnero oppido deferret, duabus horis peruenit illò, percurrens tam exiguo tempore pluſquam triginta leucas.

De grato canum animo recens exemplum nunc miratur Belgium. Mortuo filio Marchionis Cannetenſis, canis eius ſe in ſepulturam vnâ immiſit terra condendum. Violenter extracta, numquam ex loco ſepulchri recedere voluit. Ibi hactenus perſeuerare, accepi ex teſtibus oculatis.

CAPVT XLI.
De animali canicida.

VVlgaris eſt noue Hiſpaniæ fera canum interſectrix, *yzcainchini* dicitur vernaculis. Improbum eſt animal, & noxium indigenis. Accedit noctu oppidorum ſuburbia, vlulatúque euocat ad ſe canes, trucidat, vorat, atque ita interemptis probis fidiſque domiciliorum cuſtodibus, vitæ ſuæ parato victu proſpicit. Sunt qui idem animal cum *cuitlamizli* eſſe putent.

MAPACH.

C.I.

CAPVT XLII.
De animali Mapach.

ANtra & cauitates montium atque collium Tzozocolci hoſpitatur animal peregrinum, quod cuncta manibus prætentat. *Mapach* ab Indis dicitur, ſed non firmo

nomine; alij *illamaton* ſeu *vetulam* appellát, alij *maxtle* ſeu *goßypinum cingulum*, alij *cioatlamacazqui* ſeu *ſacerdotißam*. Exæquat magnitudinem quauhpecotli, imò cane Melitenſi paulò eſt maius, humile, teres, pilis veſtitum, nigro albóque colore promiſcuè variatis, magno capite, paruis auriculis, roſtro
canis

P 4

Figure 4.7. Mapach, in Nieremberg's *Historia naturae, maxime peregrinae* [. . .] (1635). Courtesy of the John Carter Brown Library at Brown University.

a "healthy and delicious food, similar to the meat of eels." The Mexicans fry or grill it and traditionally season it with pepper, "a common spice that they really like." The creature's name originated from the way people perceived it, that is, "from the peculiar and amusing" form it possessed.[108]

In the commentaries to Pliny, immediately before discussing the Tower of Babel, humankind's linguistic diversity, and the question of the primordial tongue, Hernández inserted a comment on the softness, richness, and descriptive prowess of Mesoamerican languages. Responding to Pliny's description of the astounding multiplicity of speeches and tongues, Hernández proposed a few observations of his own. Nowhere in the world did languages abound as in the West Indies, "where within a distance of a few leagues languages often vary." This was "remarkable, because each of these [languages] could be very perfect and copious if were propagated by wise Indians who could know and impose in their way names onto things." The variety of letters and sounds, "so natural and soft to its natives and so hard and difficult to foreigners," could provide for very rich, variegated, and even perfect languages.[109]

In his *Antiquities of New Spain*, Hernández repeated his assessment of Indigenous languages and linked the Mexicans' ability to "impose names in their own way" to their ancient customs and history. He compared the Americas' enormous natural variety with its equally impressive linguistic wealth: "What will I say of the admirable natures of so many plants, animals and minerals, of so many differences of languages, Mexican, Tezconquense, Tlaxcalan, Huaxtecan, Michoacano, Chichimeca and many others that can barely be enumerated; of so many customs and rituals of men, of so many dresses with which they cover themselves . . . and other ornaments that can barely be followed by human intelligence."[110] The depth and richness of these languages demonstrated that New Spain's people possessed an ancient relationship with their natural surroundings. The complex set of experiences embedded in the words they used to designate things bespoke this antiquity.[111] More broadly, Hernández impressed upon his readers the richness and diversity of Mesoamerican cultures and the immense challenges of describing them. In the Plinian commentaries he made this latter point repeatedly. When describing the flower called *ornithogalum*, Hernández marveled that Spaniards still called it by its Latin name without assigning it one of their own. In the Indies, he claimed, "where people are so uncultured and barbarous, there is such a great number of herbs that the virtues of some are known and some are not known." Despite this great variety of plants, "there are almost none that are not known and named with a particular name." He went on to describe the language of Mexico as one "of very ancient stock" that was "never soiled by commerce with other nations."[112]

"Old and wise Indian" name-makers learned the intricacies of their language from an early age. Bernardino de Sahagún—perhaps Hernandez's main source for the *Antiquities*—described the institution known as the *calmecac*, where Indigenous boys would learn the art of public speaking, in detail.[113] Repeating Sahagún's description almost verbatim, Hernández explained that in those institutions, young boys were "taught to tell the truth and speak with eloquence." They had "to pay reverence to their elders" and "were also instructed to learn their divine songs that were preserved written in papers with hieroglyphics (that they are also taught to draw)."[114] Language was cultivated with care, as the large corpus of surviving *huehuetlatolli*, or "words of the elder[s]," further demonstrated. An important source for Hernández, the *huehuetlatolli* were the sayings of the ancients, or didactic speeches that children memorized to learn moral lessons and proper conduct. As we saw in chapter 3, the Franciscan friars who progressively recorded them from 1546 until about 1600 thought they had a comparable role to orations in the Greco-Roman world. For Hernández, ceremonial addresses and ornate forms of expression, could, if faithfully memorized, transmit much about the history, traditions, and values of past times.[115]

Hernández further observed that the Mexica were such "uncultivated and barbarous people . . . one scarcely comes across a word that does not have a considered significance and etymology, for almost all their words have been adapted to things with such precision and care that the name alone is enough to indicate the nature of any important thing."[116] Nahuatl fit the description of what Vives considered an ideal language, as it was deliberate, soft, and prolific, capable of faithfully transmitting, in spite of its speakers' lack of alphabetic writing, their community's knowledge of the natural world.

Even if Indigenous physicians had not, in Hernández's view, understood the causes of diseases and were merely empirical doctors, they nonetheless possessed a rich and reliable corpus of oral natural knowledge that could be logically organized and communicated to a distant reader using conventions only slightly modified from those of European natural knowledge.[117] In this respect, the conjectural knowledge of the peasants he interviewed in the Castilian countryside before setting out to the Americas differed significantly from the knowledge he encountered in New Spain. In Mexico, Hernández found an elaborate and ancient oral tradition, formally and deliberately transmitted in schools.[118]

Indigenous Antiquities and the Origin of the Mexican Languages

Hernández's reliance on Indigenous etymologies is consistent with other strategies he employed for translating knowledge from Mesoamerican sources into a natural history for a European readership. The *Antiquities of New Spain* was conceived as an integral part of his natural history, a response to Pliny's book 7. As Mary Beagon has argued, book 7 is devoted to showing the extraordinary and atypical aspects of the human condition, such as monstrous births, children born with strange abilities, and wonderful machines.[119] Hernández equally compiled what is extraordinary about the people of the Americas, including their schools, childbirth practices, their botanical gardens, and aviaries. As we would expect, his study of Indigenous antiquities relied on the dissection of etymologies in a fashion that resembled their historical use. Every entry on a Mexican ritual or habit began with an explanation of its Nahuatl name. When describing Mexican dances, Hernández recalled that their many names vividly expressed the characteristics of each type of performance. Their dances were "worthy of being seen." The Mexicans gave them the general name of *nitoteliztli*, which Hernández identified as the *areitos* of the Haitians. For instance, the *nenahuayzcuicatl*, or "song of the embraces," would take place on an afternoon before the celebrations offered to their many gods.[120]

Hernández repeated some of the attributes of the Mesoamericans' linguistic abilities in a section devoted to the Indigenous American's possible origins. "There are those who assure that all of these peoples came from Palestine." Even if this appeared to be untrue, Hernández cautiously stated that the conjecture should not be blindly rejected. Contrary to the Franciscan Jerónimo Mendieta, or to other authors who defended the theory of the ten lost tribes in the Americas strictly on biblical interpretations or on a particular understandings of Indigenous moral character, Hernández, like Alonso de Zorita, was willing to consider the supposition on account of their linguistic aptitudes: "Their names, not unlike those of the Hebrews, were imposed by deliberation and council and not without etymology [*ethimo*]."[121] Hernández concluded by narrating the Mexica's own understanding of their origins as described in Sahagún. Even though in his Latin writings Hernández often was dismissive of Indigenous culture, and their medical science, he always praised their language.[122] The natives' rich etymologies, with their expressive and visual capacities, could help "to put before the eyes of the absent" images that can be fully understood only "by those who are present and by experience itself."[123]

The correspondence among Hernández, president of the Council of Indies Juan de Ovando, and Philip II reveals that by 1575 the king had already become impatient because Hernández had not sent him any writings; the expedition, after all, was expensive, and Philip was eager for new plants to grow in his botanical gardens and distill in his alchemical laboratories.[124] The main reason for Hernández's delay, as he revealed in a letter that accompanied his natural history the following year, was that his work required more exhaustive editing; he expected to continue his labors once he returned to Spain. The doctor finally surrendered his writings to his patron, dispatching a Latin copy of the treatises to Philip II from Mexico in 1576. The volumes comprised about 893 pages of text accompanied by 2,250 pages of paintings of plants and 179 pages with images of animals. In his will, Hernández also mentioned five books (*adminiculativos*) meant to assist in the reading and interpretation. The works were first received by the king's personal treasury, then transferred to the Council of the Indies in Seville, from where they made their way to the Escorial.[125] In 1671, alas, the king's volumes perished in the devastating fire that engulfed the monastery for fifteen days. Even so, Hernández's own copies survived; they found their way to the Jesuit college in Madrid, eventually becoming the basis of Juan Eusebio Nieremberg's writings on American animals in his *Historia naturae, maxime peregrinae* of 1635.[126] A 1626 copy of the original index of the destroyed volumes, the *Index alphabeticus plantarum Novae Hispaniae*, survives in Montpellier.

Why did Hernández never publish his books? Providing a definitive answer is challenging on account of the scant evidence. Several historians have pointed to possible censorship of his writings, yet no document confirms this theory. A *consulta* (formal consultation) of 1578 regarding a possible edition of Pliny's *Natural History* and other works in the Council of the Indies stated that although members of the council recognized the "public utility" of Hernández's writings, they thought that the works' length and their many illustrations would make them expensive to print and hard to sell. The member of the council in charge of answering the query regarding the publication advised caution and that perhaps it would be better to make a clean manuscript copy with its paintings. The rest of the contents, the council member averred, could be reduced to summaries (*sumarios*) in cheaper pocket editions.[127]

In 1580, for unknown reasons, Philip II commissioned the Neapolitan physician Nardo Antonio Recchi (1540–1594) to organize Hernández's voluminous works into a synthesis that could be used to train physicians and apothecaries working in the palace complex of the Escorial.[128] Recchi first encountered the manuscripts in the Escorial, copying and reorganizing about 30 percent of the original entries. The Neapolitan physician, who had never been to the Americas, had access to all of Hernández's works, including the auxiliary books mentioned in the will that do not seem to have any extant copies. Recchi's writings survive in manuscript. They also served as the basis of a 1651 edition of Hernández's works published by the Accademia dei Lincei in Rome. The *Treasury of the Medical Things of New Spain* (*Rerum medicarum Novae Hispaniae thesaurus*) reproduced Recchi's prologue to Hernández's writings.

Recchi's introductory chapters, which either derived directly from Hernández or else represent Recchi's reactions to the *protomédico*'s works, contain one of the most explicit discussions in all of Hernández's writings of etymology's role in natural historical knowledge. Recchi outlined the main logic behind the imposition of New World plant names: "And so, of the plants, whose names [*nomina*] have an etymological origin, certain ones have taken their name [*appellationem*] from their chief faculty, or the excellency of their effect,"[129] like the *tlahuelilocaquahuitl*, or "madness tree," which healed the sick by expelling demons.[130] Some plants took their names not from their effects but from their peculiar properties: the leaves of the *pinahuihuitztli*, or "embarrassed herb," retreated when touched, as if recoiling from human contact. Sometimes the name reflected the plant's flavor, its color, or its resemblances to other objects or drugs. Sometimes a single plant had multiple names because in one region it was known mainly for one of its qualities, whereas in another region, a different quality had proven more useful, like the *cohayellim*, or "stinky snake," named for its odor; also known as the *tlipoton*, or "black and stinky plant," on account of its color and smell; and elsewhere called *chichic*, or "bitter herb of closed leaves," because of its taste and form.[131]

Even if the multiple denominations might have seemed confusing, Recchi understood this variety not as a problem that needed to be remedied but rather as useful plant data. He believed that the inhabitants of New Spain themselves would gain much from seeing the multiple names side by side, so that they would more easily learn to identify the same plant across languages and regions. Recchi concluded with the hope that for those plants that seemed to lack an etymology, one might either be discovered in the future or else other names might be imposed on them, so that their designations would reflect their qualities.[132]

The Challenges of Organizing

Organizing such a massive addition to European materia medica and classifying it so that it would be intelligible to a reader unfamiliar with the Americas' geography, history, or languages presented an enormous challenge. The paradigms European herbalists used to arrange their materials—alphabetical organization according to the four types of plants defined by Theophrastus (herbs, shrubs, "undershrubs," and trees)—did not provide sufficient flexibility, in Hernández's view, to accommodate all the varieties of New Spain. When Recchi confronted his difficult editorial task, he found the manuscripts to be "confusing, incomplete and disorganized" and opted to follow the traditional European arrangement for herbals.[133] Thus, he divided the works into the typical four books, corresponding to the four types of plants, and each book into smaller sections dedicated to plants with morphological similarities or common effects. Within each plant entry, Recchi also emulated Dioscorides by introducing first the name and its meanings, second the physical description and geography (loci), and finally the plant's virtues (main qualities) and its applications.[134]

Recchi's work was typical, for few Renaissance naturalists formulated their own classificatory systems.[135] In his *De plantis libri* of 1583, Andrea Cesalpino proposed a botanical nomenclature in harmony with the premises of Aristotelian logic. Aristotle had asserted that it was impossible to categorize plants using the canons of logic; because no single feature of a plant could be selected as the fundamental characteristic on which all others depended, any choice would inevitably be subjective. Cesalpino attempted to solve this problem by positing that the plant's manner of reproduction would serve as the essential characteristic to which all other features were subordinated.[136] Although Hernández's endeavor preceded Cesalpino's, the comparison serves to highlight the challenge inherent in establishing an organizational scheme for such a massive quantity of newly described animals and plants.

In fact, Recchi misunderstood Hernández's organizational criteria, as José María López Piñero, José Pardo-Tomás, and Jesús Bustamante García have demonstrated. Hernández had arranged the plants in their traditional alphabetical order but kept their names in the original Nahuatl and other Indigenous languages.[137] This choice differed from the practice of contemporary European botanists, who usually related new species of plants to those described by Dioscorides, Pliny, or Theophrastus. The decision created multiple challenges. First, as Bustamante García explains, Hernández had to establish a standard spelling for Nahuatl words

when there was still no consistency. The phonology of this language was also problematic, given that Roman letters like [c] and [t] represented more than one Nahuatl
phoneme. Second, Hernández grouped similar plants according to their Nahuatl
root into genera, assigned them a common name in Nahuatl, and defined these
groups in accordance with the European concept of botanical virtue. These genera became the bases for the alphabetical organization of the book.[138] For instance, the group of plants categorized under the Nahuatl root *metl* (types of agave)
are those that share features of "the century plant."[139]

Through his study of etymology, Hernández discovered that the Nahua people
described nature using composite words that made a plant or animal readily iden-
tifiable from the descriptions embedded in its name. Toponyms, names of geo-
logical phenomena, and those of mineral and animal products, all likewise
encoded such meanings. As Francisco del Paso y Troncoso explained in his clas-
sic monograph, Nahua taxonomy grouped plants into three natural orders similar to those recognizable to Europeans: trees, bushes, and herbs. It also afforded
more detailed anthropocentric taxa, or "artificial classes," depending on the plants'
uses and whether they were edible, medicinal, or "ornamental."[140]

In making the leap to this nomenclature, Hernández adopted the Franciscans'
understanding of the Nahuatl language as well as developments in how Europeans
understood and deployed etymologies. As the sixteenth-century grammarian Molina explained, Nahuatl attaches multiple suffixes and prefixes to a core that can
be either verbal or nominal, to modify the meaning of a central unit.[141] Molina attempted to write a grammar that accounted for the properties of the Mexican
language and invoked the precedents that seemed most appropriate. The Franciscan grammarian did not rely exclusively on metalinguistic Latin terms but rather
applied a more eclectic method and compared the Mesoamerican language to Castilian, Hebrew, and Latin, when relevant.[142] Increasingly influenced by the study
of Hebrew, scholars in the sixteenth century, as Demonet explains, began to
organize etymologies not alphabetically but by a nucleus or "root" that contained
an important clue to the word's meaning. Readers of Johann Reuchlin's 1512 Hebrew
grammar, for instance, could identify the root as the semantic core of the (Hebrew)
word.[143] This conceptualization, reminiscent of Aristotelian "essence" and "accidents," provided a system for Hernández to begin to decipher the Americas' flora
and fauna.[144] Because a full understanding of the "morpheme" as a unit did not
emerge until the nineteenth century, Hernández referred to this semantic core as
the *ethimo*, or "origin," when explaining the organizational paradigm in a letter
to Juan de Ovando. Underlying this organizational choice was the assumption

that the Indigenous language is sufficiently adept at describing and highlighting the common features that unite classes of plants and animals.[145]

Conclusion

In his study of the fauna and flora of the Indies, Hernández deployed etymologies in ways that resemble the exegetical or historical uses of his Spanish contemporaries. In the Americas as in Europe, the attributes that names encoded might provide the natural historian with insight into the properties of things themselves, always, if possible, corroborated by firsthand observations or reliable witness reports.

At the same time, Hernández's work also fulfilled different functions than in Old World scholarship. When describing a world distant from and unfamiliar to European audiences, etymologies served a mnemonic purpose. Hernández often pointed to names as a fundamental descriptive aid in conveying information to a reader unfamiliar with the object of study, through etymology's powers of association. In this respect, etymology resembles the analogies and comparisons with Old World plants, animals, and landscapes that the early chroniclers of the Indies commonly used to make their material intelligible to readers who might be unfamiliar with what they were describing.

Hernández's use of Indigenous names also buttressed the authority of Indigenous knowledge systems, and his own. His propensity to maintain the names of plants and animals in their original languages, and to provide translations of the local designations, suggests that he believed that only Indigenous languages could fully convey the properties of the flora and fauna of the Americas. By maintaining the names in their original languages, he also provided the reader with additional proof of his credibility: he had been in the American viceroyalties, he had seen the plants and animals with his own eyes, and he had learned their true names.

In 1790, more than 200 years after Hernández's death, the botanist Casimiro Gómez Ortega first published Hernández's complete works. A member of the newly created Royal Academy of the Sciences, Gómez Ortega recognized the linguistic peculiarity of the doctor's writings. He valued the treatises because "the Mexican names, sometimes of plants, sometimes of animals, sometimes also of minerals, diligently explored by Hernández, and taken from the from the language of the Indians, express their form, or something singular or health-giving, or, for the most part, a noxious quality" before they had been irretrievably altered by the

Indigenous inhabitants' contact with Spanish.[146] For Gómez Ortega, Hernández's botanical and zoological entries were a testament to a bygone world.

In the eighteenth century, Spanish scholars would begin to abandon the method of first and foremost seeking out the Indigenous names of plants. Instead, as Antonio Lafuente and Nuria Valverde have shown, royal botanical institutions attempted to introduce the Linnean system of naming and classifying plants. This was part of a broader set of reforms that aspired to centralize the production of botanical and cartographic knowledge by employing a single descriptive language across all regions of the Spanish Empire. Ideally, standardized recording local phenomena would facilitate the transfer of information across regions.

Like all other fiscal, administrative, and cultural aspects of the Bourbon reforms, this initiative met with mixed responses among local authorities and elites. Some posed serious challenges and objections, while others embraced the new scientific tenets. In the particular case of Linnean classification, the scholar José Antonio de Alzate y Ramírez (1738–1799) and others rejected it on the grounds that "a plant's name should not express a logical, but a functional order."[147] Like Aristotelian classifications, Nahua practices of naming and organizing plants above all expressed functional orders. Yet what Hernández considered the greatest asset to the Indigenous knowledge of materia medica had come to appear to some eighteenth-century botanists as an improper language of science. Vicente Cervantes (1755–1829) the director of the botanical garden of Mexico, criticized Nahuatl in 1788 more broadly: he considered it worthy only of being "spoken in public places and small groups, with Indian women selling herbs and vegetables, but not in the academies of the learned."[148] His assessment stands in stark contrast to Hernández's praise of Nahuatl as an ecumenical language that played the role of a lingua franca like Latin and possessed the expressive capabilities and semantic abundance of Hebrew.

With its imperial and universalizing ambitions, the Plinian model emphasized compilation as the fundamental activity of knowledge production. Thus, Pliny's model could accommodate non-European modes of producing knowledge that operated under broad categories of classifying the natural world. More broadly, the Bourbon reforms envisioned information gathering in a way that Hernández and his contemporaries would not have considered possible or even desirable. Hernández's emphasis on antiquities and his approach to studying the natural world not only through direct observation but also through the local testimony of Indigenous doctors show that, in the late sixteenth century, the most cutting-edge natural knowledge of the Americas was also in many ways the most ancient.

CHAPTER FIVE

The Rudiments of All Languages

Only one Indigenous American word appears in Bartolomé Valverde's *Treatise on Castilian Etymologies*: *cacique*. Valverde credited its presence in Castilian as an Arabic loanword and then traced it to Hebrew. According to Valverde, in Hebrew *cacique* meant "great prince."[1] In the late sixteenth century, when Valverde was writing, the Taíno word *cacique* had become widespread enough in Iberia to merit his attention as a Castilian word.[2] Yet Valverde made no significant mention of its New World origin. To be sure, he knew that all languages adopt words from neighboring nations through trade, as was observable in "mercantile lands" where no pure, single language predominated. Soldiers also borrowed language. He even observed that the propensity to adopt new words from travel and trade was visible in "those who come from the Indies." Even though "they speak their own languages," they mix "many foreign words from the lands where they have been with their own."[3]

Despite Valverde's understanding of language borrowing, he insisted that *cacique* had Hebrew origins. If he could not find the etymology of a Castilian word in Arabic or Latin, he proposed that the "safest and most correct" strategy was to turn to Hebrew. Spain's former Jewish inhabitants had left countless Hebrew words, and even though the "Catholic Kings expelled the Jews from Spain they were unable to do [the same] to the words that they had introduced."[4] Like a doctor, the etymologist had to employ purgatives and medicines to strip words from the changes introduced by the passing of time and the mistakes of ignorant speakers. For Valverde, since the rudiments of all languages came from Hebrew, the "substance" or "root" of a word could be found by removing all vowels, because in

Hebrew "the vowels are some extraordinary dots that are not part of the word's substance." Then, the etymologist could move around letters and syllables according to Valverde's guidelines until he reconstructed a word's original form.[5]

Valverde's instructions for identifying the etymologies of Castilian words combined his reading of Plato's *Cratylus*, his knowledge of Hebrew and sacred scriptures, and a particular interpretation of Spain's past. Given Spain's history, the origins of Castilian words, when not apparent in Latin, could be found in Arabic, Greek, or Hebrew. Valverde's recognition of the need to turn to the languages formerly spoken in Spain was not new; Juan de Valdés had developed a similar understanding of the relationship between language borrowing and the histories of speech communities. Unlike Valdés, who in the 1530s had stressed that Latin was the main pillar of the Castilian language, Valverde, working fifty years later, as Francisco Perea Siller has noted, primarily emphasized the Hebrew origins of all words.[6] A teacher of Hebrew and Greek in the Escorial monastery in the 1570s and 1580s, Valverde shared a belief in the primacy of Hebrew with other scholars in his circle, among them Benito Arias Montano and the Augustinian friar Luis de León.[7] His treatise on Castilian etymologies suggests the importance that the study of Hebrew had attained among scholars under Philip II's patronage: beyond biblical studies, it even informed Castilian lexicography.

Through his etymologies, Valverde sought to ennoble the Castilian vernacular. By treating *cacique* as a Hebrew word, Valverde also implicitly adopted a stance about the Americas and its languages, and their structural relationship to Hebrew. Like their predecessors who debated Spain's original language (see chap. 2), Arias Montano and Luis de León argued about New World toponyms and their relationship to Noah's progeny. For these scholars, the Americas were not newly discovered lands but had already appeared in the Hebrew Bible. Luis de León believed that the sacred scripture also expressed the imperative to convert the Indigenous inhabitants of the Americas—if one knew where to look and how to read. Thus, through their etymologies, these scholars reaffirmed the idea that the Bible contained all knowledge, past and present.

The previous chapters have considered the varied uses of the etymological method to study the human and natural history of the Iberian kingdoms and the American viceroyalties. Scholars relied on the tradition of name study to harness local knowledge and to make arguments about the Spanish Empire's linguistic diversity. Etymologies, however, were also central to biblical interpretation. This was the highest-stakes field for early modern European scholars because its goal was ultimately to establish a correct understanding of the word of God.

The Antwerp Polyglot Bible, Arias Montano's most significant editorial project, was published in the workshop of Christophe Platin between 1568 and 1573.[8]
The last volume of the eight that make up the Antwerp Polyglot, part of the apparatus, contains a number of treatises to aid interpretation of the biblical text.[9]
These works, especially their prefaces, expressed Arias Montano's linguistic ideas,
for instance, that the language of the Bible contained numerous levels of meaning
that attested to Hebrew words' unique forcefulness. Montano argued that in this
original tongue, objects corresponded to the words that signified them. That is,
words in Hebrew signified naturally and not conventionally. Moreover, like Valverde, he stressed that features of Hebrew underlay all other languages.

While the Americas barely appear in the treatises that make up the apparatus,
its interpretative principles had consequential implications for understanding
the New World. One of the most significant works of the late sixteenth century,
the Antwerp Polyglot was not only read and debated by peninsular Spanish
and European scholars but also made its way to the libraries of the American
viceroyalties soon after its publication. As discussed in chapter 3, the treatise entitled *Phaleg* in the apparatus, for instance, was a definitive source for Miguel Cabello Valboa, who probably consulted it in Quito or Lima less than ten years
after its publication to develop his account of how the descendants of Ophir had
sailed to the Americas and then spread, mutating their languages willingly as a
result of conflict. Cabello Valboa embraced the *Phaleg* as an authoritative source
on the earliest ages of the Americas' habitation before he turned to the genealogies of Inca rulers as attested by the *quipucamayocs* (Inca administrative officials
who specialized in the use of the quipu). However, Cabello Valboa rejected the
idea that there were Hebrew vestiges in the languages of the Americas, arguing
that the descendants of Ophir had created countless new languages unrelated to
the first language of humankind.

This chapter focuses on the linguistic beliefs of Spanish Hebraists such as Valverde, Arias Montano, and Luis de León. It puts their etymological exercises in
dialogue with scholars concerned with the history of the Americas and its people.
It shows that debates about the proper form of biblical exegesis in Spain also ended
up encompassing questions about the origins of Indigenous Americans and their
place in the history of humankind. In his influential *Natural and Moral History
of the Indies* (1590), the Jesuit scholar José de Acosta not only rejected Arias Montano's and Luis de León's interpretations about the ancient inhabitants of the
Americas but also intervened in long-standing debates about linguistic knowledge
and its validity for historical reconstruction.

By relying on expert interpreters of Indigenous languages, Indigenous histories, and information about the natural world, Acosta distinguished between different modes of etymologizing. He relied on words and examples of language change to discover local traditions and understandings of the natural world, but like other scholars in the 1590s, Acosta refuted the notion that linguistic vestiges provided a way to reconstruct origins. Acosta believed that, in general, languages were conventional and did not contain vestiges of ancient wisdom, beyond what was known to their name makers. Despite Acosta's refutation of the Indigenous Americans' Hebrew origins, the search for Hebrew rudiments continued to be a significant pursuit among lexicographers who sought to connect Indigenous lexical items to Hebrew, and thus account for the linguistic diversity of their broadened world.

Benito Arias Montano and the Antwerp Polyglot

The Antwerp Polyglot was one of the most controversial projects in sixteenth-century Spain and the most important work of its editor, the humanist scholar Benito Arias Montano.[10] Whether translating the Bible, commenting on specific passages, seeking out works for the Escorial's library, or collecting curious antiquities, Montano abided by a linguistic logic in which Hebrew was the primordial language, as it was replete with arcane meanings, and the foremost tool for understanding the order of the world.[11]

The question of how Montano developed his method of investigating and reading the Bible—that is, of the intellectual influences that shaped his work and his scholarly goals—has received a fair amount of attention in the context of broader debates on the nature and ultimate fate of the Spanish Renaissance.[12] Baldomero Macías Rosendo argues that upon his arrival to Antwerp, Montano had probably already composed some of his most controversial works like the *Book of Joseph, or On the Arcane Language* (1571, *Liber Ioseph, sive De arcano sermone*), and that he had written some of the other treatises of the apparatus when he was still a student of theology at Alcalá de Henares. Consequently, Macías Rosendo maintains that Montano's ideas cannot be the sole product of his experiences in Antwerp but more likely the result of his studies in Spain and of the particularities of the Spanish academic context, with its strong tradition of Hebrew studies on account of its converso inheritance.[13] In his works, Montano displayed numerous affinities with Erasmus's approach to biblical studies.[14] Montano's insistence on reading the texts in their original languages and his willingness to distance himself from the Vulgate when the Latin text contradicted the Hebrew are among the many similarities.[15] However, Montano's interest in Hebrew, and consequently in the Hebrew Bible, and his investment in the study of the arcane and symbolic meanings of words separated him from other aspects of Erasmus's thought.[16]

Montano's inclination toward the study of Hebrew and the Hebrew Bible can be seen, rather, as part of broader trends particular to Spanish biblical humanism. Emilia Fernández Tejero and Natalio Fernández Marcos have pointed out that the confluence of two circumstances, "the Christian coexistence in the Middle Ages with Jewish communities and the contribution of the conversos, who gave access to the interpretation of the Spanish Rabbis, in particular Moses Maimonides, Abraham ibn Ezra, and Moses Nahmanides as well as to the Rabbinic Bibles of Felix Pratensis (1517) and Jacob ben Hayyim (1525)," significantly shaped biblical scholarship in sixteenth-century Spain. Early Spanish humanists confronted questions over the authenticity of the Vulgate, the differences between the Hebrew and the Greek text, and the many meanings of Scripture and their "hierarchy in the practice of exegesis," with a knowledge of the Hebrew language and rabbinic exegesis that allowed them to develop distinct and creative approaches to the Bible.[17]

The influence of these exegetical traditions was present even in the edition of Cardinal Francisco Jiménez de Cisneros's Polyglot. At the end of the fifteenth century, in an attempt to improve the intellectual and theological preparation of the Spanish clergy, Cardinal Cisneros coordinated an ambitious program of religious reform that included the founding of the University of Alcalá de Henares (1499) and the printing of a Polyglot Bible to replace the numerous Bibles in *romance* (*biblias romanceadas*) that had proliferated in Spain. Cisneros founded the Trilingual College of San Ildefonso in 1510 and instituted chairs in Hebrew, Greek, and Latin at the university, fulfilling a necessity that the church had recognized since the Council of Vienne's 1312 decree *Inter sollicitudines*, but had not been able to achieve because of the lack of teachers. The cardinal himself sought out native Greek and Hebrew speakers to work on editing the different parts of the text. The conversos Alonso de Zamora (c. 1474–c. 1544), Pablo Coronel (c. 1480–1534), and Alfonso de Alcalá (c. 1465–c. 1540) edited the Hebrew text. Zamora later became the first professor of Hebrew of the college. Among Zamora's most famous students was the Cistercian Cipriano de la Huerga (c. 1509/1510–1560), who in his turn taught Arias Montano, among others.[18]

Cipriano de la Huerga connected the scholars of the Complutensian Polyglot with Arias Montano and the scholars of the Antwerp Polyglot. During his professorship at Alcalá de Henares, de la Huerga defended the need to read Scripture in its original language and, like Erasmus, emphasized the literal sense of scripture over the four senses of medieval exegesis. To compose his biblical commentaries on the Song of Songs, the books of Job and the prophet Nahum, and Psalms 38 and 130, de la Huerga drew on an astounding variety of Greco-Latin authors, "ancient Egyptian theologians, and the ancient Kabbalah." Fernández Tejero and Fernández Marcos point out that Cipriano de la Huerga, like his contemporaries, had an imprecise understanding of the Kabbalah. De la Huerga believed it to be part of a set of secret traditions passed down from God to Moses on Sinai, and then passed down orally through a chain of chosen people.[19] Most Christian Hebraists at the time learned about the nature and transmission of the kabbalistic tradition in Giovanni Pico della Mirandola's (1463–1494) *Oration on the Dignity of Man* (1496) and *Apologia* (1487).[20] Sixtus of Siena quoted Pico at length on these issues in his standard reference work the *Bibliotheca sancta* (1566).[21]

These traditions, according to Cipriano de la Huerga, were an important source for studying the etymology "and the true form of proper names in relation to their meanings," since their integrity "were respected in the Old Testament up to the coming of Jesus."[22] The works of de la Huerga had an important influence on Montano's concern with deciphering the symbolism inherent in Hebrew words and perhaps were the inspiration for the treatise dealing with these matters in the apparatus. Cirpiano de la Huerga also left his imprint on the works of his student Luis de León.[23] In his work devoted to the names of Christ, *De los nombres de Cristo* (1583, 1586, 1587, 1595), for instance, in a preliminary section dedicated to names and the ways they signified, Luis de León stated that although it was true that the correspondence between words and the objects they defined was not "always maintained in [all] languages," in Hebrew, "the first language[,] it is almost always kept. God, at least kept it so in the names that he imposed, as is seen in scripture." Luis de León further elaborated on the nature of this correspondence by invoking Adam's naming practices in Genesis:

> Because if it is not this, what is it that is said in Genesis that Adam, inspired by God, imposed upon each thing its name, and what he named them became the name of each? This means that to each [thing] its name was appropriate, and that it was its own for some particular and secret reason, and if it had been imposed on something else it would not suit it as well. But as I said, this similitude and conformity is seen in three things: in the figure, in the sound, and pointedly in the origin of its derivation and meaning.[24]

Thus, for Luis de León, Hebrew names were suited to the objects they signified in three ways: in the "figure" or the letters that composed them, in the sounds of the words, and in their etymology.[25]

Drawing from a variety of similar spiritual and intellectual influences as Luis de León, Montano developed the notion that the Bible contained "the absolute compendium of all of the [forms of] knowledge that interest men."[26] The Bible openly teaches humankind what the world is, what man is, "what was his beginning, what his nature is like and how he participates [in the world]." It tells human beings about "their creator and moderator, for what cause they came to be, the finality of their lives," and guides them toward their well-being.[27] These beliefs led Montano to embark on the study of humankind and nature by relying on the exegetical principles that he devised to study the Bible. He would develop this program in a trilogy that he entitled the *Magnum opus*. Montano completed the first two parts. The first volume, the *Book of the Generation and Regeneration of Man, or History of Humankind* (*Liber generationis et regenerationis adam, sive De historia generis humani*), appeared in 1593. The second volume, *The History of Nature* (*Naturae historia*), was published posthumously in 1601. The third part of the work never appeared in print. It was either lost or never completed.[28]

Montano believed that all truths about the world could be found in the Scriptures. From the principles governing nature to the ideal forms of government and laws that people should adopt to live in happy and harmonious societies, all could be observed or derived from proper interpretations of Scripture. The Hebrew Bible itself was so complete in its contents that it predicted even the discovery of the Americas.[29] Given the richness of the Bible, its multiple meanings, and infinite uses, Montano considered that it was not enough to read the sacred texts in translation but that it was necessary to engage with them in their most authentic form, in the languages in which they were dictated and written in other times. Simultaneously, as a translator, he struggled with the question of making this knowledge accessible to the majority who had no mastery of the ancient languages.

No language was more noble, beautiful, or worthy of study than Hebrew. This was the language through which God revealed his word to human beings and created the world, and as such, each letter was rich with meaning, much of which was incomprehensible to human reason. In the *Preface to the Latin Interpretation of the Old Testament from the Hebrew Text* (or the *Preface to the text of the Italian Dominican scholar Santes Pagnini*), one of the most controversial components of the *Biblia Regia*, Montano wrote, "Indeed, this fecundity can be found and observed nowhere more than in that primordial language. It is clear that, in this kind, that that tongue, which has been called holy, precedes all others and whose nearly infinite expressions doubtless contain within themselves manifold meanings, that you will not find in any other genre of languages. And we must believe that it is for this reason above all others that it has been chosen by God so that it would be like a certain shrine [*sacrarium*] of divine mysteries."[30]

Each word was like a "shrine [*sacrarium*]" enclosing many meanings. Not only was Montano a defender of the need to learn Hebrew, but he also believed in the authenticity of Hebrew sources.[31] Most humanists upheld the notion that Hebrew was the primordial language. Montano differed from other Spanish scholars in his fixation on unveiling the arcane meanings of words (this aspect in particular he shared with Luis de León) and in the ways in which he envisioned translation.[32]

Soon after completing the works on the Antwerp Polyglot and submitting all of its parts for review to the censors of the University of Louvain, Montano began to receive letters that expressed a general discomfort with two specific parts of the edition: the *Book of Joseph*, dedicated to explaining the arcane meanings of the words of the Bible, and the inclusion in the main text of the Bible of the Latin translation from the Hebrew of the Dominican scholar Santes Pagnini (1470–1536).

In 1570, after reading a draft of the *Book of Joseph*, University of Louvain theologians Augustinus Hunnaeus (1521–1577/1578) and Cornelius Reyneri (1525–1609) wrote to Montano expressing their surprise. Having Montano's best interests in mind, they informed him that after submitting the work to many learned men, none thought it should be included in the apparatus. They claimed that if educated men had the comments of the Holy Fathers, they did not require this treatise at all, and that readers would be confused by it. They emphasized this point to Montano: "Its contents are treated with such obscurity, that we have barely been able to find anyone who is capable of understanding this, or what it contributes."[33] Furthermore, it worried them that the meanings of the words were not supported by the authority of any author and that the mere comparison of passages from the Holy Scriptures would do nothing but generate confusion because this source alone could not help ascertain a concrete meaning of the words. They very forcefully suggested to Montano that he not include the work in the apparatus and advised him to seek the opinion of other censors or theologians if he was not satisfied with their assessment.[34]

The issue of the inclusion of Pagnini's translation alongside the Vulgate in the Antwerp Polyglot was also contentious. Pagnini's Latin translation of the Bible, which he worked on for thirty years, endeavored to offer a line-by-line rendering of the original Hebrew text. Published in Lyon in 1528, Spanish scholars like

Cipriano de la Huerga, Martín Martínez Cantalapiedra (1519–1577?), and Montano, held the translation in high esteem. Since the late Middle Ages, however, the Vulgate had become the basis of biblical exegesis, and as such corrections or changes to the text of Jerome were extremely polemical. In response to mounting criticisms against the Vulgate, the Council of Trent promulgated in 1564 a decree that established it as the reference for all Catholics, enshrined by an authoritative interpretative tradition. Heated debates about the authority of Vulgate became commonplace in Spanish universities in the middle decades of the sixteenth century.[35]

In this context, Montano was denounced to the Inquisition in Spain for allegedly overlooking the Council of Trent's decrees on the authenticity of the Vulgate in 1568 while the printing of the Antwerp Polyglot was in its earliest stages. His most fierce opponent was the Salamanca Greek scholar León de Castro, the student of Hernán Núñez, who published his book of sayings posthumously. Castro's rivalry with Montano stretched back to their university days. Castro's denunciation indeed had as its backdrop the Tridentine Vulgate decree of 1546. However, Castro's accusations, at first, were based on rumors because the printed text of the Antwerp Polyglot had not yet become available.[36] This signaled a broader opposition to Montano's exegetical practices. As Theodor Dunkelgrün has argued, Castro developed the idea that "adherence to the literal reading of the Hebrew text went against the rich and deep traditions of Christian allegorical and topological exegesis."[37] Castro's denunciations of Christian Hebraists would also lead, for instance, to the arrest, imprisonment, and later exoneration of Luis de León.

Although Montano was dissuaded from including Pagnini's text in the main edition of the Bible and ended up demoting it to the pages of the apparatus, he was more defiant about the *Book of Joseph*. This work was an essential component of translating the arcane dimension of the words of the Hebrew Bible, and its lack of citations of the Holy Fathers was an important choice on the part of its author. Baldomero Macías Rosendo suggests that the work was perhaps even a tribute to Cipriano de la Huerga. Soon after receiving the criticism from Louvain, Montano expressed his intention to submit the work to other theologians for their opinion. In the matter of the *Book of Joseph*, the censors and Montano did not reach an agreement, and, as a consequence, he had to seek the support of the bishop of Antwerp and the censors of that city.[38]

The *Book of Joseph* and Pagnini's translation also became the main source of much contention when papal approval was sought. After an initial evaluation, the papacy decided that the Antwerp Polyglot was not yet ready for publication and created a committee to further examine the text. Cardinal Guglielmo Sirleto (1514–

1585) was among the members of the committee who conducted the investigation and presented many objections in a 1572 letter to the Spanish ambassador Juan de Zúñiga. Their fourth point specifically concerned the *Book of Joseph* because it contained "many things that are uncertain and not well researched . . . and because it is not certain that it is not cabbalistic."[39] A long confrontation between the papacy and the Spanish Crown ensued over the question of who possessed the privilege to control the publication of liturgical books. The approval finally came in September 1572 when Pope Gregory XIII signed a *Motu proprio* granting his ecclesiastical acceptance. The Antwerp Polyglot sold poorly and did not achieve the wide diffusion that Montano had hoped. Plantin was able to make up for the costs only when the Crown granted him the privilege of printing liturgical books such as missals and breviaries.[40]

Although Montano was exonerated in Spain 1577, he chose to retire from his former duties. In the interim, he taught "Oriental" languages at the Hieronymite Monastery at the Escorial and worked as the palace's first librarian. There, he cataloged numerous works in Hebrew and Arabic and furthered the monarch's collection. He also devoted himself to composing the volumes of the *Magnum opus*. In the 1590s, he left the Escorial and spent the last years of his life between the convent of Santiago de Sevilla, where he was prior, and the retreat of La Peña de Aracena (Huelva).[41] Controversies over the Antwerp Polyglot and the writings of its editor continued even after Montano's death.

Montano's Translation Practices

Montano's views on the power and clarity of the Hebrew language shaped his ideas about the act and meaning of translation, as well as his predilection for certain subjects or works that he deemed worthy of translation. In Montano's view, the apparatus was central to the translator's attempts "to soften the asperity or difficulty that might prevent readers from obtaining a simple understanding of the message," that is, to convey to a reader not versed in Hebrew or Greek the contents of scripture in a language much less clear and eloquent than the original tongue.[42] Consequently, translation for Montano encompassed a broader range of activities than simply rendering accurately the contents of the text from one language to another. Instead, translation of the sacred scriptures required nothing less than a full-fledged investigation of all aspects of biblical antiquity and of the sacred language, but also of the materiality, measures, colors, gestures, geographies, chronologies, and cityscapes described in the texts.

Reconstructing even the quotidian aspects of biblical antiquity could fulfill an essential pedagogical task. It could aid beginning and advanced students, missionaries, or priests in recreating in their imaginations some of the more symbolic or elusive properties of the original writings not immediately available in translation, and perhaps even guide them to the knowledge of eternal happiness. To accomplish this task, Montano immersed himself in a wide variety of Hebrew sources, ranging from rabbinical commentaries to Hebrew maps, and even produced a Latin translation of the *Itinerary of Benjamin of Tudela* (1575, *Itinerarium Beniamini Tudelensis*).[43] In doing so, he defended the use of Hebrew sources as a valuable and authentic means of reconstructing antiquity.

Perhaps precisely because of his admiration for Hebrew, as a translator Montano struggled with the question of making the Bible and its meanings accessible to those who did not know its original languages. He and the scholars of the Antwerp Polyglot advocated for a word-for-word translation of the Bible. This approach had a twofold purpose. The first was to preserve the order and sense of the text as faithfully as possible. When dealing with a sacred text, religious respect was to be preferred over the beauty of words, and the translator should attempt, above all, to reproduce "[rhetorical] figure for [rhetorical] figure, idiom for idiom."[44] This was a particularity of the biblical translator, undesirable in the translation of works of literature. In his Castilian *Paraphrase of the Song of Songs* (1553–1554), for instance, given the nature of the work as poetry, Montano took the liberty of inserting adjectives and abstract nouns to enhance the biblical verse.[45] The second goal of the word-for-word method of translation was to allow readers to learn the sacred languages. Montano optimistically claimed that his translations were arranged in such a way that they "could teach Hebrew and Greek to those who know Latin, and Latin to those who know Greek without the need of employing a teacher and almost effortlessly."[46]

When translating a text that held the key to human history, the order of nature, and salvation from a language as meaningful as Hebrew, a translator faced problems transcending grammar. Hebrew's more arcane properties could never be replicated in translation. As Montano observed, "It occurs often that what has a unique and certain meaning in the first tongue, the translators render with ambiguity. And in turn what is rendered in a second or third tongue has an even more uncertain meaning." Montano further clarified, "A second translation, which takes into consideration the first [translation], does not have the same purity and elegance as the first, not even that first [translation] which is expressed from the original version, has the force and properties of the vocabulary that you can recognize in the original."[47]

For Montano, all biblical scholars had to learn Hebrew, at least enough to understand verbal forms and their properties. The same was true for Greek because translations were never able to capture the range of meaning and the forcefulness of the language through which God revealed his word. Sacred words in their original form were full of meaning. In them, the properties of nature appear as if "before the eyes."[48] They had the energy and forcefulness that etymologies tried to render for other less forceful languages. The more words were transformed in a chain of translations, the more ambiguous their meaning became.

Montano believed that it was necessary to reconstruct some of these more difficult or subtle aspects of Hebrew for the sake of his readers. To this end, he sought nothing less than to restore "antiquity itself," using "the ancient arts of the most evolved and complete disciplines," and "readings of ancient and recent authors to explain the difficult passages of many books."[49] Thus, he included in the apparatus a set of treatises that would allow the reader to approach the multidimensional meanings of the text progressively. The first treatise dealt with idiomatic expressions of the Hebrew language, the second with the arcane meaning of words, and the third with the gestures of the prophets, the language of gestures. The goal of these treatises was to unravel each word's power of signification. Since etymologies were ornaments of brevity, the texts of the apparatus were meant to unpack grammatical signification with great philological precision and then to present to the reader the historical and arcane meanings of words.

A description of the gestures and expressions of biblical figures followed the work on the arcane language. The *Book of Jeremiah* meant to fill an important gap in biblical scholarship. Montano claimed that many wise men wrote grammars of the ancient languages in which the sacred texts were written. Some wrote dictionaries and took great care in explaining the metaphors, idioms, tropes, and figures of these tongues. However, Montano claimed that he was unable to locate any works on *actio*, or gestures, that more universal form of expression that is common to all men, even to those who speak unintelligible languages.[50] In this book, Montano aimed to explore how things were said, "because it is one thing to consider . . . the meaning of a passage or be led by someone's commentary. It is another to observe them tearful, burning, or smiling."[51] Montano collected passages from the Bible that represented a set of attitudes or gestures in order to arrive at a more exact definition of the words spoken in scripture.

Montano, then, desired to expound on the meaning of many passages that could not be understood without a proper appreciation of the gestures that accompanied their elocution. In doing so, Montano attempted to devise a sacred rhetoric in the tradition of Erasmus's *Ecclesiastes, or The Method of Preaching* (1535), an exposition on how to preach effectively by drawing from scripture; the writings of the Church Fathers; and traditions of classical rhetoric.[52] More broadly, Montano's objectives resemble Erasmus's rhetorical hermeneutics. Kathy Eden has argued that for Erasmus, "like rhetorical imitation, then, biblical hermeneutics . . . is an act of accommodation, one in which both the writer and the reader come to feel at home." Through "historical reconstruction" the interpreter of texts can seek to understand beyond the literal meaning of the words and, rather, grasp the writer's intention. In this way an interpreter could conjecture how a writer "would accommodate his meaning to new and unforeseen circumstances."[53] For Montano, gestures were part of the reconstruction of antiquity, since they possessed historical specificity. In defending his work on gestures, Montano invoked ancient masters of oratory who placed great care in teaching their students the proper gestures to support their speech. More importantly, a precise awareness of the emotional undertones of specific passages could also bridge some of the inevitable gaps caused by the translation. The reader could thus be readily moved by the passages and truly comprehend what it meant for the biblical figures to cry, to raise an arm, or to smile. They would achieve this by means of an artificial recreation of the forcefulness of the original text.

If gestures were important in elucidating the emotional meaning of certain passages of the Bible, a proper understanding of the measurements and weights of things, or the distances between places, would contribute to rendering more exactly the dimensions of the physical setting in which the actions of the Bible took place. *Thubal-Cain, or On Measurements* is dedicated to this purpose.[54] Sacred measurements revealed "sacred mysteries."[55] Moreover, since God introduced measurements and they were the basis of all human interaction and harmony, like commerce and government, their origins and first applications were worthy of study in their own right. In the preface to the investigation on measurements, Montano complained that scholars of Greek and Roman antiquities, on account of their ignorance of "Oriental" languages, projected their conclusions on measurements onto the other nations of the world, when the characteristics of each nation should be correctly determined. It is for this reason that he had gathered through his study of numismatics, using sources like Nahmanides's commentary on the Pentateuch[56] and "the book that the Hebrews call

Mishnayoth and St. Jerome," all of the information that he could on the subject.[57] While many Spanish antiquarians focused on discovering remains of the Greco-Roman past in the Iberian Peninsula, Montano was interested in applying anti-quarian methodologies to the people and places in the Bible.

Books on sacred geography, architecture, and the chronologies of the Bible fol-lowed these initial treatises in the reading aid. In the preface of the *Book of Pha-leg*, Montano urged his readers to seek in the Bible all the information they needed to know on all aspects of culture: "Almost infinite is that which refers to the cus-toms, rites, the religions of peoples, the rules to administer the private and the public, for war, and for commercial uses, and really for all customs and habits of life, even for the ways of dressing, for [all] of this knowledge if you seek in sacred geography, you will find without a doubt a perfect and absolute knowledge of all of these things."[58]

The volume included elaborate maps of the places mentioned in the Bible, iden-tified with their current names, "before the eyes of each," to imagine how the world was when the biblical scenes unfolded and to contemplate how all that was and is to be can be recognized in the description and proper interpretation of the sacred text. The *Book on the Construction of the Objects Mentioned in the Bible* and the *Book of Chaleb* follow the same logic. The *Book of Chaleb* presented a study of sacred topography because "if they narrate to you the things happened without noting where they take place, or if you read the histories without a knowledge of the topography, everything is so confusing and mixed that you would not be able to decant anything that is not dark and difficult."[59] Montano also believed that ge-ography was the eye of history. Topography was essential to visualizing all the events of the Bible, of imagining their immediacy and full significance.

In the *Book on the Construction of Objects*, Montano used his knowledge of Hebrew and architecture to reconstruct the appearance of biblical buildings. A passage on Noah and the construction of the Ark is representative of the way Mon-tano used etymologies to unravel the biblical words' power of signifying. Montano argued that God indicated to Noah the use or utility of the Ark in the word that he used to designate the object that was to be constructed. Montano explained that the "name itself expressed the way of preparing and building the thing." God commanded Noah to build not a general "ark" but one that in Hebrew was called *THEBAH*. "The Hebrews," Montano explained, "have two names in which the object signified and meaning differ in its very use. One is *ARON*, and the other *THEBAH*."[60] The Latin translators did not distinguish between these two terms and rendered them both as "ark." However, *ARON* is used to store things that are not alive. Montano clarified that there were examples of this kind of construction in the Hebrew Bible, for instance, "the very famous ark of the Old Testament dedicated to God, built to keep the tablets of the Law, the volumes of Moses." The ark called *THEBAH* is prepared with the goal of safekeeping living beings, specially to safeguard humans who are exposed to the waters. It is the same term used to describe the "little basket" that Moses's mother prepared "to expose him in the riverbank." It was for this reason, Montano contended, that once Noah heard the word *THEBAH*, he knew with clarity what he was supposed to make, the materials he was to use, and the structure that the ark had to have for the preservation of life. Montano then clarified that this word "did not only fulfill this function" of explaining the mode of construction and utility of the ark "but it also exposed its mystical meaning so that at another time it could explained by me or someone else to whom God revealed it to."[61]

Hebrew contained various levels of meaning that even a reader who understood this language could not grasp. They could not even be accessed through regular etymologies. Montano dedicated the controversial *Book of Joseph, or On the Arcane Language* to unveiling this mysterious form of signification. In the introduction, he explained that in the sacred scriptures, names had been imposed onto things after an attentive and minute observation of nature. The *Book of Joseph* did not pursue the translation of simple words, but rather the "[path] of those who desire to comprehend the authentic meaning, typical of the divine oracles that are contained in these words replete with meaning, [will be taken]."[62] Montano believed that the more arcane level of signification, which can be termed metaphorical, facilitated an understanding of numerous passages of scripture that would otherwise remain obscure.

For instance, if a scriptural passage mentioned a lion and the context did not suggest that an actual animal was involved in the action, the reader then could choose from a variety of metaphorical meanings that might have been presented more clearly elsewhere in the sacred text. To support this kind of reading, Montano sought to bring all biblical symbols together in a reference work of arcana. There was a long tradition of this type of writing in other languages. Plutarch attempted to decode Pythagorean symbols. Horapollo's *On the Images of the Egyptians* (found in 1419) sought to present the symbolic meaning of ancient Egyptian writings, sparking the Renaissance interest in symbolism and allegorical representations. As Don Cameron Allen has argued, "One of the immediate effects of the enormous popularity of the *Hieroglyphica* of Hor Apollo was the compiling of sym-

bolic lexicons, which attempted to bring together a world of interpreted representations."[63] One of these works was Pierio Valeriano's (1477–1558) *Hieroglyphica* of
1556, a compendium of signs and symbols. Valeriano collected the secret meanings of plants, animals, metals, and the human body, among many other things.
The work appeared in numerous editions and translations, attesting to its popularity.[64] Valeriano provided a model on how to read symbolic languages. This
mode of reading could also be applied to interpret the allegories and symbolic
messages embedded in sacred writings.[65]

To compile his own symbolic lexicon, Montano would use as evidence for his
derivations only biblical passages and "the attentive observation of things."[66] This
decision underlay the fact that for Montano, ideally, the biblical text combined
with straightforward observations of the world was enough to gain all the knowledge available in the universe. A tree, for instance, was not simply a type of plant.
"In the arcane sense the tree represents man. In this way: every good tree yields
good fruits."[67] The same symbolic nature could be derived from gold, silver, weapons, laughter, and simple foodstuffs like bread.

To return to the example of the ark, Montano dedicated an entry to this word in
the *Book of Joseph*. Besides the distinction between *ARON* and *THEBAH* he had
indicated and that signaled the diverse uses of this object, Montano also described
how each Hebrew term for "ark" also encompassed other arcane meanings. The
ark where living beings were to be preserved, *THEBAH*, also meant "Christ." Noah's Ark was destined to bring "life, restoration and renewal to all living things
by divine gift." For this reason, Montano believed, "it enclosed an arcane, meaning Christ, who promised and brought salvation and life to all earth." *ARON* also
encapsulated a deeper arcane meaning. This term was used to signify a place to
safeguard inanimate objects, like the "tablets and the book of the Law. . . . [F]or this
reason it received the name Ark of the Covenant." Its meaning, according to Montano, was also that of "human heart," where each human being is to preserve
their allegiance to the Law of God.[68]

Like Joseph, after whom he named his treatise, Montano sought to be an interpreter of words, using only biblical passages and rabbinical commentaries to discover their true meanings. "Words mean," he wrote, "that which we think with the
mind and pronounce with the mouth." However, he continued, "they have by nature a precise sense and convey the things in themselves like the properties and
actions of those things."[69] For this reason, their translation rendered many meanings. God created everything in the world simultaneously through words. Humans

could find the sense of their own existence and the order of the world only in the correct interpretation of the divine word. Grammar, then, was essential for unveiling humanity's true sense of existence.

Sacred Geography and the Americas

Montano cast his net wide when he searched for sources to reconstruct even the minutest details of the Bible. During his participation in the Council of Trent, he obtained a copy, via a Venetian friend, of the *Itinerary of Benjamin of Tudela*. Thinking that it provided important details on sacred geography, he produced a Latin translation of the work soon after (1574). Even though Benjamin lived and traveled in the twelfth century, Montano considered him an important source in his studies of sacred geography. Zur Shalev argued in his study of this translation that for Montano "Benjamin, a Jew writing in Hebrew, the sacred tongue, presented a certain continuation of Scripture itself."[70] According to Shalev, Benjamin was a model traveler in Montano's eyes because of the detail with which he recorded everything he saw. He represented the manner in which an informed reader should approach Scripture.

Moreover, Montano dedicated the work to his friend Juan de Ovando. In the book's dedication, Montano emphasized the difficulties of governing very different territories from a great distance. In Benjamin's capacity for observations, Montano hoped that Ovando would find a model for reading cosmographical accounts and learning to discern geographical information with great care.[71] Similarly, in a missive congratulating Ovando for his newly acquired position as the president of the Council of Indies, Montano reaffirmed that "it requires a great wit [*ingenio*] that knows how to imagine lines and measures and angles and ports and fields and animals and plants and natures that many have not seen and that do not resemble those from around here."[72] From Ovando, the Spanish Crown had "demanded among other functions" to possess "the most important" and "definitive knowledge" to "govern all of the things extant in the kingdoms and jurisdictions belonging to our king throughout the Ocean." This required a great deal. So "to keep this care with prudence and wisdom," Montano advised his friend who had been "acquiring through attentive study a very useful previous knowledge of geography" to learn about Benjamin of Tudela's "representation of the terrestrial globe," which this author traced "in accordance with the knowledge of his time."[73] An eager reader, then—like Benjamin traveling the medieval world or Ovando virtually crossing the ocean with each memorandum or description of the Americas—

even if they did not possess the skills to understand the sacred tongue, could seek the information they required in the detailed antiquarian treatises of the apparatus to bridge worlds temporally and geographically distant. With these tools in hand, the reader could gain sufficient context to engage in a meaningful and fulfilling reading of the Bible that engaged all the senses and that began to capture the infinite energy of Hebrew words.

Montano's interest in the Americas is evident not only in his correspondence with Ovando but also in his collection of natural and artificial objects.[74] In his natural historical writings, a few mentions of New World plants also appear. For instance, in his discussion of roots, Montano mentioned the *batata*, or "potato."[75] However, as María Portuondo has argued, Montano's attitude toward the Americas was based on a "historically informed hermeneutics of nature that in practice worked to nullify the temporal dimensions of any discovery." His natural philosophical writings thus endeavored to demonstrate how all natural phenomena were already prefigured in the Bible.[76] This rested on an understanding of Adam's role as name-giver (*nomothete*), the one who had known and named all natural phenomena.

In the *Phaleg*, Arias Montano argued for the importance of studying sacred geography. Having a mental image of the worlds described in the Bible allowed the reader to reach a better understanding of the actions unfolding in the text. But sacred geography also facilitated the reader's apprehension of the basic unity of humankind, which was genealogical but could also be mapped onto all the lands of the earth. In sacred scriptures, Montano argued, one could find evidence of the fact that the Americas were known to the ancient Israelites and that they "traveled there with great frequency."[77]

Ancient poets and philosophers like Plato and Aristotle did not know anything about the Americas, in Montano's opinion. Moses, however, Montano argued, wrote clearly about the land of Ophir. This knowledge "the prophet Jonathan transmitted with elegance and rigor in a work about the kings of Judah." Ophir was also described with great eloquence by the author of *Paralipomenon* "who wrote this work by dictation from the Holy Spirit." This author mentioned the fleet that Solomon built to navigate to the East, to a land called Parvaim, an "expression that clearly designates, for those who know how to read Hebrew, two regions, called in another time Perú: the only one with which the same word nowadays is called Perú, and another, which navigators have now called New Spain."[78] Ancient writers described the gold that came from this region as pure and plentiful. It provided extraordinary wealth to all nations. No river in Europe or Asia, Montano averred, could have provided the ample quantities of gold that the Romans brought back from their conquests. Montano emphasized how the Tagus, the Pactolus, and the Hermus Rivers were then as they were in his day: poor in gold. In the Bible, gold circulated as the main medium of exchange. Montano believed that the quantities of gold that sustained ancient economies must have come from the Americas. This comparative outlook also relied on the strategy of using the Americas as living example of what had once been.[79] A map accompanied the description, locating Ophir in the Americas (fig. 5.1). Montano's friend, the cartographer Abraham Ortelius, also included the identification in his *Synonymia geographica* of 1578.[80]

Arias Montano's Peruvian etymological conjectures came under scrutiny once more a mere four years after Cabello Valboa completed the *Antarctic Miscellany*. José de Acosta's *Moral and Natural History of the Indies* (*Historia moral y natural de las Indias*) appeared in Seville in 1590. Like Jerónimo Román Zamora's *Repub-*

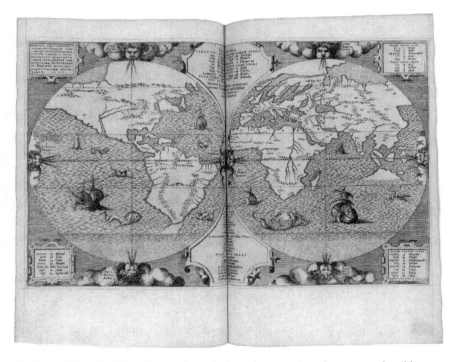

Figure 5.1. Map depicting the earth, including the Americas, being populated by Noah's descendants. Their names appear in Latin and Hebrew. In Benito Arias Montano, *Phaleg, siue De gentium sedibus primis, orbisque terrae situ, liber,* Antwerp Polyglot Bible (Antwerp: Christophe Plantin, 1572). Courtesy of the John Carter Brown Library at Brown University.

lics of the World, Acosta's book was one of the few accounts about the Americas to be published during the reign of Philip II. It became a best seller.[81] Acosta arrived in the viceroyalty of Peru in 1571. There, he traveled extensively into the interior of Peru to Cuzco, Arequipa, La Paz, and Potosí. He witnessed the extraction of mercury from the mines of Huancavelica and silver from the mines of Potosí. In 1582, he participated in the Third Council of Lima, the provincial council summoned to determine future strategies for the conversion of the Indigenous inhabitants of the viceroyalty of Peru. He then traveled to New Spain, where he gathered materials for his historical works, and returned to the Iberian Peninsula in 1587. There he would remain until his death in 1600, serving as rector of the Jesuit college of Salamanca.

As its title announces, Acosta brought together in one single work natural information about the so-called torrid zone together with the customs (*mores*) of the "Indians, ancient and natural inhabitants of this New Orb."[82] Fray Luis de León read and approved the work for publication in 1589.[83] To describe both the "natural" and the "moral," Acosta relied on the most up-to-date information about American plants, animals, and minerals, and on informants fluent in Indigenous languages to apprise him of both antiquities and customs. Among his two most important sources were his fellow Jesuit the missionary Juan de Tovar (d. 1626) in New Spain and the jurist Polo de Ondegardo (d. 1575) in Peru.

Acosta began the *Moral and Natural History* by debating the different theories of origins that European scholars had advanced about Indigenous Americans and their connection to the people of Afro-Eurasia (book 1). He then discussed the climate and nature of the region and the insufficiency of Aristotelian explanations to account for the New World's physical characteristics (book 2). Books 3 and 4 built on these geographical considerations to then discuss animals, plants, and minerals, and how the Indigenous societies of the Americas understood and exploited them, including their Indigenous names. The subsequent sections (books 5–7) analyze the religious customs, the calendars, systems of inscription, and political institutions of the Inca and the Mexica. Finally, the work culminates with the Spanish conquest of Tenochtitlan and a short exposition on miracles and the future of the Catholic conversion of the Americas.[84]

The seven books bring together human and natural knowledge, following Pliny's model for the historical study of nature, with Acosta's immediate concerns about the accomplishments of missionaries and the outlook of conversion. Unlike Hernández's translation of Pliny, however, Acosta's account culminated in specific proposals about the future of conversion efforts in the Americas. All seven of

Acosta's books alluded to the discussion of origins that occupied the first section. They also mobilized evidence embedded in local languages to both counteract existing interpretations and support Acosta's own theory on the early habitation of the Americas.

Acosta's theory was essentially that people had migrated to the Americas by crossing on foot a still undiscovered land bridge. He discussed the Ophir hypothesis in detail. Acosta traced back Montano's Ophir conjecture from the *Phaleg* to the Bible of Robert Estienne (1503–1559). Published in 1545, Estienne's Bible contained the scholia of François Vatable, "a man learned in the Hebrew tongue." This work located Ophir on the island of Hispaniola. Subsequently, Arias Montano shifted its location to Peru and New Spain.[85] Acosta was aware of the longer tradition of this identification and wanted to dispel all its forms. Acosta took issue with various parts of Montano's argument. He first addressed the etymological evidence and then approached broader issues about the status of navigation in ancient times. Cabello Valboa, as we have seen, had anticipated the objection over the identification of Ophir with Peru on the basis of his own experiences in the Indies but ultimately accepted the linguistic association. For Acosta, "the etymology of the name Ophir and its reduction to the name Peru" was not in the least convincing. He argued that "the name Peru is not that ancient and not general to this land." Acosta explained how it had been common for Spaniards to impose on the kingdoms of the Americas names that were convenient, that is, that they had learned in the ports or from what they could understand from their informants.

Acosta expanded on the implications of this statement to debunk other hypotheses that relied on etymological evidence alone. The general opinion in the viceroyalty of Peru was that the Spaniards had named the whole territory after a river they believed was called Piru. Since the Indigenous Peruvians did not use this name, or have any memory of it, Acosta argued that those who defended the Ophir conjecture must also accept that the Indigenous Americans "do not use and do not know this name for their land."[86] Hence, there was no local knowledge to support Montano's etymology. Likewise baseless were Montano's conjectures that Sefer was the Andes Mountains or that the Yucatán Peninsula was settled by Joktan.[87]

Acosta also called into question the notion that the Americas had been in close exchange with the fleets of Solomon. If ancient Israelites had reached the Americas as many times as the *Phaleg* presumed, Acosta argued, there would be evidence of these journeys somewhere in the West Indies. Yet he could find no such traces in local traditions. From Ophir, sailors brought Solomon gold, silver, ivory, and monkeys, among other commodities. Acosta believed that this assortment of goods could be found only in the East Indies. There was no ivory in Peru or any local memory of elephants, although monkeys, gold, and silver were indeed plentiful. Acosta then proposed his own interpretation: in the Hebrew Bible, Ophir did not mean a specific place but was used in the same way that the Spaniards use the term Indies, grouping lands as distant as China, Mexico, and Peru under one general appellation. The name India was given to the Americas after 1492 because of their perceived remoteness. Acosta believed that the toponym Ophir similarly signified a general place in Hebrew from which these commodities came.

Acosta called into doubt the connections between ancient Hebrews and the Americas on technological grounds as well. Solomon's fleets would have had to cross the seas past China and the East Indies to search for gold in the Americas. This was a long oceanic voyage that Acosta considered impossible for ancient people. He had asked himself many times, "Is it only in our time . . . and only our men, who have achieved the secret of sailing on the ocean?" The Holy Scriptures described how Solomon received from the people of "Tyre and Sidon master mariners and pilots who were very skilled at sea and that with them he made a three-year voyage." Yet Acosta still concluded that the ancients did not attain the degree of skill in navigation with which men crossed the ocean in his time. He rejected notions of *prisca sapientia* (ancient philosophy) that underlay the "Mosaic" natural philosophy of Arias Montano and other contemporaries.[88] Moreover, no ancient text mentioned the use of the compass or the lodestone to sail, which were fundamental to measuring latitude and finding one's direction on the high seas. To anyone who had made the transatlantic crossing, the centrality of the compass was clear. Even though ancient authorities like Pliny and Saint Augustine mentioned the loadstone, none described its "strange quality of always turning to the north." Moreover, "there is no word in Latin, Greek, or Hebrew for such a remarkable object as the compass. If they had known it, such an important thing would inevitably have had a name in these languages."[89]

Many scholars shared Acosta's views about the modern origin of the compass. Lists of ancient and modern (postclassical) inventions circulated in this period. Jean Bodin included the compass in his *Methodus* in 1566 as a modern invention. Bodin stated that the "ancients were not aware of its use, clearly divine, and whereas they lived entirely in the Mediterranean basin, our men traverse the whole earth every year."[90] The idea that the compass was a recent innovation can be traced back at least to Lorenzo Valla, who compiled his own list of modern inventions for which there were no words in classical Latin.[91] The Medici court artist Stradanus (1523–1605) also included the compass among his images of modern inventions (fig. 5.2).[92] Finally, Acosta pointed to the fact that in the ancient world, pilots sat in the prow of ships. The Greeks called their pilots *proritas* because they steered their ships from the prow. In Acosta's day, pilots had to sit on the highest part of the stern to observe the movements of the compass. They sat in the prow only when entering or leaving ports or sailing close to land.[93]

As for the supposed prophecy in the book of Obadiah that announced Spain's providential role in the discovery and conversion of the Americas, here too Acosta expressed his skepticism. Obadiah was one of the twelve minor prophets. In 1589, Fray Luis de León argued that Obadiah foretold both the discovery of the West Indies and the conversion to Christianity of Indigenous Americans by the Spanish. Luis de León's reading of Obadiah was part of a broader effort to address the problem of the "New World" and its presence in the sacred scriptures. Like Arias Montano, Luis de León found it incredible that the most "important and memorable event to have happened to the Church" was not foretold in the Bible.[94] He argued that Obadiah predicted the spread of the Gospels in three stages, first to

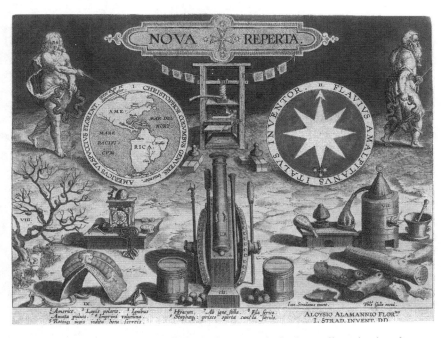

Figure 5.2. Jan van der Straet, called Stradanus, and Thedor Galle, title plate from the *New Inventions and Discoveries of Modern Times* (*Nova reperta*), including the compass on the right (Antwerp, c. 1600). Courtesy of the John Carter Brown Library at Brown University.

the Jews, and second to those who inhabited the Roman realms. Lastly, the verses, "and those deported from Jerusalem which is in the Bosphorus will inherit the lands of the south," portended the third stage. Luis de León interpreted this verse to mean that the "faithful who inhabited Spain, who descend in part from all of those Jews who emigrated to Spain after the destruction of Jerusalem, would possesses the cities of the South, that is they would possess the New World, which is south of Spain, situated in the southern hemisphere."[95] For Bosphurus, Fray Luis argued, the word that appeared in the Hebrew Bible was Sefarad.[96] And Sefarad was the Hebrew name for Spain, as Arias Montano had also demonstrated.

Acosta's explanation of Obadiah differs from that of Fray Luis de León.[97] In this instance, he chose to leave the long linguistic disputes about Sefarad aside and focus instead on the part of the verse that discussed the "cities to the south." It was not clear to him that this necessarily meant "people of the New World," although it was possible, because "most of this New World is toward the south and faces the South Pole." Acosta conceded that "it seems like a very reasonable supposition that there would be some mention in the Sacred Scriptures of a matter as great as the discovery and conversion to the faith of Christ of the New World." However, he believed that the "kingdom" was not to be for Spaniards or Europeans but for "Christ Our Lord." While not dismissing the foresight of prophecy, Acosta reaffirmed his belief that "much was unknown to the ancients," and that even in his day "a good part of the world is still to be discovered."[98]

For sixteenth-century European scholars, all of humankind descended from Adam, and Acosta still had to account for the monogenetic descent of the people of the Americas. He believed that the Atlantis hypothesis, which both Agustín de Zárate and Francisco López de Gómara embraced, could easily be dismissed. Acosta chastised these scholars for taking Plato's writings as literal truth.[99] He dismissed the popular notion that Indigenous Americans descended from the lost tribes of the book of Esdras as "idle conjectures." One of the arguments against this reading of Esdras was the fact that the "Hebrews used writing, but among the Indians there is no trace of this." Moreover, Acosta remarked, "the Jews have been so assiduous in preserving their language and ancient traditions," in other parts of the world that it was difficult to believe that they would forget them in the Americas.[100]

Like other New World chroniclers before him, Acosta lamented how difficult it was to investigate the Americas' ancient past because there were no writings. The question to which he could find no answer, however, was where the Indigenous inhabitants had come from, because the Christian tradition claimed that all people descended from one human being. These were the "obscure" first ages of the world that had forced Iberian scholars like Ocampo and Garibay to turn to the writings of Annius. Acosta had judged that Indigenous peoples' accounts of origins were "dreams rather than history." Acosta claimed this situation resulted from their lack of "books and writing."[101] Notwithstanding this seemingly categorical assessment, Acosta relied on Indigenous histories to make his conjectures about origins, and on etymologies and oral traditions, to relate all aspects of the Americas' natural history.

To explain Indigenous Americans' genealogical connection to the inhabitants of Afro-Eurasia, Acosta proposed that they must have crossed on foot over some land bridge that was still undiscovered and that connected the Americas to Europe. This theory could account for the particularities of American fauna. It also did not require Indigenous Americans to have known about the compass. Moreover, Acosta believed that evidence both in Indigenous histories and in the nature of the Americas supported this hypothesis. His reading of Juan de Tovar's historical account of how the Mexica related their origins gave him evidence to support the idea that Indigenous Americans had migrated on foot from the north and populated the Americas slowly.

Based on Tovar's account, Acosta related how the Nauatlaca, or "people who speak well and express themselves clearly," related their most ancient history. Their ancestors slowly made their way to what would become the city of Tenochtitlan by walking from the north. In the sections relating this migration, Acosta reproduces Tovar's glosses of Indigenous toponyms and ethnonyms. Aztlan, the land in the north where the Nauatlaca originated, means "place of herons." The Nauatlaca called the first inhabitants of the land, who lived only from hunting and in cliffs and had no permanent abodes, Chichimecas, "which means people who hunt . . . and who are known by another name as Otomies."[102] The Nauatlacas, Acosta continued, "depict their origins and descendance as a cave and say that they came to settle the land of Mexico from seven caves."[103] Acosta described how all the lineages descended from the north, settled, and slowly made their dwellings and cities. The ancient Chichimecas also began to adopt some of the traits of the other lineages but continued to live separately from them. Acosta believed that this account, which he took to be reliable, was evidence for the general way the settlement of the Americas had taken place. "I am convinced," he explained, "that most of the provinces and nations of the Indies have developed in the same way: the first of them were barbarians, and in order to maintain themselves by hunting, they little by little penetrated inhospitable lands and discovered a new world, living in it almost like beasts; they had no dwelling places nor roofs nor cultivated fields, nor livestock nor king, nor law, nor God, nor the use of reason." Later, Acosta explained, people who were more powerful and energetic began to subdue others until they formed their great kingdoms. He was thus persuaded "that the settlers of the West Indies came by land, and that in consequence all of the territory of the Indies is contiguous with that of Asia, Europe, and Africa, and the New World with the Old."[104]

Acosta's general point was to show that from what he took to be more rudimentary forms of social organization, cities and empires slowly developed. Tovar clarified that the "seven caves" was a figure of speech of sorts. He explained that although the Mexicans say that they left from seven caves, it is not because "they lived in them, because they had their houses and fields with much order and civility [*policía*] of a republic, their gods, rites and ceremonies, since they are very political people." In that northern province, Tovar further contended, "it was accustomed that each lineage would have their known place, which they used to signal with a cave, saying the cave of such lineage or such descent, like in Spain it is said the house of the Velasco or Mendoça."[105] According to Tovar, the lineages that descended from the north already cultivated fields and possessed *policía* before their southern migration. Acosta reproduced this explanation and added that in their "libraries they make a history of this, painting seven caves with their descendants" (fig. 5.3).[106]

Acosta's reliance on Tovar's etymologies to represent the early history of the Indigenous inhabitants of New Spain was replicated in other parts of his works. In the sections that deal with animals and plants, Acosta attempted to reproduce local nomenclature as faithfully as he could. For instance, in book 3, he described the flowers of New Spain. Acosta believed that above all other people Indigenous Mexicans esteemed flowers. They specially esteemed a flower called *yolosuchil*, "which means flower of heart because it has the same making as a heart."[107] In these instances, as in his use of ethnonyms, Acosta demonstrated an Aristotelian understanding of language. Words were conventional, and they contained the knowledge of their name makers.

This same approach to spoken language underlies Acosta's understanding of writing systems. The categories he used to understand various forms of Indigenous American writing systems were limited by his Aristotelian outlook. Acosta's assessment of Indigenous forms of transmitting history is not always seamless. For instance, in a chapter dedicated to the memory aids of the Peruvians, Acosta argued that even though the Indigenous people of Peru had no "kind of writing at

Figure 5.3. "Cave of the seven lineages that populated Mexico and its surroundings." Tovar Codex (c. 1585), fol. 85r, Codex Ind 2. Courtesy of the John Carter Brown Library at Brown University.

all . . . this did not prevent them from preserving the memory of their antiquities, nor did they fail to keep a reckoning for all their affairs, whether of peace or war or government." Specialized chord keepers produced and interpreted these records, "for whatever books can tell of histories and laws and ceremonies and accounts of business all is supplied by the quipus so accurately that the result is astonishing."[108] Despite the order maintained in these devices and the fact that mail connected imperial domains, Acosta still considered alphabetic inscription superior. Nonetheless, Acosta followed Tovar's elaborate arguments for the reliability of oral traditions in Mexico. On the authority of local experts, like Polo de Ondegardo, and his own experiences in Peru, Acosta would also accept the reliability of quipus and other forms of oral commemoration in the Andean region.

Tovar had assured Acosta in a letter that the histories he compiled from interviewing "old and wise" Indigenous men from Tula fully agreed with what the Indigenous authorities in Mexico and Texcoco had related to him. Moreover, despite not possessing alphabetic writing, Tovar described having seen all the histories he transcribed for Acosta written "with characters and hieroglyphs." Importantly, Tovar admits that "he did not understand" these characters and that for this reason the viceroy Martín Enríquez (d. 1583) "commanded the elders of Mexico, Tezcuco, and Tulla, to come with me" and explain the histories as inscribed in these characters. Only then was Tovar able to compose "an accomplished" account that was based on the Indigenous elders' testimonies, not on his own reading of their books.[109] Tovar reaffirmed once more that these accounts had great authority not only because had he seen that their "books" existed, but, more importantly, because he had collected the accounts of the elders before the *cocoliztli* epidemic, and none of the "elderly men disagreed" among each other.[110]

Tovar then explained, as he understood it, how "hieroglyphs" worked. The Mexicans would "paint" objects that could be rendered as images. For "things that did not have images of their own they had other meaningful characters, and, in this manner, they could paint what they wanted." Although interpreters could differ in the exact words they used to translate these pictographs, they agreed on the "concepts" the images conveyed. In this way Tovar conceded that this form of writing was "not enough like ours" to allow words to be rendered in the same order without disagreement. To complement pictographs, however, Tovar explained to Acosta, and to remember ancient speeches and songs exactly as they had been once uttered, the Mexicans would commit them to memory in special schools and assiduously teach them to young boys, generation after generation.[111] Tovar collated enough testimonies from Indigenous elders to assure Acosta that this form of transmission was authoritative and truthful.

The chapters in which Acosta compared Chinese and Indigenous American forms of writing drew from informants like Tovar and the writing of the Jesuits who had been to China. The sections began with a summary of Aristotle's understanding of language. "Letters were invented," Acosta explained, "to refer to and immediately signify the words that we pronounce." Words, as Aristotle taught, "were signs of the concepts and thoughts of men." An image like a "painted sun" cannot be regarded as a "letter" because it depicts an object; it is merely a "painting." There were three main forms of inscribing and transmitting memory to those who are absent, according to Acosta: ciphers, letters, and paintings. Acosta classified Chinese writing and Indigenous Mexican characters as paintings. He claimed that their "writing is to paint or cipher, and their letters do not mean parts of words, like ours, but are figures of things, like the sun, fire, man, the sea."[112] In doing so, Acosta also argued for inferiority of Chinese characters to alphabetic writing. He misunderstood this writing system as well.

Acosta did not compare the Mexican writing system to Egyptian hieroglyphs. However, there were precedents for doing so. As described in chapter 3, the chronicler Francisco Cervantes de Salazar compared Mexican "paintings" to "Egyptian ones." The scholar and collector of art and antiquities Felipe de Guevara (ca. 1500–ca. 1563), who composed his unpublished *Commentaries on Painting and Ancient Painters* (*Comentarios de la pintura y pintores antigüos*) in 1560, also elaborated on the similarities.[113] In this early work, Guevara described how the Egyptians practiced two types of painting. The first was also common to the Greeks and the Romans and included figures "and natural things of history and poetry"; the second type "they used in place of letters, for arcane things of great veneration." These were hieroglyphs, "or so-called letters which consisted of paintings and varieties of animals and similar things." Herodotus, Guevara claimed, called these "sacred letters," since they were replete with arcane meanings and were understood only by a select few who were initiated into their contents.[114]

Guevara drew on a long tradition that, as Erik Iversen has shown, regarded hieroglyphs as a "divinely inspired Egyptian invention, a unique form of symbolic writing." This belief originated in Marsilio Ficino's translation of Plotinus and remained accepted until the eighteenth century.[115] Guevara then discussed how certain types of painting that he had observed from the New World codices that had come to Spain from the Americas resembled the Egyptian images of Horapollo. He argued that "this type of painting," or glyphs, "which express concepts, the Western Indians seem to have imitated, and those of the New World, especially those of New Spain." The causes for this were unknown, and Guevara contemplated two possibilities. The first was that they were perhaps able to produce these arcane letters "because of an ancient tradition that they inherited from the Egyptians, which could have been." The second explanation Guevara devised from this comparative exercise was that it was also feasible that the "natives of these two nations concurred in their imaginings."[116]

Because Acosta did not believe in notions of an ancient wisdom that had been passed down in these "paintings," as his assessment of ancient knowledge in the sections dedicated to the origins of Indigenous Americans reveal, he reaffirmed in these chapters about writing the conventional approach to spoken and written language. Hence, Acosta did not draw from traditions that compared Mexican writing systems to Egyptian hieroglyphs. He also did not understand the linguistic or compositional principles of Mesoamerican writing systems; in fact, very few Europeans did at the time, and what they knew was rudimentary.[117] In his assessment of Mexican script, as in his treatment of etymologies, Acosta disentangled the uses that words and images could provide. They could only transmit the immediate knowledge of their makers.

The Hebrew Rudiments of Castilian Lexicography

In a passage describing the events of the Tower of Babel, Montano described how the intention to defy God was not assuaged by this occurrence but only thwarted by the inconveniences generated by linguistic confusion. The inability to communicate led humankind to disperse across the earth. Montano expressed that in another volume dedicated to the "doctrine common to all languages" he would expound on the "confusion of speech." However, the dispersal was an apt reminder of how nothing opposes the will of God.[118] This treatise on the doctrine common to all languages was not completed. However, a manuscript entitled *Adam, or About Language, Interpreter of Human Thought and of the Principles Common to All Languages*, written around 1571, might correspond to this work.

In this treatise, Montano sought to present the principles common to all languages, as derived from Hebrew. As Navarro Antolín and Gómez Canseco have argued, Montano believed that from this natural language, other languages emerged. The analogies that existed between languages were evidence of divinity, "while the anomalies that existed between languages were caused by human intervention."[119] Montano began the work by reminding his readers that "the faculty of [using] words and language, the habit of conversation and communication, were given to men together with the other gifts of nature." Men had the capacity to speak since the beginning of creation, and later they developed writing. Furthermore, all existing languages, if interpreted correctly, could be seen to function in the same manner as Hebrew and contain the same essential parts of speech.[120]

Montano contributed to a tradition of linguistic speculation that those like the Swiss Reformed scholar Theodor Bibliander had already explored. Bibliander's *De ratione communi omnium linguarum et literarum commentarius* (1548) treated the principles common to all languages as derived from Hebrew, the pre-Babelian tongue. Bibliander also compiled written evidence from many sources to show the interrelationships between all the languages he could gather evidence for and Hebrew.[121] The confusion of tongues, according to Bibliander, occurred suddenly.[122] Reducing tongues to their rudiments had, for the Swiss scholar, a very explicit pedagogical goal: it could allow humankind to communicate better and to recognize the essence of Christianity.

In the early modern period, as Umberto Eco explained, there was no clear distinction between attempts to establish a universal grammar or a set of principles common to all languages and the search for a primordial tongue.[123] In his 1555 comparative work on languages, Conràd Gessner, for instance, insisted that all languages retained words that possessed Hebrew origins, even though these words were now corrupted.[124] Likewise, Luis de León argued in the prologue of his translation of the Song of Songs that Castilian resembled Hebrew in many aspects, including its phrases and ways of speaking. Luis de León described how he strove for his translation to "correspond with the original not only in the expressions and words, but even still in the agreement and air of them, imitating their figures and ways of speaking as much as is possible in our tongue, which in all truth, corresponds with the Hebrew language in many things."[125] Comparisons to Hebrew, as in this case, served to ennoble vernaculars, to promote a particular understanding of the Bible, and to reconcile the diversity of the world to a common history.

Acosta's refutation of the connection between Ophir and Peru appears cited in Sebastian de Covarrubias's *Treasure of the Castilian or Spanish Language* (1611, *Tesoro de la lengua castellana o española*), the first published etymological dictionary of Castilian. Covarrubias may have studied some Hebrew at the University of Salamanca between 1565 and 1571. At the time, as Dominique Reyre has shown based on internal evidence in the *Treasure*, Covarrubias learned rudimentary Hebrew from Castilian. At Salamanca, in the late sixteenth century, classical languages were beginning to be taught with "the aid of vulgar tongues." This might have encouraged Covarrubias's tendency to establish analogies between Hebrew and Castilian lexical items, sometimes on the basis of phonetic similarity alone, since most of his lessons were based on oral readings of the Hebrew Bible. Although his knowledge of Hebrew was limited, Covarrubias relied on sources like the grammar and dictionary of Pagnini (Antwerp, 1572).[126] The ample diffusion of Covarrubias's work allowed for the translation and broader dissemination of some of the central principles that inspired Arias Montano's treatises in the Antwerp Polyglot but were now rendered in the form of encyclopedic lexicography.[127] Moreover, Covarrubias's collection of Indigenous American words represents an attempt to integrate these lexical objects into a broader account of the history of the Castilian language that was in harmony with the biblical paradigm, despite almost fifty years of debates about the origins of Indigenous American words.

Covarrubias contrasted Acosta's interpretation with Arias Montano's Ophir hypothesis in his entry dedicated to Peru, which began by relating that the Spanish kings had conquered this province, and that this very famous territory was the source of much gold and silver. Covarrubias then turned to varying "opinions" of the word's etymology. "Some claim," Covarrubias speculated, that Peru was so called because of a river that ran through that province called Peru. Acosta, Covarrubias asserted, in "his book de natura *novi orbis*," explained the etymology as the result of a misunderstanding between an Indigenous informant and a Spaniard. Here, Covarrubias repeated an anecdote that was not actually in Acosta. This story appeared, rather, in the recently published *Royal Commentaries of the Inca* (Lisbon, 1609) of Garcilaso de la Vega "El Inca." Bernardo de Aldrete cited Acosta's and Garcilaso's explanations of the Peruvian toponym in his work on the origins of Castilian (1606); Garcilaso had communicated this explanation to Aldrete informally. Aldrete was probably Covarrubias's source, hence the conflation of the two accounts.[128]

Nonetheless, Covarrubias correctly attributed to Acosta a skeptical position about the origin of the toponym Peru. Covarrubias explained, later, that many "grave authors believed that this was the land of Ophir." He cited other reference works like Ortelius's *Synonymia* as evidence for this interpretation. He then included the relevant biblical passages the supporters of Ophir used to explain their theory. Then Covarrubias related another opinion about the possible origin of the toponym: that this name was Hebrew and came from the verb *parad*. This verb meant "to divide," and it was applied to Peru "because its land was so apart and divided from the rest."[129] Covarrubias routinely applied Hebrew phonetics and morphology to analyze certain Castilian verbs, toponyms, and nouns the lexicographer thought possessed roots with three radicals.

Covarrubias applied this mechanism of reducing words to Hebrew roots to many of the Indigenous American words the etymological dictionary collected. *Cacique*, which had already inspired Valverde's etymological derivations, also merited Covarrubias's attention. He believed this word meant "as much in the Mexican language as lord of vassals, and amongst those barbarians the lord that has the most strength to subject the others." While many chroniclers had attributed this lexical item to the Caribbean languages, Covarrubias took it to be "Mexican." Although the current form of the word was attested to in the Mexican language, it had a much more ancient history that could be traced by reducing the name to its constitutive parts.[130]

In this entry, Covarrubias explained his justification for his etymological method. After the flood, "those who populated the world" divided themselves "in the confusion of the tower of Babel or Bablyon." Despite adopting different languages, Covarrubias believed, "each nation that set themselves apart took with them some trace of the original language, in which they had all spoken." The original language "remained with Heber, his family, from where the Hebrews proceeded." It was for this reason that Covarrubias proposed, on the basis of phonetic similarity, that *cacique* originated from the Hebrew verb *chizek*, meaning "to strengthen" and the noun *chazak*, meaning "strong."[131] In this way, *cacique* proved how all languages, even those in the Americas, descended from Hebrew. All languages retained some vestiges of the original language. Thus, *cacique* contained the common origins of humanity and upheld the history of linguistic descent put forth in the Bible.

Covarrubias admitted when he could not provide an etymology for certain words. The word *cayman*, a "fish that grows in the rivers of the Indies and that eats men who swim in the waters," stumped him. Covarrubias claimed, however, that given that this name came "from a barbarous tongue," its etymology could not be located.[132] Other Indigenous American words, he proposed, had clear Hebrew roots. *Hamaca* (hammock), for instance, was one. A hammock was "a bed of Indians." Covarrubias specified that it was a "big blanket made of cotton" they would hang from trees to remain cool in the heat. They would also wrap themselves in them and thus deter the insects, "which are much more annoying in those lands than here." Covarrubias then suggested that the word might come from the Hebrew verb *hhamaK*, meaning "to turn," because Indigenous Americans "stir themselves in them."[133]

In several instances, Covarrubias relied on López de Gómara, whom he did not explicitly cite, and perhaps even on Pedro Cieza de León, to determine the meaning of Indigenous American words. The entry dedicated to Moctezuma is an example of this. Echoing López de Gómara almost word for word, Covarrubias defined Moctezuma as "the name of that very powerful Indian king" whose name means "sullen and grave." The "Indians call him Motezumacin: the *cin* is the equivalent of our *don*, and they only used it at the end of the name of a king or great lord."[134] His sources demonstrate that information about New World lexical items continued to circulate mainly embedded in the early chronicles of the Indies, more than fifty years after their initial publication.

Conclusion

For Arias Montano, translation, with all the dimensions that it encompassed, could provide for a more faithful rendition of the infinite contents of the Bible, the source that contained all knowledge, and allow for a better understanding of history, nature, and humankind. Etymology was among the most important tools for extracting the chains of signification embedded in sacred words.

The Antwerp Polyglot was influential in debates about the Americas and their place in ancient and biblical history. Despite having access to Acosta's condemnation of the many theories of origins that had proliferated since Columbus's landing in Hispaniola, Covarrubias, for instance, continued to uphold a linguistic genealogy that equated the search for principles common to all languages with a return to the original, first tongue of humankind, which he took to be Hebrew. Covarrubias, and other etymologists before him, seamlessly integrated the Americas into this model of understanding linguistic diversity and proposed this genealogy as the organizing principle of the first etymological dictionary of Castilian. In so doing, Covarrubias echoed the strategy that scholars like the jurist Alonso de Zorita and the natural historian Francisco Hernández had employed in the previous decades in their efforts to praise Nahuatl, especially the way words in that language seemed to convey the properties of the objects that they signified.

Epilogue

In 1629, in Madrid, the jurist Antonio León Pinelo published a catalog of all the works in manuscript and print about the East and West Indies. The *Epitome of the Oriental and Occidental, Nautical, and Geographic Library* collected titles in forty-four languages. More than twenty of these languages were indigenous to the Americas. To assemble the *Epitome*, León Pinelo consulted more than five hundred manuscripts buried in archives in New Spain, Peru, and Castile, discovering numerous works that dealt with the cities, customs, provinces, and nature of a world that was once unfamiliar to Spain and still remained so for most.[1] The compilation meant to bring to light forgotten authors and their histories, for many Spaniards, Léon Pinelo believed, had been more interested in the Americas' gold and silver than in its knowledge. His book, he affirmed, could serve as a library.

In the entry dedicated to Castilian, León Pinelo explained that this language was now spoken in many parts of the world. In Spain, Castilian was widely spoken for communication and business, even though each kingdom had its own language. This could also be observed in many parts of Europe. In the Indies, Castilian could be heard from the Straits of Magellan to the most western part of New Spain and all the islands of the Caribbean. To the "Indians it is ordered," Pinelo claimed, "that they learn it." He suggested that among other efforts, the "sons of Indian *caciques* should be sent to the cities of Spain," thus echoing the strategy, recorded by the prophet Daniel: "King Nebuchadnezzar commanded so that the [people] of his kingdom would learn the Chaldean language."

After this counsel, León Pinelo briefly reconstructed his understanding of the history of the languages of Spain. Many authors related that, among the pre-

Roman languages, Basque, brought to the Iberian Peninsula by Tubal, was the first, or at least the only one still alive as an island of antiquity in a sea of *romances*. It had been preserved for many centuries, a feat because it is "notably difficult to preserve something as corruptible as the language of kingdoms."[2] Hebrew, Greek, Latin, Arabic, Gothic, and the Punic language had all been spoken in Spain at some point. León Pinelo concluded that seventeenth-century Castilian took most of its vocabulary and structure from Latin, but it also contained traces of all of the languages that were once spoken in Iberia. For León Pinelo, as for his predecessors, lexical borrowing was still one of the primary ways of understanding historical change in languages, as was the concept of "corruption"—the idea that purer languages had decayed to their present form.

Thus, León Pinelo wove together the relationship between language and political power as he understood it in the early decades of the seventeenth century: various languages could coexist in a kingdom, but they were organized hierarchically depending on the linguistic preferences of rulers and in accordance with languages' specific domain of use. This language flexibility seemed foregone by the eighteenth century, when the new Bourbon Crown sought to impose Castilian and eradicate Indigenous languages. Yet in the seventeenth century, when León Pinelo produced his catalog, the power of languages, and the language of power—specifically, the possibility of monolingual empires—could not have been more uncertain.

The previous chapters have revealed the many ways Spanish and early Latin American scholars seized the opportunity posed by the Americas to expand the comparative linguistic research they had long practiced in Europe. While Europeans developed no consistent set of principles to study linguistic change until the eighteenth century, well before then linguistic speculation served to write the history of the Americas and to make political and religious claims that were not linguistic in nature. To do so, writers relied on a common textual tradition and bound them to causal explanations to account for language change. These causes could be political, environmental, or temporal.

Immersions in the archives of language could bridge various forms of cognitive distance, thus linking the ancient world to the sixteenth century, European readers to American landscapes, the arcane meaning of the divinity to plain human understanding, and the natural wealth of distant places to the knowledge of European botanists. Using the Americas as a living example, scholars often debated linguistic corrosion, borrowing, and contact. Ultimately, they generated a rich harvest of theories linking these processes to general historical patterns.

Unwittingly, such conflicting and ever more detailed explanations eventually rendered the etymological artistry of the first generations of Spanish humanists unreliable as testimony or proof. They were surpassed by newer theories explaining the evolution and genealogy of languages.

The emphasis on the power of words and on unveiling their histories as sources of local knowledge and authority reveals an unexpected relationship between imperial domination and the production of knowledge about subject territories and people. Yet Spanish administrators understood local knowledge as necessary to their efforts to govern, exploit, and transform the Americas and their Indigenous inhabitants. They relied on oral knowledge in all its manifestations and, thus, developed techniques to parse, classify, and organize it.

Projects, like León Pinelo's, to compile and catalog languages emerged later in the eighteenth century. The most substantial effort of that period is the work of the Jesuit scholar Lorenzo Hervás y Panduro (d. 1805). Hervás y Panduro studied more than three hundred languages from all over the world. After the Bourbon monarch Charles III expelled the Jesuits from the Spanish Americas and Spain in 1767, Hervás y Panduro found himself in Italy surrounded by fellow Jesuits and their books. He took the opportunity to study all the tongues that his multilingual colleagues had learned in their efforts to convert the people of the world. Hervás organized languages into families drawing from Joseph Justus Scaliger's (d. 1609) method of using similarities in the word used to designate "god" to establish genetic filiation. However, unlike his scholarly predecessors and even though he accepted the linguistic history postulated in the book of Genesis, Hervás y Panduro abandoned the search for one common language from which all others originated.[3]

Within the Spanish realms, a parallel initiative was incited by Catherine II of Russia. In 1785, the empress requested from Charles III information about all the languages spoken in his realms in order to compose a universal dictionary. The request included a list of words like "god, father, mother, son and daughter," which the architects of the dictionary considered universal enough to serve as global categories of comparison. The Spanish monarch forwarded the request to his viceroys, who in turn reached out to missionaries and local experts to provide the required information.

Several surviving manuscripts demonstrate the many inquiries that were carried out from the Philippines to Chile in response to Catherine's query. The manuscripts also contain reflections about linguistic change almost three hundred years after Columbus had settled Hispaniola. The botanist José Celestino Mutis

(1732–1808), writing from New Granada in 1788, for instance, declared that among his most prized possessions were "the two only original manuscripts known of the Chibcha or Mosca language, which was once general to this New Kingdom, and that now its memory seems extinguished." Mutis would share the manuscripts but only after they had been copied, because he would not risk their loss at sea. In the letter, he referred to these works as "precious antiquities."[4] He attributed the loss of this language to the "extinction" of its community of speakers, highlighting, but only tacitly, the devastating consequences of Spanish rule for the speakers of Muisca. Mutis called for preserving these books, as even printed grammars of the Muisca language had become scarce. In this, Mutis echoed León Pinelo's concern with assembling a library of works on the Americas since, even in the eighteenth century, many, if not most, still remained in manuscript form and vulnerable to loss.

The nineteenth-century wars of independence brought about commitment among the new nations' elites to the Castilian language in the creation of their new republics, radically departing from long-standing views on language in the Americas. Except for José Gaspar Rodríguez de Francia (1766–1840), who ruled over Paraguay in the early decades of the nineteenth century and made Guaraní one of Paraguay's official languages together with Spanish, the rest of the new republics accepted Castilian Spanish as a de facto national language.[5] Yet, as this book has shown, the new nations' emphasis on forging a homogenous linguistic identity based on local varieties of Spanish represented a break with earlier practices. It constitutes yet another way the colonial past was recast in contested attempts to create foundational national mythologies based on the model of the nation-state.[6] Grammarians and language academies would seek to standardize and regulate language usage. Although their linguistic ideas differed, from Mexico to Chile to Argentina to Colombia, grammarians believed that it was through linguistic unity that the bonds of the national community were best sustained.[7]

In this spirit, the Colombian philologist Rufino José Cuervo (1844–1911) published in 1886 the first volume of a work dedicated to the history and proper usage of Castilian. Cuervo would complete only the first two volumes of this dictionary, merely reaching the end of the letter *D* before his death in 1911.[8] The *Dictionary on the Construction and Regime of the Castilian Language* (*Diccionario de construcción y régimen de la lengua castellana*) included a section describing etymological practice. To correctly arrive at the origin of a word, Cuervo averred, one had to not only understand the history of a language and the patterns that dictated its changes across time. It was also necessary to "reestablish historically or ideally the bond that unites the present and the past." Cuervo believed that "even in the midst of the transformation of elements that constitute the nucleus of a language, as it passes from mouth to mouth, and from generation to generation, there is in words, no less than in human beings, a continuity in body and spirit that bears witness to its identity across time."[9] This continuity between past and present, which he believed to be detectable in words, equally bound speakers in a transhistorical community that possessed a natural identity. Language, thus, had to be preserved, because it "symbolized . . . the Fatherland." It was for this reason, Cuervo claimed, that "after those who worked to conserve the unity of religious beliefs no one does as much for the brotherhood of Hispano-American nations" than those who work to "conserve the purity of its language."[10] In Cuervo's writings etymology transformed into a method for reasserting the identity of a community, iterated in each spoken and written word, safeguarded by the erudite grammarian. This community was simultaneously formed by each nation's specific rendering of the Spanish language and by the Spanish-speaking world at large as a sort of Iberian brotherhood.

Despite the eventual discursive successes of Castilian Spanish, Ibero-America's multilingual reality bears witness to the resilience of Indigenous language speakers. It is also the result of centuries of power relations, scholarship, and antiquarian research centered on language, which ironically led to both the extinction of various Indigenous languages and the survival of others. This polyglossia, thus, challenges the notion that the spread of a single language, in the wake of empire, was always the goal of an early modern monarchy that sought, at one point, to convert, rule, and know the whole world.

Notes

ABBREVIATIONS

AGI   Archivo General de las Indias—Seville
Antwerp   *Biblia Sacra Hebraice, Chaldaice, Graece, et Latine: Philippi II. Reg.*
  Polyglot   *Cathol. pietate, et studio ad Sacrosanctae Ecclesiae usum.* Edited by
               Benito Arias Montano. 8 vols. Antwerp: Christophe Plantin, 1569–1573.
BCS   Biblioteca Capitular y Colombina Sevillana—Seville
BME   Real Biblioteca del Monasterio de San Lorenzo de El Escorial—El
               Escorial
BNE   Biblioteca Nacional de España—Madrid
BNF   Bibliothèque nationale de France—Paris
BRP   Biblioteca Real del Palacio—Madrid
JCB   John Carter Brown Library—Providence, RI
JGI   Joaquín García Icazbalceta Manuscript Collection at the Benson Latin
               American Collection—Austin, TX
HL   Huntington Library—San Marino, CA
Prefacios   *Prefacios de Benito Arias Montano a la Biblia Regia de Felipe II.* Edited
               and translated by María Asunción Sánchez Manzano. Humanistas
               españoles 32. Salamanca: Junta de Castilla y León, Consejería de
               educación y cultura, 2006.
PUL   Special Collections, Princeton University Library—Princeton, NJ
RAH   Real Academia de la Historia—Madrid
UTA   Nettie Lee Benson Library, University of Texas at Austin—Austin, TX

INTRODUCTION

1. *Instructio[n], y memoria, de las relaciones que se han de hazer, para la descripcion de las Indias, que Su Magestad manda hazer para el buen gouierno y ennoblescimiento dellas,* 1577, JCB, BB S7333 1577 1, 2.

2. "Ameca, 1579," in *Relaciones Geográficas of Mexico and Guatemala, 1577–1585,* Nettie Lee Benson Latin American Collection, University of Texas Libraries, University of Texas at Austin, box 1, folder XXIII-10, 14 folios, fols. 1r–3r. For a critical edition of the *relación,* see René Acuña, *Relaciones geográficas del siglo XVI,* 10 vols. (México, 1988), vol. 10, 23–50. Acuña suggests that the "Cazcan language" was perhaps related to Nahuatl, as the toponym seems to indicate. Acuña conjectures that the Spanish interlocutor probably rendered the toponym inexactly. In Nahuatl it would have derived from Amieccan, meaning "place of much water" or "water in many parts." For an introductory studies on the *relaciones geográficas de Indias,* see Howard F. Cline, "The *Relaciones Geográficas* of the Spanish Indies, 1577–1586," *Hispanic American Historical Review* 44, no. 3 (August 1964): 341–374. For a study of the natural

historical sections of the questionnaire, see Raquel Álvarez Peláez, *La conquista de la naturaleza americana* (Madrid, 1993).

3. "Barajas," in *Relaciones topográficas de Felipe II: Madrid*, ed. Alfredo Alvar Ezquerra, transcribed by María Elena García Guerra and María de los Angeles Vicioso Rodríguez, 4 vols. (Madrid, 1993), vol. 1, 121–28, 121. In his *Compilation of Some Arabic Names*, the Franciscan lexicographer Diego de Guadix (d. 1615) provided a completely different etymology. He claimed that the toponym is entirely in Arabic and that it could be rendered as place "hope or that of trust." See Diego de Guadix, *Recopilación de algunos nombres arábigos que los árabes pusieron a algunas ciudades y a otras muchas cosas* [1593], ed. E. Bajo Pérez and F. Maíllo Salgado (Gijón, 2005), 380.

4. The bibliography on the *relaciones topográficas* carried out in Spain is extensive. For a set of representative studies, see F. Javier Campos y Fernández de Sevilla, *Las relaciones topográficas de Felipe II: Índices, fuentes y bibliografía, Separata del Anuario Jurídico y Económico Escurialense* (San Lorenzo de El Escorial: XXXVI-2003); Campos y Fernández de Sevilla, *La mentalidad en Castilla la nueva en el siglo XVI: Religión, economía y sociedad, según las "relaciones topográficas" de Felipe II* (Madrid, 1986); and Campos y Fernández de Sevilla, "Los moriscos en las 'relaciones topográficas,' de Felipe II," *Anuario jurídico y económico escurialense* 43 (2010): 413–30.

5. Arndt Brendecke, *The Empirical Empire: Spanish Colonial Rule and the Politics of Knowledge* trans. Jeremiah Riemer (Berlin, 2016), chap. 1.

6. The term "speech community" is used in linguistic anthropology to describe groups of people who "share values and attitudes about language use, varieties and practices." These communities develop over prolonged interactions. On the concept and a synthesis of its history, see Marcyliena H. Morgan, *Speech Communities: Key Topics in Linguistic Anthropology* (Cambridge, 2014), 1–17.

7. On Philip II's finances, see Mauricio Drelichman and Hans-Joachim Voth, *Lending to the Borrower from Hell: Debt, Taxes, and Default in the Age of Philip II* (Princeton, NJ, 2014); On Philip II's collection of relics, see Guy Lazure, "Possessing the Sacred: Monarchy and Identity in Philip II's Relic Collection at the Escorial," *Renaissance Quarterly* 60, no. 1 (Spring 2007): 58–93; and Adam Beaver, "Sticks, Stones, and Ancient Bones: Relics and the Historical Record," in "A Holy Land for the Spanish Monarchy: Palestine in the Making of Modern Spain, 1469–1598" (PhD diss., Harvard University, 2008), 61–107.

8. Noteworthy studies on the relationship between royal power and information include David Goodman, *Power and Penury: Government, Technology, and Science in Philip II's Spain* (Cambridge, 1988); and Geoffrey Parker, *The Grand Strategy of Philip II* (New Haven, CT, 2000); and Brendecke, *The Empirical Empire*, 1–16.

9. The revised stance on the Crown's management of linguistic diversity appears, for instance, in Bruce Mannheim's work on the transformations of Southern Peruvian Quechua after the Spanish conquest of the Inca Empire. Bruce Mannheim, *The Language of the Inka State since the European Invasion* (Austin, TX, 1991), 1–28.

10. Peter Burke, *Languages and Communities in Early Modern Europe* (Cambridge, 2004), 63. John H. Elliott has also warned against projecting the current understanding of language and its relationship to membership in a political community onto the early modern period. While language, for instance, has always been an expression of group identity, it is not clear that it possessed the significance in earlier periods that it came to enjoy after Johann Gottfried von Herder and the Romantics developed their organic concept of nation and *Volk*. See J. H. Elliott, *History in the Making* (New Haven, CT, 2012), 54.

11. Sabina Collet Sedola, "La castellanización de los Indios (S. XVI–XVII): Conquista del Nuevo Mundo y conquista lingüística," in *Actas del Congreso Internacional de Historiografía*

Lingüística: Nebrija V centenario, vol. 2 (Murcia, 1992), 85. See also Otto Zwartjes, introduction to *Las gramáticas misioneras de tradición hispánica (siglos XVI–XVII)*, ed. Otto Zwartjes (Amsterdam, 2000), 1.

12. Juan Fernando Cobo Betacourt, "Colonialism in the Periphery: Spanish Linguistic Policy in New Granada c. 1574–1625," *Colonial Latin American Review* 23, no. 2 (2014): 118–142, 120.

13. Daniel Wasserman-Soler, *Truth in Many Tongues: Religious Conversion and the Languages of the Spanish Empire* (University Park, PA, 2020), 154–155; and Wasserman-Soler, "*Lengua de los Indios, Lengua Española*: Religious Conversion and the Languages of New Spain, ca. 1520–1585," *Church History* 85, no. 4 (December 2016): 690–723. On the influence of religious orders and the secular clergy in determining contrasting language policies, see Stafford Poole, *Pedro Moya de Contreras: Catholic Reform and Royal Power in New Spain, 1571–1591* (Berkeley, CA, 1987), 69.

14. Nancy Farriss, *Tongues of Fire: Language and Evangelization in Colonial Mexico* (Oxford, 2018). See also Klaus Zimmerman, *La descripción de la lenguas amerindias en la época colonial* (Madrid, 1994); and Zimmerman, "Translation for Colonization and Christianization: The Practice of the Bilingual Edition of Bernardino de Sahagún (1499–1590)," in *Missionary Linguistics V / Lingüística Misionera V: Translation Theories and Practices*, ed. Otto Zwartjes, Klaus Zimmermann, and Martina Scharder-Kniffi (Amsterdam, 2014), 85–112.

15. César Itier, "Lengua general y quechua cuzqueño en los siglos XVI y XVII," in *Desde afuera y desde adentro: Ensayos de etnografía e historia del Cuzco y Apurímac*, ed. Luis Millones, Hiroyasu Tomoeda, and Tatsuhiko Fujii (Osaka, 2000), 47–59.

16. Alan Durston, "Standard Colonial Quechua," in *Iberian Imperialism and Language Evolution in the Americas*, ed. Salikoko Mufwene (Chicago, 2014), 225–243; *Pastoral Quechua: The History of Christian Translation in Colonial Peru, 1550–1650* (Notre Dame, IN, 2007).

17. Durston, "Standard Colonial Quechua," 230.

18. In Santa Fé, for instance, local authorities selected the Muisca (Chibcha) language spoken in the vicinity of Bogotá. This language, however, was not understood widely, and it was very dissimilar to other languages in the territory, creating numerous communication problems for the missionaries. See Cobo Betacourt, "Colonialism in the Periphery," 134. For the Mayan-speaking regions, see William F. Hanks, *Converting Words: Maya in the Age of the Cross* (Berkeley, CA, 2010).

19. Vera Candiani, *Dreaming of Dry Land: Environmental Transformation in Colonial Mexico City* (Stanford, CA, 2014), xxv–xxxii.

20. In support of this chronology, Ángel María Garibay Kintana in *Historia de la literatura náhuatl* shows that high literary production in alphabetized Nahuatl subsisted well until the eighteenth century. Only then did increased pressures to Castilianize begin to modify the contents of this tradition. Ángel María Garibay Kintana, *Historia de la literatura náhuatl* [1953–1954] (México, 2000), 24. See also James Lockhart, "Postconquest Nahua Society and Culture Seen Through Nahuatl Sources," in *Nahuas and Spaniards: Postconquest Central Mexican History and Philology* (Stanford, CA, 1991), 2–22.

21. Durston, "Standard Colonial Quechua," 230.

22. For a representative bibliography on Indigenous chroniclers, see Camilla Townsend, *Annals of Native America: How the Nahuas of Colonial Mexico Kept Their History Alive* (Oxford, 2017); Domingo de San Antón Muñon Chimalpahin Quauhtlehuanitzin, *Annals of His Time*, ed. and trans. James Lockhart, Susan Schroeder, and Doris Namala (Stanford, CA, 2006); Rolena Adorno, *Guamán Poma: Writing and Resistance in Colonial Peru* (Austin, 2000); Sabine MacCormack, *On the Wings of Time: Rome, the Incas, Spain, and Peru* (Princeton, NJ, 2007), 1–28, 170–201; and Margarita Zamora, *Language, Authority, and*

Indigenous History in the Comentarios Reales de los Incas (Cambridge, 1988). For the important role of Latin education in the introduction of alphabetic writing, see Andrew Laird, "Colonial Grammatology: The Versatility and Transformation of European Letters in Sixteenth-Century Spanish America," *Language & History* 61, nos. 1/2 (2018): 52–59; and Laird, "The Teaching of Latin to the Native Nobility in Mexico in the Mid-1500s: Contexts, Methods, and Results," in *Learning Latin and Greek from Antiquity to the Present*, ed. Elizabeth P. Archibald, William Brockliss, and Jonathan Gnoza (Cambridge, 2015), 118–135.

23. The bibliography on Indigenous, mestizo, and Spanish and Portuguese intermediaries is enormous, as intermediation has become a subfield in colonial Latin American Studies. See Yanna Yannakakis and Gabriela Ramos, introduction to *Indigenous Intellectuals: Knowledge, Power, and Colonial Culture in Mexico and the Andes*, ed. Yanna Yannakakis and Gabriela Ramos (Durham, NC, 2014), 1–19.

24. Durston, "Standard Colonial Quechua," 230.

25. Joan Rappaport and Tom Cummins, *Beyond the Lettered City: Indigenous Literacies in the Andes* (Durham, NC, 2012), 22.

26. As Mannheim and Durston have argued, "Indigenous languages expanded into other arenas at the wake of the Iberian invasions and when they did lose ground, the gains often went to other indigenous languages." In addition, like all languages, Indigenous tongues were neither "monoethnic" nor "ahistorical." They continued to change because of people's mobility and the incorporation of new speakers. Alan Durston and Bruce Mannheim, introduction to *Indigenous Languages, Politics, and Authority in Latin America*, ed. Alan Durston and Bruce Mannheim (Notre Dame, IN, 2018), 1–17.

27. Otto Zwartjes, *Portuguese Missionary Grammars in Asia, Africa, and Brazil, 1550–1800*, ed. Otto Zwartjes (Philadelphia, 2011), 205–241.

28. Larissa Brewer García, *Beyond Babel: Translations of Blackness in Colonial Peru and New Granada* (Cambridge, 2020), 1–33 and 164–206.

29. Claire Gilbert, *In Good Faith: The Arabic Voices of Imperial Spain: Arabic Translation and Translators in Early Modern Spain* (Philadelphia, 2020), 5–6. Also see Seth Kimmel, *Parables of Coercion: Conversion and Knowledge at the End of Islamic Spain* (Chicago, 2015), 18–42; and Mercedes García-Arenal and Fernando Rodríguez Mediano, *Un oriente español: Los moriscos y el Sacromonte en tiempos de Contrarreforma* (Madrid, 2010).

30. Mercedes García Arenal, "The Religious Identity of the Arabic Language and the Affair of the Lead Books of Sacromonte of Granada," *Arabica* 56 (2009): 495–528.

31. See James Lockhart, *Nahuas and Spaniards: Postconquest Central Mexican History and Philology* (Stanford, CA, 1991), xi. For the most representative works and a brief history of the group of scholars working on Indigenous language documents and their methods, see Alan Durston, "Indigenous Languages and the Historiography on Latin America," *Storia della storiografia* 67, no. 1 (2015): 51–65. For a history and bibliography of the "New Philology," see Matthew Restall, "A History of the New Philology and the New Philology in History," *Latin American Research Review* 38, no. 1 (February 2023): 113–134.

32. Burke, *Languages and Communities in Early Modern Europe*, 38.

33. On the works of Florentine humanists and their attempts develop tools to study the Florentine vernacular, see Ann Moyer, *The Intellectual World of Sixteenth-Century Florence: Humanists and Culture in the Age of Cosimo I* (Cambridge, 2020), 123–174.

34. Toon Van Hal, "Early Modern Views on Language and Languages (ca. 1450–1800)," in *Oxford Research Encyclopedia of Linguistics* (Oxford, April 2019); and Van Hal, "Linguistics 'ante Litteram': Compiling and Transmitting Views on the Diversity and Kinship of Languages before the Nineteenth Century," in *The Making of the Humanities: From Early*

Modern to Modern Disciplines, ed. Rens Bod, Jaap Maat, and Thijs Weststeijn (Amsterdam, 2012), vol. 2, 37–53.

35. Miguel Martínez, "Language, Nation, and Empire in Early Modern Iberia," in *A Political History of Spanish: The Making of a Language*, ed. José del Valle (Cambridge, 2013), 44–60; Eugenio Asensio, "La lengua compañera del imperio: Historia de una idea de Nebrija en España y Portugal," *Revista de filología española* 43 (1960): 399–413.

36. Byron Ellsworth Hammann studied the production of dictionaries of Indigenous languages based on the word lists that Nebrija devised for his bilingual Latin-Castilian dictionary, showing how early lexicographical fieldwork was deeply inflected by the grammatical and cultural world of the ancient Mediterranean. See Bryan Ellsworth Hamann, *The Translations of Nebrija: Language, Culture, and Circulation in the Early Modern World* (Boston, 2015), 1–11.

37. "Ambrosio de Morales, sobrino del Maestro Oliva al Lector," in Francisco Cervantes de Salazar, *Obras q[ue] Francisco Ceruantes de Salazar, ha hecho, glosado, y traduzido* (Alcalá de Henares: Imprimiase en esta casa de Alcalá: [Juan de Brocar], 1546), B2v.

38. "Ambrosio de Morales, sobrino del Maestro Oliva al Lector," B3r–v.

39. "Ambrosio de Morales, sobrino del Maestro Oliva al Lector," B5v.

40. "Ambrosio de Morales, sobrino del Maestro Oliva al Lector," B6r.

41. Likewise, Antonio Feros explains that the "Castilianization" of Spain was a gradual process, as "Spanish" was a language in process of being created in the sixteenth and seventeenth centuries. Antonio Feros, *Speaking of Spain: The Evolution of Race and Nation in the Iberian World* (Cambridge, MA, 2017), 38. On the ways Iberian imperialism transformed the many languages of the New World, see *Iberian Imperialism and Language Evolution in Latin America*, ed. Salikoko S. Mufwene (Chicago, 2014). Also see Lucia Binotti, "La 'Lengua compañera del imperio': Discursos peninsulares sobre la hipanización de América," in *Actas del III Congreso Internacional de Historia de la Lengua Española*, ed. Alegría Alonso González (Salamanca, 1996), vol. 1, 621–632; and Binotti, *La teoria del castellano primitivo: Nacionalismo y reflexión lingüistica en el Renacimiento español* (Münster, 1995).

42. Theodor Bibliander, *De ratione communi omnium linguarum et literarum commentarius*, trans. and ed. Hagit Amirav and Hans-Martin Kim (Geneva, 2011), 430–431; Anthony Arthur Long "Stoic Linguistics, Plato's *Cratylus*, and Augustine's *De Dialectica*," in *Language and Learning Philosophy of Language in the Hellenistic Age*, ed. Dorothea Frede and Brad Inwood (Cambridge, 2005), 36–55.

43. Isidore of Seville, *The Etymologies of Isidore of Seville*, trans. Stephen A. Barney, W. J. Lewis, J. A. Beach, and Oliver Berghof (Cambridge, 2010), book 1, chap. 29, 54–55. Also see Frank L. Borchhardt, "Etymology in Tradition and in the Northern Renaissance," *Journal of the History of Ideas* 29 (1968): 415–429.

44. This idea of strength drew from a long-standing mnemonic tradition. See Paolo Rossi, *Logic and the Art of Memory: The Quest for a Universal Language*, trans. Stephen Clucas (Chicago, 2000), 1–28; Mary J. Carruthers, "Etymology: The Energy of a Word Unleashed," In *The Craft of Thought: Meditation, Rhetoric, and the Making of Images, 400–1200* (Cambridge, 1998), 155–60. Also see Ernst Robert Curtius, "Etymology as a Category of Thought," in *European Literature and the Latin Middle Ages*, trans. Willard R. Trask (Princeton, NJ, 1952), 496–500.

45. Rhodri Lewis, *Language, Mind and Nature: Artificial Languages in England from Bacon to Locke* (Cambridge, 2007), 10–19.

46. See Allison Coudert, "Some Theories of a Natural Language from the Renaissance to the Seventeenth Century," in *Magia naturalis und die Entstehung der modernen*

Naturwissenschaften: Symposion der Leibniz-Gesellschaft, Hannover, 14. und 15. November 1975, ed. Albert Heinekamp Redaktion and Dieter Mettler (Wiesbaden, 1978). Also see Clara Auvray-Assayas, Christian Bernier, Barbara Cassin, André Paul, and Irène Rosier-Catach, "To Translate," in *Dictionary of Untranslatables: A Philosophical Lexicon*, trans. Steven Rendall, Christian Hubert, Jeffrey Mehlman, Nathaniel Stein, and Michael Syrotinski, ed. Barbara Cassin, and trans. and ed. Emily Apter, Jacques Lezra, and Michael Wood (Princeton, NJ, 2014), 1139–1151.

47. Isidore of Seville, *The Etymologies of Isidore of Seville*, book 20, xiii.

48. Plato, *Cratylus*, 390a, in *Cratylus, Parmenides, Greater Hippias, Lesser Hippias*, trans. Harold North Fowler, Loeb Classical Library (Cambridge, MA, 1939).

49. In his work *On the Names of Christ* (1582), the Augustinian friar Luis de León described how names in Castilian worked. For instance, *corregidor* (magistrate) "is a name that is born and taken from *corregir* [correct], because their function is to correct that which is bad." Luis de León, *De los nombres de Cristo*, ed. Cristobál Cuevas (Madrid, 1997), book 1, 160.

50. Ralph Bauer and Jaime Marroquín Arrendondo reframe the history of procuring the knowledge of the new people and territories under the rule of Spain as one of translation rather than discovery. See Ralph Bauer and Jaime Marroquín Arredondo, "Introduction: An Age of Translation," in *Translating Nature: Cross-Cultural Histories of Early Modern Science*, ed. Ralph Bauer and Jaime Marroquín Arredondo (Philadelphia, 2019), 1–23. Allison Bigelow's study of mining practices in the sixteenth- and seventeenth-century Spanish Empire explains how Indigenous American, African, and South Asian knowledge of minerals and mining technologies, encoded in language, shaped technical and scientific vocabularies for the exploitation of metals. See Allison Bigelow, *Mining Language: Racial Thinking, Indigenous Knowledge, and Colonial Metallurgy in the Early Modern Iberian World* (Chapel Hill, NC, 2020).

51. "Vocabula barbara," in *De orbe novo decades* [. . .] *cura & diligentia.* [. . .] *Antonii Nebrissensis* [. . .] (Alcalá de Henares: Arnaldi Guillelmi de Brocario, 1516), i5r–i7v. For a study of the Indigenous words in this list, see José G. Moreno de Alba, "Indigenismos en las *Décadas del Nuevo Mundo* de Pedro Mártir de Anglería," in *Nueva revista de filología hispánica* 44, no. 1 (1996): 1–26.

52. Nardo Recchi, *De materia medica Novae Hispaniae Philippi Secundi Hispaniarum ac Indiarum regis invictissimi iussu*, Codex Lat 5, JCB, fol. 7r. I have also consulted the Spanish translation and critical edition of the Recchi manuscript of Florentino Fernández González and Raquel Álvarez Peláez, *De materia medica Novae Hispaniae: Manuscrito de Recchi*, trans. and ed. Florentino Fernández González and Raquel Álvarez Peláez, 2 vols. (Madrid, 1998), vol. 1, prologue, chap. 4, 162–165.

53. *De materia medica Novae Hispaniae* [1998], prologue, chap. 4, 162–165.

54. Philip II's desire for books and manuscripts of all kinds was well known. When preparing his edition of Raymond Lull's *Blanquerna*, Fray José de Andreu recorded that the monarch's passion for Raymond Lull's works was legendary. See Raymond Lull, *Blanquerna*, trans. José Andreu, BNE, Ms. 5611, fol. 4.

55. See Angel Riesco Terrero, "Colaboración del Obispo y Cabildo central de Málaga a la empresa real de selección y edición de las obras de San Isidoro de Sevilla (Edic. Regia 1597–1599) y al enriquecimiento de dos grandes centros documentales: El Archivo General de Simancas y el Archivo del Escorial," *Baetica* 11 (1988): 301–321; Guy Lazure and Antonio Dávila Pérez, "Un catálogo de las obras de Isidoro de Sevilla conservadas en diversas bibliotecas españolas en el siglo XVI," *Excerpta philologica: Revista de filología griega y latina de la Universidad de Cádiz*, nos. 10–12 (2000–2002): 267–290; and Carmen Codoñer, "La edición de Juan de Grial de las *Etymologiae* de Isidoro de Sevilla, un informe de Juan de Mariana y el trabajo de Alvar Gómez de Castro," *Faventia* 31, nos. 1/2 (2009): 213–225.

56. Isidore of Seville, *Diui Isidori Hispal. Episcopi opera* [. . .]: *E vetustis exemplaribus emendata* (Madrid: ex Typographia Regia, 1599).

57. Stephen A. Barney, W. J. Lewis, J. A. Beach, and Oliver Berghof, introduction to *The Etymologies of Isidore of Seville*, 3–28.

58. John Henderson, *Truth from Words: The Medieval World of Isidore of Seville* (Cambridge, 2007), 7.

59. Anthony Grafton, *New Worlds, Ancient Texts: The Power of Tradition and the Shock of Discovery* (Cambridge, MA, 1992), 1–11.

60. Antonio Alatorre, *Los 1001 años de la lengua española* (México, 1979), 258–298.

61. See, for instance, Walter Mignolo, "On the Colonization of Amerindian Languages and Memories: Renaissance Theories of Writing and the Discontinuity of the Classical Tradition," *Comparative Studies in Society and History* 34, no. 2 (April 1992): 301–330; and Mignolo, *The Darker Side of the Renaissance: Literacy, Territoriality, and Colonization* (Ann Arbor, MI, 1995), 21–67. David Rojinksky, *Companion to Empire: A Genealogy of the Written Word in Spain and New Spain* (Amsterdam, 2002).

62. Fernando Bouza, *Communication, Knowledge, and Memory in Early Modern Spain*, trans. Sonia López and Michael Agnew (Philadelphia, 2004), 1–15; and Daniel Woolf, *The Social Circulation of the Past: English Historical Culture, 1500–1730* (Oxford, 2003), 259–391.

63. Juvenal, *Satira* 10.168, in Juvenal and Persius, *The Satires of Persius / The Satires of Juvenal* trans. Susanna Morton Braund, Loeb Classical Library 91 (Cambridge, MA, 2004), 380–381; Geoffrey Parker, *The World Is Not Enough: The Imperial Vision of Philip II of Spain*, *Charles Edmondson Historical Lectures* (Waco, TX, 2001), 10–13. The Jesuits modified the saying in the seventeenth century to "unus non sufficit orbis." See José Luis Betrán Moya, "Unus non sufficit orbis: La literatura misional Jesuita del Nuevo Mundo," in *Historia social*, no. 65 (2009), 167–185.

CHAPTER ONE: The World in the Library

1. See Valeria López Fadul, "Juan Páez de Castro and the Project of a Universal Library," in Wars of Knowledge: Imperial Hegemony and the Assembling of Libraries Forum, *Pacific Coast Philology* 52, no. 2 (October 2017): 173–183, https://doi.org/10.5325/pacicoasphil.52.2.0173. Used with permission from Penn State University Press.

"En este gran año se descubren nuevas gentes, y se pierden otras. Pierdense los imperios y señorios, y las noblezas, y linajes antiguos, y alçansé otros de nuevo. Vencensé las memorias humanas con dilubios, pestilencias, grras, y fuego gral. De todas estas cosas ay rastro no solo en estas nras partes, pero tambien entre los Indios sy queremos leer sus historias." Juan Páez de Castro, "Notations and additions to the dedication of an anonymous history of the Emperor Charles V. In Spanish," BNE, Ms. 23083/4, fol. 3v.

2. Páez de Castro, "Notations," fol. 3v.

3. "De manera que a la historia podemos atribuir la invencion de las artes y el bivir politivo de que gozamos." Páez de Castro, "Notations," fol. 1v.

4. "Mas copiosamente pudieramos tratar esta parte si vinieran a nuestras manos diez libros que Porfirio filosofo escribio del provecho que los Reyes podrian sacar de Homero. Pero por negligencia de los principes como se han perdido otras muy preciosas reliquias de la antiguedad se perdieron tambien estos libros y se perderan muchos de los que aun no son publicados, si Dios no pone en corazon de tan gran Principe como V.M. que ponga algun remedio en esto, haziendo libreria publica real en parte segura como seria en España." Juan Páez de Castro, PL, Ms. 174, fol. 8v. Scholars have debated the purpose of this autographed text. In it, Páez reflects on the first Castilian translation of Homer's *Odyssey* produced by his friend and royal secretary Gonzalo Pérez. He also claims to be working on a biography of

Homer that has yet to be located. See Luis Arturo Guichard, "Un autógrafo de la traducción de Homero de Gonzalo Pérez (*Ulyxea* XIV–XXIV) anotado por Juan Páez de Castro y el Cardenal Mendoza y Bovadilla," *International Journal of the Classical Tradition* 15, no. 4 (December 2008): 525–557.

5. Páez described to Philip's father, Emperor Charles V, in this manner the history he endeavored to write in a text known as the *Memorandum of the Things Necessary to Write History* (c. 1555, *Memorial de las cosas necesarias para escribir historia*): "En esta historia que sera continua y perpetua." Esteban Eustacio, "De las cosas necesarias para escribir historia: Memorial inédito del Dr. Juan Páez de Castro al Emperador Carlos V," *La ciudad de Dios* 29 (1892): 34. The original copy of this manuscript in the Escorial library has been lost. Two eighteenth-century copies survive in the national library of Madrid, both copied from the original. See Arantxa Domingo Malvadi, *Bibliofilia humanista en tiempos de Felipe II: La biblioteca de Juan Páez de Castro* (Salamanca, 2011), 48; and Juan Páez de Castro, *Método para escrivir la historia por el doctor Juan Páez de Castro, Chronista del Emperador Carlos Quinto á quien le dirige*, BNE, Ms. 5578, fols. 77r–131v. Throughout the chapter I will cite the Eustacio edition that appeared in *La ciudad de Dios*. The first part of the manuscript appeared as Esteban Eustacio, "De las cosas necesarias para escribir historia: Memorial inédito del Dr. Juan Paéz de Castro al Emperador Carlos V," *La ciudad de Dios* 28 (1892): 604–610.

6. On humanists and their concern with the preservation and dissemination of knowledge, see Ann Blair, "The 2016 Josephine Waters Bennett Lecture: Humanism and Printing in the Work of Conrad Gessner," *Renaissance Quarterly* 60, no. 1 (Spring 2017): 1–37, 8. In the particular case of Páez and his horror at the dispersal of library collections after the death of their owners, see Javier Patiño Loira, "Imagining Public Libraries in Sixteenth-Century Spain: Juan Páez de Castro and Juan Bautista Cardona," *Pacific Coast Philology* 52, no. 2 (2017): 184–194.

7. Pliny, *Natural History*, trans. Horace Rackham, Loeb Classical Library 352 (Cambridge, MA, 1942), 7.33.115; and Pliny, *Natural History*, trans. Horace Rackham. Loeb Classical Library 394 (Cambridge, MA: 1952), 35.2.10. On Pliny and the *Natural History* as a catalog of empire, see Sorcha Carey, *Pliny's Catalogue of Culture: Art and Empire in the Natural History* (Oxford, 2003).

8. J. H. Elliott, "A Europe of Composite Monarchies," *Past & Present*, no. 137 (November 1992): 48–71. On the history of this manuscript and its many copies, see Domingo Malvadi, *Bibliofilia humanista*, 51.

9. One of his most enduring interests, for instance, was commenting on the works of Aristotle in Greek.

10. Alejandra B. Osorio succinctly calls attention to this point by claiming "that forms of rulership, laws, urban structures and practices, and ceremonial performances—in other words, the entire political culture under the Spanish Habsburgs—were more broadly devised coevally in this new worldwide context and developed over time through mutual influences. Such a reframing suggests that familiar dichotomies—East-West, center-periphery, colonial-metropolitan—are later elaborations, as they did not pertain to the seventeenth-century Spanish imperial geopolitics." See Alejandra B. Osorio, "Of National Boundaries and Imperial Geographies: A New Radical History of the Spanish Empire," *Radical History Review*, no. 130 (January 2018): 100–130, 106.

11. Jesús Bustamante, "Los círculos intelectuales y las empresas culturales de Felipe II: Tiempos, lugares, y ritmos del humanismo en el siglo XVI," in *Élites intelectuales y modelos colectivos: Mundo ibérico (siglos XVI–XIX)*, ed. Jesús Bustamante and Monica Quijada (Madrid, 2002), 33–58.

12. For a summary on the royal decrees concerning the publication of books about the Americas, see Stephen C. Mohler, "Publishing in Colonial Spanish America: An Overview," *Inter-American Review of Bibliography* 28, no. 3 (1978): 259–273, 262.

13. About the circulation of Indigenous grammars in Europe, see W. Keith Percival, *Studies in Renaissance Grammar* (Burlington, 2004), 15–19.

14. See Jesús Bustamante García, "Las lenguas amerindias: Una tradición española olvidada," *Histoire épistémologie langage* 9, no. 2 (1987): 75–97.

15. See Juan Francisco A. de Uztarroz and Diego J. Dormer, *Progresos de la historia en Aragon y vidas de sus cronistas, desde que se instituyó este cargo hasta su extinción: Primera parte que comprende la biografía este de Gerónimo Zurita* (Zaragoza, Impr. del Hospicio, 1878), 100.

16. Páez's only published work appeared in the early seventeenth century, a short treatise dedicated to defending Zurita's *Annals of the Crown of Aragón* (1562–1580, *Anales de la Corona de Aragón*) from the attacks of an injurious censor. Páez's defense, entitled *Parecer de Juan Páez sobre los Anales de la corona de Aragón de Zurita*, appeared in print in 1610. The original manuscript is in the Royal Academy of History in Madrid. A copy also survives in the Princeton MS. 174. See Jerónimo Zurita, *Anales de la Corona de Aragón. Va añadida de nuevo, en esta impresión, en el ultimo tomo una Apología de Ambrosio de Morales, con un parecer del Doctor Juan Páez de Castro, todo en defensa de estos anales*, 6 vols. (Zaragosa: Luis de Robles, 1610), vol. 6.

17. Throughout the chapter I follow Arantxa Domingo Malvadi's account of Páez's life. In her pioneering study Domingo Malvadi reconstructed the contents of his scattered library through an examination of his letters, inventories, and surviving books and manuscripts.

18. Domingo Malvadi, *Bibliofilia humanista*, 17–58. These were intertwined positions. Domingo Malvadi believes that Páez must have been ordained at some point during his stay in Italy, probably in Rome.

19. On the concept of "public libraries" in ancient Rome, see T. Keith Deix, "'Public' Libraries in Ancient Rome: Ideology and Reality," *Libraries & Culture* 29, no. 3 (Summer 1994): 282–296.

20. Juan Páez de Castro, "Memorial al Rey Don Felipe II, sobre las librerías por el Doctor Juan Páez de Castro," in *Carta del doctor Juan Páez de Castro al secretario Matheo Vázquez, sobre el precio de libros manuscritos*, copia digital (Valladolid, 2009–2010), 9–50, 13. Also see Teodoro Martín Martín, "Juan Páez de Castro: Aproximación a su vida y obra," *Ciudad de Dios* 201, no. 1 (1988): 35–55; and Martín Martín, *Vida y Obra de Juan Páez de Castro* (Guadalajara, 1990). On chained libraries, see Anthony Hobson, *Great Libraries* (New York, 1970), 12; and Burnett Hillman Streeter, *The Chained Library: A Survey of Four Centuries in the Evolution of the English Library* [1935] (Cambridge, 2011), 1–76.

21. Páez de Castro, "Memorial al Rey Don Felipe II, sobre las librerías," 27.

22. Páez de Castro, "Memorial al Rey Don Felipe II, sobre las librerías," 26.

23. Páez de Castro, "Memorial al Rey Don Felipe II, sobre las librerías," 30. Hobson, *Great Libraries*, 150.

24. Páez de Castro, "Memorial al Rey Don Felipe II, sobre las librerías," 26. Páez might have been thinking here of the ancient Greek texts from the French royal library printed by Robert Estienne in the 1540s in the Greek types known as the *Grecs du roi*. See John Considine, *Dictionaries in Early Modern Europe: Lexicography and the Making of Heritage* (Cambridge, 2008), 13. In a 1547 letter from Trent addressed to Zurita, Páez commented on desiring to procure these works. See Domingo Malvadi, *Bibliofilia humanista*, 359.

25. Uztarroz and Dormer, *Progresos de la historia en Aragon*, 478. On the Vatican Library's layout and history, see Hobson, *Great Libraries*, 76–103.

26. Páez de Castro, "Memorial al Rey Don Felipe II, sobre las librerías," 36.

27. Páez de Castro, "Memorial al Rey Don Felipe II, sobre las librerías," 35–36. Descriptions of golden animals and plants had circulated in early reports on the conquest of Mexico and Peru. For instance, in the Inca temple of the sun of Q'oricancha, in Cuzco, several histories described a golden garden with metal cornstalks, llamas, and shepherds. Here, Páez might be referring to descriptions that appear in Cieza and Pizarro. See Emily C. Floyd, "Tears of the Sun: The Naturalistic and Anthropomorphic in Inca Metalwork," Medium Study, *Conversations: An Online Journal of the Center for the Study of Material and Visual Cultures of Religion* (2016), https://mavcor.yale.edu/conversations/medium-studies/tears-sun -naturalistic-and-anthropomorphic-inca-metalwork; Jose Toribio Medina, "Algunas piezas notables del rescate de Atahualpa," *Publicaciones del Museo de Etnología y Antropología de Chile* 4 (1927): 293–296; and Pedro Cieza de León, *Crónica del Perú: Segunda parte*, ed. Francesca Cantù (Lima, 1986), chap. 27, 79–80. Agustín de Zárate also describes these figures in his 1555 *Historia del descubrimiento y conquista del Peru, con las cosas naturales que señaladamente allí se hallan y los successos que ha avido. La cual escrivia Agustin de Çarate, exerciendo el cargo de Contador de cuentas por su Magestad en aquella provinvia, y en Tierra firme* (Antwerp: En casa de Martin Nuncio, a las dos Cigueñas, 1555), 27r.

28. Páez de Castro, "Memorial al Rey Don Felipe II, sobre las librerías," 37.

29. Páez de Castro, "Memorial al Rey Don Felipe II, sobre las librerías," 39.

30. Arndt Brendecke, *The Empirical Empire: Spanish Colonial Rule and the Politics of Knowledge*, trans. Jeremiah Riemer (Berlin, 2016), 280.

31. Richard Kagan, *Clio and the Crown* (Baltimore, 2009), 96–100.

32. In a letter of September 15, 1548, Páez asked Zurita whether the two missing volumes of Oviedo's history had been published and if anything of Hernán Cortés's was available for purchase. Domingo Malvadi, *Bibliofila humanista*, 383.

33. Domingo Malvadi, *Bibliofila humanista*, 354. On the idea of new worlds in the Renaissance and the struggle to reconcile the existence of a "plurality of worlds," see Joan-Pau Rubiés, "New Worlds and Renaissance Ethnology," *History and Anthropology* 6, nos. 2/3 (1993): 157–197.

34. Domingo Malvadi, *Bibliofila humanista*, 357.

35. Páez de Castro's interest in medicine and works of natural history is well attested in his correspondence. See Elisa Andretta, "The Medical Cultures of the 'Spaniards of Italy: Scientific Communication, Learned Practices, and Medicine in the Correspondence on Juan Páez de Castro (1545–1552)," in *Medical Cultures of the Early Modern Spanish Empire*, ed. John Slater, Maríaluz López-Terrada, and José Pardo-Tomás (Farnham, 2014), 129–148. Also see José Pardo-Tomás, "Andrés Laguna y la medicina europea del Renacimiento," *Los orígenes de la ciencia moderna: Seminario Orotava, Actas* 11/12 (2002): 45–68. Andrés Laguna included an acknowledgment of Páez's role in procuring the manuscript of Dioscorides. See Andrés de Laguna, *Pedacio Dioscorides Anazarbeo, Acerca de la materia medicinal, y de los venenose mortiferos, traduzido de lengua Griega, en la vulgar Castellana, & illustrado con claras y substantiales annotationes, y con las figuras de innumeras plantas exquisitas y raras, por Doctos Andres de Laguna Medico de Iulio III Pont. Maxi.* (Salamanca: Mathias, 1570), epistola nuncupatoria.

36. Uztarroz and Dormer, *Progresos de la historia en Aragon*, 544.

37. Domingo Malvadi, *Bibliofilia humanista*, 368.

38. José Miguel Morán Turina and Fernando Checa Cremades, *El coleccionismo en España: De la cámara de maravillas a la galería de pinturas* (Madrid, 1985), 94–95. On European perceptions of the relationship between collecting and understanding, see Paula

Findlen, *Possessing Nature: Museums, Collecting, and Scientific Culture in Early Modern Italy* (Berkeley, 1994), 17–47.

39. On the presence of American objects in Spanish sixteenth-century collections, see Morán Turina and Checa Cremades, *El coleccionismo en España*, 132–137; Alessandra Russo, "Cortés's Objects and the Idea of New Spain: Inventories as Spatial Narratives," *Journal of the History of Collections* 23, no. 2 (2011): 229–252; Antonio García-Abásolo, "Cristos mexicanos de caña y su devoción," in *Imaginería indígena mexicana: Una catequesis en caña de maíz* (Córdoba, 2001), 307–367; Rafael López Guzmán and Gloria Espinosa Spínola, "Memoria artística de los virreinatos americanos," in *Tornaviaje: Arte iberoamericano en España* (Madrid, 2021), 21–49; Brendecke, *The Empirical Empire*, 74–75; Domingo Malvadi, *Bibliofilia humanista*, 60–62. On the attempts to create comprehensive library collections and their analogies in equally ambitious cosmographical works, see Seth Kimmel's study of the private library of Hernando Colón and his indexes. Seth Kimmel, "Early Modern Iberia, Indexed: Hernando Colón's *Cosmography*," *Journal of the History of Ideas* 82, no. 1 (January 2021): 1–28.

40. "Leyendo esse libro de V.M. me viene al pensamiento la encadenacion que dios n.s. puso en todas las cosas, y como han venido desde el principio del mundo perfeccionandose las cosas de los hombres hasta estos tiempos de v.m. en esta nra. Parte habitable." BME, Esc. & IV.22., 126r.

41. This would be reminiscent of other Habsburg collections in Vienna. See Christian F. Feest, "The Collecting of American Indian Artifacts in Europe, 1493–1750," in *America in European Consciousness 1493–1750* (Chapel Hill, NC, 1995), 324–360.

42. Domingo Malvadi, *Bibliofilia humanista*, 149–299.

43. Domingo Malvadi identified these annotations, along with Páez's ex libris in her reconstruction of Páez's library. Domingo Malvadi, *Bibliofilia humanista*, 225.

44. F. Javier Campos y Fernández de Sevilla, *Catálogo del fondo manuscrito americano de la Real Biblioteca del Escorial* (San Lorenzo de El Escorial, 1993), 343–372; Carmelo Sáenz de Santa María, "Los Manuscritos de Pedro Cieza de León," *Revista de Indias*, no. 36 (January 1, 1976): 81–215.

45. His inventories also reveal that he possessed a copy of Fernaõ Lopes de Castanheda's *History of the Discovery and Conquest of the Indies by the Portuguese* (1552–1561, *Ho livro primeiro dos dez da historia do descobrimento & conquista da India pelos Portugueses*) and the published writings of Alvar Nuñez Cabeza de Vaca (1542, 1555). As chronicler of the king, Páez would also have had access to manuscripts about the Americas that circulated in the royal court. Eustacio, "De las cosas necesarias para escribir historia," 28 (1892): 604–610, 608. The Sevillian philosopher and royal tutor Sebastian Fox Morcillo also described in his *De historiae institutione dialogus* (1557) what may have been the Mixtec Codex Nuttall from the region of Oaxaca. See Antonio Cortijo Ocaña, *Teoría de la historia y teoría política en Sebastián Fox Morcillo* (Alcalá de Henares, 2000), 205.

46. The bibliography on the "chronicles of the Indies" is vast. For a representative selection, see Rolena Adorno, "The Discursive Encounter of Spain and America: The Authority of Eyewitness Testimony in the Writing of History," *William and Mary Quarterly* 49, no. 2 (1992): 210–228.

47. I adopt the term "lexical object" from John Considine. Considine, *Dictionaries in Early Modern Europe*, 13.

48. Antonio Alatorre, "Sobre americanismos en general y mexicanismos en especial," *Nueva revista de filología Hispánica* 49, no. 1 (2001): 1–51.

49. Bryan Ellsworth Hamann, *The Translations of Nebrija: Language, Culture, and Circulation in the Early Modern World* (Boston, 2015), 1–11 and 43–84; John Considine, "The

History of the Concept of Lexicography," in *The History of Linguistics 2014: Selected papers from the 13th International Conference on the History of the Language Sciences (ICHoLS XIII), Vila Real, Portugal, 25–29 August 2014*, ed. Carlos Assunção, Gonçalo Fernandes, and Rolf Kemmler (Philadelphia, 2016), 31–42.

50. Domingo Malvadi, *Bibliofilia humanista*, 225.

51. W. Keith Percival, "Renaissance Linguistics: The Old and the New," in *Studies in the History of Western Linguistics in Honour of R. H. Robins*, ed. Theodora Bynon and F. R. Palmer (Cambridge, 1986), 56–68, esp. 62. Juan de Ovando (c. 1515–1575), who served as president of the Council of Indies, for instance, owned a copy of a *Vocabulario en lengua mexicana*. See Fernando Bouza and Alfredo Alvar Ezquerra, "Apuntes biográficos y analisis de la biblioteca de un gran estadista hispano del siglo XVI: El presidente Juan de Ovando," *Revista de Indias* 44, no. 173 (1984): 81–139, 125.

52. Elio Antonio de Nebrija, *Dictionarium hispano latinum* (Salamanca: Juan de Porras, 1495), fol. XXIIIv. Also see María Águeda Moreno Moreno, "Las voces americanas de los diccionarios de los diccionarios generales del español (siglos XV–XVII)," *RAHL: Revista argentina de historiografía lingüística* 3, no. 2 (2011): 133–15.

53. Toon Van Hal, "Early Modern Views on Language and Languages (ca.1450–1800)," *Oxford Research Encyclopedia of Linguistics* (Oxford, 2019), 1–22, esp.7.

54. Eustacio, "De las cosas necesarias para escribir historia," 28 (1892): 609.

55. Considine, *Dictionaries in Early Modern Europe*, 15.

56. I will cite throughout the chapter Páez's copy of Francisco López de Gómara, *Historia general de las Indias y todo lo acaecido en ellas dende que se ganaron hasta ahora y de la conquista de Mexico y de la nueva España* (Antwerp: Martin Nuncio, 1554), BME, Esc. 60. IV.29., 3r.

57. Antonello Gerbi, *Nature in the New World: From Christopher Columbus to Gonzalo Fernández de Oviedo*, trans. Jeremy Moyle (Pittsburgh, 1985), 50–75.

58. Pietro Martire d'Anghiera, *Decadas del Nuevo Mundo por Pedro Martir de Angleria, primer cronista de Indias*, 2 vols., trans. Agustín Millares Carlo, study and appendixes by Edmundo O'Gorman (México: J. Porrúa, 1964–1965), vol. 1, decade 3, book 7, 351.

59. Martire, *Decadas del Nuevo Mundo*, decade 3, book 7, 351.

60. According to the *Glosario etimológico taíno-español histórico y etnográfico*, the word *areito* is related to the Arawak verb *aritin*, which means "to summon" or "to call," and the verb *aiintunnua*, which means "to sing." See Esther Hernández, *Vocabulario en lengua castellana y Mexicana de Fray Alonso de Molina: Estudio de los indigenismos léxicos y registros de las voces españoles internas* (Madrid, 1996), 47–48.

61. See Kathleen Ann Myers, *Fernández de Oviedo's Chronicle of America: A New History for a New World*, trans. Nina M. Scott (Austin, 2007).

62. Gerbi, *Nature in the New World*, 136; Nicolás Wey Gómez, "Memorias de la zona tórrida: El naturalismo clásico y la 'tropicalidad' americana en el *Sumario de la natural historia de las Indias de Gonzalo Fernández de Oviedo* (1526)," *Revista de Indias* 73, no. 259 (2013): 609–632.

63. Sixteenth-century authors used the term *romance* to refer to Romance languages like Castilian. The term also refers to a traditional poetic composition in romance. Gonzalo Fernández de Oviedo, *Coronica de las Indias: La hystoria general de las Indias agora nueuamente impressa corregida y emendada* (Salamanca: Juan de Junta, 1547), iiiv–ivr. See Jesús Carillo Castillo, "Naming Difference: The Politics of Naming in Fernández de Oviedo's *Historia general y natural de las Indias*," *Science in Context* 16, no. 4 (2003): 489–504.

64. Gonzalo Fernández de Oviedo, *La historia general y natural de las Indias, islas y tierra-firme del mar océano*, ed. D. José Amador de los Ríos, 4 vols. (Madrid: Imprenta de la

Real Academia de la Historia, 1851–1855), vol. 1, book 12, chap. 7, 392–396, 393. On the crocodile, see Isidore of Seville, *The Etymologies of Isidore of Seville*, trans. Stephen A. Barney, W. J. Lewis, J. A. Beach, and Oliver Berghof (Cambridge, 2010), book 12, 19, 260–261. Oviedo revised the placement of the animal. In the printed editions of 1535 and 1547, the iguana appears among the aquatic animals, while in a subsequent manuscript it appears among the land animals.

65. Antonio Barrera Osorio, *Experiencing Nature: The Spanish American Empire and the Early Scientific Revolution* (Austin, 2006), 1–12; Jeremy Paden, "The Iguana and the Barrel of Mud: Memory, Natural History, Hermeneutics in Oviedo's *Sumario de la natural historia de las Indias*," *Colonial Latin American Review* 16, no. 2 (2007): 203–226.

66. Juan Cano is supposed to have written an account of the conquest that has never been located. The jurist Alonso de Zorita provides a short biographical vignette in in his 1585 *Account of New Spain*. See Alonso de Zorita, *Relación de la Nueva España*, ed. Ethelia Ruiz Medrano and José Mariano Leyva, 2 vols. (México, 2011), vol. 1, 112.

67. Gonzalo Fernández de Oviedo, *Historia general y natural de las Indias* [1851–1855], vol. 3, book 33, chap. 54, 545–549.

68. Oviedo, *Historia general y natural de las Indias* [1851–1855], vol. 1, book 6, chap. 1, 463–464.

69. Oviedo, *Historia general y natural de las Indias* [1851–1855], vol. 1, book 6, chap. 1, 463–464. The images appear as woodcuts in the editions of 1535, 1547, and 1557. See Oviedo, *Historia general* [1547], 58v–59r.

70. As one of the earliest and most important chroniclers of Peru, the bibliography on Cieza is enormous. For a representative selection, see Franklin Pease García-Yrigoyen, introduction to Cieza de León, *Crónica del Perú*, xxi–xxii. Also see Sabine MacCormack, *On the Wings of Time: Rome, the Incas, Spain, and Peru* (Princeton, NJ, 2007), chap. 3; and Luis Millones Figueroa, *Pedro Cieza de León y su crónica de Indias: La entrada de los Incas en la historia universal* (Lima, 2001). For early modern translations of Cieza and other chronicles of the Americas published before 1556, see Roberto A. Valdeón, *Translation and the Spanish Empire in the Americas* (Philadelphia, 2020), 153–207.

71. Domingo de Santo Tomás, *Gramática o Arte de la lengua general de los Indios de los Reinos del Perú. Nuevamente compuesta por el maestro Fray Domingo morador en los dichos reinos* (Valladolid: Francisco Fernández de Cordova, 1560); Santo Tomás, *Lexicón o Vocabulario de la lengua general del Perú llamada quichua* (Valladolid: Francisco Fernández de Cordova, 1560).

72. This town was located in the Cauca valley of modern-day Colombia. Throughout the chapter I cite Páez's copy of Cieza. Pedro Cieza de León, *Primera parte de la cronica del Peru* (Antwerp, 1554), BME, Esc. 20.VI.17, chap. 16, 40r.

73. Páez's annotation: "No se conquistaron por esto, si no por la division de los dos hermanos y por la pasion de Atabalipa." Cieza de León, *Primera parte*, chap. 36, 94v.

74. Jeremy Mumford, *Vertical Empire: The General Resettlement of the Indians in the Colonial Andes* (Durham, NC, 2012), 14.

75. Zárate, *Historia del descubrimiento y conquista del Peru*, chap. 15, 29v–30r.

76. Cieza de León, *Primera parte*, chap. 40, 105v.

77. Páez's annotation: "La ciudad de Cuzco esta en ladera, una parte, y otra en llano. Los de arriba se llaman Hanacuzco. Q Hana quiere dezir arriba. Los otros se dize Huricuzco por Hura quiere dezir abaxo. Pero los de arrivba eran mas principales." Cieza de León, *Primera parte*, chap. 54, 143r. Rodolfo Cerrón-Palomino has argued that "Hurin" is actually incorrect, and the term is *lurin*; however, Hurin widely circulated in the early chronicles of Peru as the opposite of *hanan*. See Rodolfo Cerrón-Palomino, "Hurin: Un espejismo léxico opuesto a

hanan," in *El hombre y los Andes: Homenaje a Franklin Pease G.Y.*, ed. Javier Flores Espinoza and Rafael Varón Gabai (Lima, 2002), 219–235.

78. Cieza de León, *Primera parte*, book end.

79. Páez's annotation: "Guava—Pacay; Aguacates—Paltas; Tunas—Caxareros." Cieza de León, *Primera parte*, chap. 46, 127r.

80. In numerous letters, one from as early as 1546, Páez becomes aware of Gómara through Zurita. See Domingo Malvadi, *Bibliofilia humanista*, 496, 401, 435.

81. López de Gómara, *La historia general de las Indias*, 42r.

82. Eustacio, "De las cosas necesarias para escribir historia," 28 (1892): 608.

83. Eustacio, "De las cosas necesarias para escribir historia," 28 (1892): 608–609.

84. Scholars have greatly debated the purposes of these two texts. A comparison between them demonstrates a concern with similar topics and even the repetition of various phrases and formulations, regardless of their final objectives. The Escorial version BME, Esc. & IV.22. appears to be an earlier draft of the more polished prologue in Mss. 23083 (4) of the BNE.

85. "Bastante exemplo es, sin q digamos las antiguedades de este nuestro mundo, lo q los indios occidentales hazian en el mundo Nuevo en sus bayles cantando las cosas de sus reyes pasados sin tener otra historia mas cierta. Despues q por grandisima merced De dios supieron los de estas partes escrivir lo primero q se puso en escritura fue historia. Esta fue al principio grosera y sin arte como acontece en las invenciones humanas, q poco a poco llegan a perfeccion." BME, Esc. & IV.22., 126v.

86. Exodus 15; 1 Samuel 18:6.

87. "Por eso fue el gran trabajo que pasaron los sabios para hallar manera con que dexassen memorias de sus invenciones y hazañas a los del porvenir. Contentanse al principio con que las cosas memorables se cantassen publicamente y se celebrassen con instrumentos musicos como canto Moyses la victoria del mar bermejo, y las otras mugeres las proezas de David. Esto hacian muchas naciones antes de que se hallasen las letras, como se ha visto claramente en las Indias de Vuestra Magestad [crossed out: en aquellas canciones que llamavan areytos] que dezian canciones que duravan muchos dias. La poesia Antigua era celebrar semejantes cosas, y aun nuestros romances viejos tomaron origen de esto. Assy es creyble que despues que se hallaron las letras por gradissima magestad de Dios, que lo primero que se escribio fueron aquellos cantares y historias groseramente compuestas como suele acontecer en todas las invenciones humanas que poco a poco se perfeccionan." BNE, Ms. 23083 (4), fol. 1v.

88. Domingo Malvadi, *Bibliofilia humanista*, 51.

89. Although Pané never uses the term *areito*, he describes how the inhabitants of Hispaniola passed down their laws in ancient songs. Pané compares this to the practices of the "Moors." Pané's original manuscript is lost. Martire and Bartolomé de las Casas used Pané's unpublished work as a source. See *Relación acerca de las antigüedades de los indios, las cuales, con diligencia, como hombre que sabe el idioma de estos, recogió por mandato del Almirante*, ed. J. J. Arrom (México, 1988).

90. In the early modern sources, the word appears spelled alternatively as *areyto, arreito*, and *areito*, etc. I have chosen to use in the main text the modern rendition of the word *areito*, although I reproduce in quotations how the various authors spelled the term in their own writings.

91. Martire, *Decadas del Nuevo Mundo*, decade 3, book 7, 351. The *areito* also appears briefly in the decade 1, book 9.

92. Oviedo, *Historia general y natural de las Indias* [1851–1855], vol. 1, book 5, chap. 1, 125–130.

93. Oviedo, *Historia general y natural de las Indias* [1851–1855], vol. 1, book 5, chap. 1, 125–130.

94. For this passage in Livy's Roman history (7.2), see Tomas Habinek, *The World of Roman Song: From Ritualized Speech to Social Order* (Baltimore, 2009), 106–107.

95. Oviedo, *Historia general y natural de las Indias* [1851–1855], vol. 1, book 17, chap. 8.

96. Oviedo, *Historia general y natural de las Indias* [1851–1855], vol. 1, book 5, chap. 1, 125–130.

97. Oviedo, *Historia general y natural de las Indias* [1851–1855], vol. 1, book 5, chap. 1, 125–130.

98. Oviedo, *Historia general y natural de las Indias* [1851–1855], vol. 1, book 16, chap. 5.

99. Oviedo, *Historia general y natural de las Indias* [1851–1855], vol. 4, book 42, chap. 11

100. Nicolás Wey Gómez, "Memorias de la zona tórrida," 628–629. In his book-length study about Columbus's four voyages, Wey Gómez reconstructs this cosmological paradigm in great detail. See Nicolás Wey Gómez, *The Tropics of Empire: Why Columbus Sailed South to the Indies* (Cambridge, MA, 2008).

101. Oviedo, *Historia general y natural de las Indias* [1851–1855], vol. 4, book 42, chap. 11.

102. Oviedo, *Historia general y natural de las Indias* [1851–1855], vol. 4, book 42, chap. 11.

103. Alonso de Molina's 1555 *Vocabulario en lengua Castellana y Mexicana*, for instance, included the word *areito* to clarify Mexican words, signaling that *areito* had become a part of the lexicon of Castilian speakers in the Americas. See Hernández, *Vocabulario*, 48.

104. Galen Brokaw, "Ambivalence, Mimicry, and Stereotype in Fernández de Oviedo's *Historia General y Natural de las Indias*: Colonial Discourse and the Caribbean Areíto," in *CR: The New Centennial Review* 5, no. 3 (2005):143–165.

105. Paul Scolieri, *Dancing the New World: Aztecs, Spaniards, and the Choreography of Conquest* (Austin, 2013), 24–43, esp. 28.

106. Ralph Bauer and Jaime Marroquín Arredondo, "Introduction: An Age of Translation," in *Translating Nature: Cross-Cultural Histories of Early Modern Science*, ed. Ralph Bauer and Jaime Marroquín Arredondo (Philadelphia, 2019), 9–25.

107. John Elliott summarizes the formative impact of the Caribbean here: John H. Elliott, "The Spanish Conquest and the Settlement of America," in *The Cambridge History of Latin America*, ed. Leslie Bethell (Cambridge, 1984), vol. 1, 147–206, 164–166. On the demographic impact of early Spanish settlement on the island of Hispaniola, see Massimo Livi Bacci, "Return to Hispaniola: Reassessing a Demographic Catastrophe," *Hispanic American Historical Review* 83, no. 1 (February 2003): 3–51. Also see Lauren MacDonald, "The Cemí and the Cross: Hispaniola Indians and the Regular Clergy, 1494–1517," in *The Spanish Caribbean and the Atlantic in the Sixteenth Century*, ed. Ida Altman and David Wheat (Lincoln, NE, 2019), 3–24, 17.

108. Cieza de León, *Primera parte*, 110r.

109. Zárate, *Historia del descubrimiento y conquista del Peru*, 22r.

110. On the challenges and possibilities of cross-cultural comparison, see Caroline Bynum Walker, "Avoiding the Tyranny of Morphology: Or: Why Compare," in *Dissimilar Similitudes: Devotional Objects in Late Medieval Europe* (New York, 2020), 183–220.

111. Oviedo, *Historia general y natural de las Indias* [1851–1855], vol. 1, book 7, chap. 13, and book 8, chap. 8; vol. 4: book 42, chap. 11.

112. Oviedo, *Historia general y natural de las Indias* [1851–1855], vol. 1, book 5, chap. 1, 125–130.

113. Scholars like Desiderius Erasmus (d. 1536) and the Valencian pedagogue Juan Luis Vives (d. 1540) were critical of *romances*. Vives, for instance, believed that epics and medieval *romances*, on account of their violence and eroticism, were harmful to the education of children. See Carlos G. Noreña, *Juan Luis Vives* (The Hague, 1970), 188.

114. For the history and origin of *romance* poetry and its similarities with other European popular ballads, see Ramón Menéndez Pidal, "Estudios sobre el Romancero," *Obras completas de Ramón Menéndez Pidal* (Madrid, 1973), vol. 11, 11–84.

115. Dorothy Clotelle Clark, "Romance," in *The Princeton Encyclopedia of Poetry and Poetics*, ed. Roland Greene, Stephen Cushman, Clare Cavanagh, Janah Ramazani, and Paul F. Rozer, 4th ed. (Princeton, NJ, 2012), 1204–1205.

116. Gonzalo Fernández de Oviedo, *Ouiedo de la natural hystoria de las Indias* (Toledo: Ramón de Petras, 1526); Oviedo, *La historia general de las Indias* (Seville: Juan Cromberg, 1535); and Oviedo, *Coronica de las Indias: La hystoria general de las Indias agora nueuamente impressa corregida y emendada* (Salamanca en casa de Juan de Junta, 1547), subsequently cited as *Historia general* [1547].

117. Oviedo, *Historia general y natural de las Indias* [1851–1855], vol. 1, book 5, chap. 1, 125–130.

118. On the use of the *bixa*, known in the Mesoamerican world as *achiotl*, and the painting of the body as an Indigenous technology, see Marcy Norton, "Subaltern Technologies and Early Modernity in the Atlantic World," *Colonial Latin American Review* 26 (2017): 18–38.

119. Oviedo, *Historia general y natural de las Indias* [1851–1855], vol. 1, book 8, chap. 6, 297–298. Botanists later identified the *bixa* as the plant that the Mexicans or Nahuas called the Achiotl. David Weil Baker has shown how the antiquarian William Camden derived the name "Britain" from the body painting practiced from the ancient British people. See David Weil Baker, "Etymology, Antiquarianism, and Unchanging Languages in Johannes Goropius Becanus's Origenes Antwerpiane," *Renaissance Quarterly* 72, no. 4 (2019): 1326–1361.

120. Caesar identified the plant as the *vitrum*, which makes a blue dye, not a red dye. Caesar, *The Gallic War*, trans. H. J. Edwards, Loeb Classical Library (Cambridge, MA, 1917), 5:14.

121. Hernández, *Vocabulario*, 100–101.

122. "[(]Se untan) costumbre fue, y aun es oy de muchas naciones pintarse de diversos colores, tomadas de varias cosas naturales segun lo que se acostumbrava en esta nueva Hespaña, antes de que viniessen a ella los Hespañoles, quando havian de exercitar sus contiendas o arreitos porque embixavan entonces sus cuerpos, creyendo pararse asi o mas hermosos, o a sus enemigos mas temerosos y espantables. . . . Mas para que nota esta censura las naciones barbaras y bestiales, teniendo a la mano tantos enxemplos familiares y domesti-cos? No se pintan de blanco y colorado nuestras mugeres la cara y los labios, y se tiñen o enrruvian los cabellos? No se imprimen, donde mas les agrada, lunares, y debuxan en los braços diversas formas y characteres. Han dexado alguna cosa intentada que pueda dar ayuda a la forzada moçedad, y detener o remediar las canas y arrugas de la vejez hasta deshollarla con mudas o a lo menos conservarla blanca y lustrosa, con otros afeites e invenciones que cria la mar, aire o tierra, que ellas no acomoden a su regalo y hermosura?" Francisco Hernández, *Historia natural libros xxi, xxii, xxiii, xiv, y xv, por Cayo Plinio Cecilio Segundo; traducida y declarada por el mismo autor*, BNE, Ms. 2868, book 20, prologue, 55v–57r.

123. Gerbi, *Nature in the New World*, 129–132.

124. For a discussion of this passage in the works of Francois Baudouin in 1561 and the origins of the *ars historica*, see Anthony Grafton, *What Was History?* (Cambridge, 2007), 112–116.

125. Carlo Ginzburg, "Selfhood as Otherness," in *No Island Is an Island: Four Glances at English Literature in a World Perspective* (New York, 2000), 30–33. Also see Arnaldo Momigliano, "Perizonius, Niebuhr and the Character of Early Roman Tradition," in *Journal of Roman Studies* 47, nos. 1/2 (1957): 104–114. The *areito* also appears described in the Latin epic of Giulio Cesare Stella of 1589, one of the first Latin poems to describe Columbus's role

in early American exploration. See Manuel Antonio Díaz Gito, "Conjurando—en vano—el amor de Colón: El areito de Anacaona en la Columbeis (Roma, 1589) de Giulio Cesare Stella," *Confluencia* 36, no. 1 (Fall 2020): 2–16.

126. Tamar Herzog, "Can You Tell a Spaniard When You See One? 'Us' and 'Them' in the Early Modern Iberian Atlantic," in *Polycentric Monarchies. How did Early Modern Spain and Portugal Achieve and Maintain a Global Hegemony?*, ed. Pedro Cardim, Tamar Herzog, José Javier Ruiz Ibáñez, and Gaetano Sabatini (Brighton, 2012), 147–161, 22.

127. See *Más de mil y un cuentos del Siglo de Oro*, ed. José Fradejas Lebrero (Madrid, 2008).

128. Alonso de Fuentes, *Quarenta cantos de diversas y peregrinas historias, declarados y moralizados por el maravilloso cavallero Alonso de Fuentes* (Seville: Doménico de Robertis, 1550), epistola. Vinceç Beltran has also noted the use of the term of *areito* in Sevillian writers like Fuentes, Argote de Molina, and Juan de la Cueva. See Vinceç Beltran, "Estudio introductorio," in *Cuarenta cantos de diversas y peregrinas historias* (México, 2020), 1–239.

129. Antonio Alatorre, *Los 1001 años de la lengua española* (México, 1979), 158.

130. Gonzalo Argote de Molina, "Al curioso lector," in *El Conde Luncanor compuesto por el excelentisimo principe don Juan Manuel, hijo del Infante don Manuel, y nieto del santo rey don Fernando* (Sevilla: en casa de Hernando Diaz, 1575).

131. Gonzalo Argote de Molina, "Indice de algunos vocablos antiguos que se hallan en este libro, para noticia de la lengua castellana," in *El Conde Luncanor.*

132. Gonzalo Argote de Molina, "Discurso Hecho por Gonçalo de Argote y de Molina, sobre la poesia Castellana contenida en este libro," in *El Conde Luncanor*, fols. 92r–97v. A short study of the treatise appears in Irene Rodríguez Cachón, "Nuevas y Viejas reflexiones literarias en el 'Discurso sobre la poesía castellana' (1575) de Gonzalo Argote de Molina," in *Romance Studies* 37, no. 1 (2019): 1–11.

133. Páez read and approved Esteban de Garibay's chronicle for publication in 1567.

134. Argote de Molina, "Discurso," 93r. For the humanist reception of Ablabius and the emergence of idea that the Goths memorialized the past through song, see "Ablabius: History," in *The Fragmentary Latin Histories of Late Antiquity (AD 300–620)*, ed. and trans. Lieve Van Hoof and Peter Van Nuffelen (Cambridge, 2020), 137–145, 140. John of Uppsala is Johannes Magnus the Catholic Archbishop of Sweden.

135. Argote de Molina, "Discurso," 93r.

136. Gonzalo Argote de Molina, *Nobleza de Andaluzia* (En Sevilla: Fernando Diaz, 1588), al lector.

137. Argote de Molina, *Nobleza de Andaluzia*, al lector.

138. See Bernardino de Sahagún, *Historia general de las Cosas de Nueva España*, ed. Ángel María Garibay Kintana (México, 2016), book 10, 565.

139. Juan de la Cueva de la Garoza, "Exemplar Poético," ed. Frans Gustaf Emanuel Walberg (Lund: Håkan Ohlsson, 1904), epistola 2 (124–143), 59–60. On the life and works of Juan de la Cueva, see Richard F. Glenn, *Juan de la Cueva* (New York, 1973); and José Cebrián, *Juan de la Cueva y Nueva España* (Kassel, 2001).

140. Domingo Malvadi, *Bibliofilia humanista*, 51.

141. López de Gómara, *La historia general de las Indias*, 116v.

142. "Primeramente concebir en su amino la verdadera imagen del buen Rey, y saber se despues tambien declarer, es una señal que supiera hazer lo que dezir. Ó a lo menos supiera bien aconsejar á los Reyes que se allegan. Porque como dize Dion porque los sabios no pueden ser todas vezes Reyes las mas de las naciones consituyeron que los Reyes tuviessen consejos de sabios principalmente los religiosos. Assi los Persas pusieron a los Magos. Los de Egypto a los sacerdotes que tenian la misma disciplina que los Magos. Los Indios a los

Brahmanes. Los Franceses a los Druydas, sin cuyo parecer no podian hacer nada." Princeton MS. 174, fol. 10v.

143. Eustacio, "De las cosas necesarias para escribir historia," 28 (1892): 608–609.

144. Cicero, *De oratore*, 2.15.62; 2.9.36–37. On the ancient rules for history writing, see George Nadel, "Philosophy of History before Historicism," *History & Theory* 3, no. 3 (1964): 291–315.

145. Eustacio, "De las cosas necesarias para escribir historia," 29 (1892): 30.

146. Zárate, *Historia del descubrimiento y conquista del Peru*, ad lectorem prefatio. Also see Giuiliano Gliozzi, "L'Altantide e il Nuovo Mondo," in *Adamo e il Nuovo Mondo: La nascita dell'antropologia come ideologia coloniale; Dale geneaologie bibliche alle teorie razziali* (Florence, 1976), 177–232; On Acosta's rejection of the Atlantis hypothesis, see Anthony Grafton, "José de Acosta: Renaissance Historiography and New World Humanity," in *The Renaissance World*, ed. John Jeffries Martin (New York, 2007), 166–88.

147. Eustacio, "De las cosas necesarias para escribir historia," 29 (1892): 32.

148. López de Gómara, *La historia general de las Indias*, 294r–294v. See Harold Cook, "Ancient Wisdom, the Golden Age, and Atlantis: The New World in Sixteenth-Century Cosmography," *Terrae incognitae* 10 (1978): 24–43.

149. On the life and works of Agustín de Zárate, see Teodoro Hampe-Martínez, "Agustín de Zárate, contador y cronista indiano (estudio biográfico)," *Mélanges de la casa de Velázquez* 27, no. 2 (1991): 129–154; Hampe-Martínez, "Agustín de Zárate: Precisiones en torno a la vida y obra de un cronista indiano," *Cahiers du monde hispanique et luso-brésilien*, no. 45 (1985): 21–36; "La difusión de libros e ideas en el Perú colonial: Análisis de bibliotecas particulares (siglo XVI)," *Bulletin Hispanique* 89, nos. 1–4 (1987): 55–84.

150. Zárate, *Historia del descubrimiento y conquista del Peru*, al lector.

151. Sabine MacCormack, *Religion in the Andes: Vision and Imagination in Early Colonial Peru* (Princeton, NJ, 1991).

152. Zárate, *Historia del descubrimiento y conquista del Peru*, fols. 7r–7v. For the idea of the existence of giants in Mesoamerica, see Mackenzie Cooley, "The Giant Remains: Mesoamerican Natural History, Medicine, and Cycles of Empire," *Isis* 112, no. 1 (2021): 45–67.

153. Zárate, *Historia del descubrimiento y conquista del Peru*, 7r–7v.

154. On the quipu, see John A. Yeakel, "The Accountant-Historians of the Incas," *Accounting Historians Journal* 10, no. 2 (Fall 1983): 39–51; John Murra, *Formaciones económicas y políticas del mundo andino* (Lima, 1975); Frank Solomon, *The Cord Keepers: Khipus and Cultural Life in a Peruvian Village* (Durham, NC, 2004).

155. Godefroyd de Catallaÿ, *Annus Platonicus: A Study of World Cycles in Greek, Latin, and Arabic* (Leuven, 1996), 214.

156. "Se hazen grandes imperios y se pasan de unas gentes a otras. Se pierden las noblezas y linages, y se vencen las memorias de los hombres con diluvios, pestilencias, Guerra, y fuego general, qual cuentan los antiguos el de Phaeton y los indios en este libro de V. M. el de Viracocha. y ql pensaron los antiguos q seria el fin del mundo y q dezian que el cielo todo arderia. Persuadiados creo yo de la opinion de algunos filosofos q ponian el fuego para principio de las cosas y pensavan q como elemento potentisimo convertiria todo el resto en si mesmo andando el tiempo." BME, Esc., MS. & IV.22, fol. 129v. Gregory Nagy, "Phaeton, Sappho's Phaon, and the White Rock of Leukas," *Harvard Studies in Classical Philology* 77 (1973): 137–177.

157. "La razon de la dificultad en conseguir estas dos partes es,] porq las causas parecen contrarias, q son unidad y diversidad. Lo q es uno dura y se perpetua, Lo q es vario contenta al ingenio humano, q fue siempre amigo de novedad." BNE, Ms. 23083/4, fol. 1r.

158. "La unidad en la historia es cierta encadenacion de toda la materia de principio hasta el fin, de manera que todo concuerda con una sencillez de la verdad tan desnuda y clara que se ven no hazer en ella lisonjas ni enemistades." BNE, Ms. 23083/4, fol. 4r.

159. Eustacio, "De las cosas necesarias para escribir historia," 29 (1892): 34.

160. Fernando Bouza, *Communication, Knowledge, and Memory in Early Modern Spain*, trans. Sonia López and Michael Agnew (Philadelphia, 2004), 1–15.

161. Kagan, *Clio and the Crown*, 104–123.

162. "Instititucion del oficio de Cronista del Reino de Aragon y primer nombramiento de Gerónimo Zurita," Acto de Corte sobre el cronista, fol. 85, col. 3a, [1547], quoted in Uzatarroz and Dormer, *Progresos*, 64–65.

163. Patiño Loira, "Imagining Public Libraries in Sixteenth-Century Spain," 190.

CHAPTER TWO: The Search for Spain's Most Ancient Language

An earlier and abridged version of this chapter appeared published as "Language as Archive: Etymologies and the Remote History of Spain," in *After Conversion: Iberia and the Emergence of Modernity*, ed. Mercedes García-Arenal (Leiden, 2016), 95–125.

1. Esteban de Garibay, *Los XL libros d'el compendio historial de las chronicas y vniuersal historia de todos los reynos de España* (Antwerp: Christophe Plantin, 1571), book 4, chap. 6, 95.

2. Garibay, *Los XL libros d'el compendio historial*, book 4, chap. 6, 95. On Garibay, see Julio Caro Baroja, *Los vascos y la historia a través de Garibay* (Madrid, 2002), 174–89.

3. Garibay, *Los XL libros d'el compendio historial*, book 4, chap. 3, 88.

4. Garibay, *Los XL libros d'el compendio historial*, book 6, chap. 5, 173.

5. Garibay, *Los XL libros d'el compendio historial*, book 3, chap. 3, 65.

6. Garibay, *Los XL libros d'el compendio historial*, book 2, chap. 6, 47.

7. See Barbara Fuchs, *Exotic Nation: Maurophilia and the Construction of Early Modern Spain* (Philadelphia, 2011), 1–10.

8. See Juan Manuel Lope Blanch, "La lingüistica española del Siglo de Oro," in *Actas del VIII Congreso de la Asociación Internacional de Hispanistas*, ed. A. David Kossoff, José Amor y Vázquez, Ruth H. Kossof, and Geoffrey W. Ribbans (Madrid, 1986), 37–58.

9. Theodor Bibliander, *De ratione communi omnium linguarum et literarum commentaries* [1548], trans. and ed. Hagit Amirav and Hans-Martin Kim (Geneva, 2011), tract 2, book 3, 430–431.

10. Jesús Martínez de Bujanada, *El índice de libros prohibidos y expurgados de la Inquisición española (1551–1810)* (Madrid, 2016), 340.

11. Bibliander, *De ratione communi omnium linguarum*, tract 2, book 3, 433.

12. Giacomo Ferraù, "Riflessioni teoriche e prassi storiografica in Annio da Viterbo," in *Principato ecclesiastico e riuso dei classici gli umanisti e Alessandro VI: Atti del convegno (Bari-Monte Sant'Angelo, 22–24 maggio 2000)*, ed. Davide Canfora, Mauro de Nichilo, and Myriam Chiabò (Rome, 2002), 151–193.

13. Anthony Grafton, "Annius of Viterbo as a Student of the Jews: The Sources of his Information," in *Literary Forgery in Early Modern Europe, 1450–1800*, ed. Walter Stephens, and Earle A. Havens (Baltimore, 2018), 147–169, 164. On the influence of Annius of Viterbo on Jean Bodin, see Anthony Grafton, "Traditions of Invention and Inventions of Tradition in Renaissance Italy: Annius of Viterbo," in *Defenders of the Text: The Traditions of Scholarship in the Age of Science 1450–1800* (Cambridge, MA, 1994), chap. 3, 76–103, 88; and Anthony Grafton, *What Was History?* (Cambridge, 2007), chap. 3, 167–188.

14. Florián de Ocampo, *Los cinco primeros libros de la crónica general de España* (Medina del Campo: Guillermo de Millis, 1553), book 1, chap. 4, fol. xxr.

15. Ocampo, *Los cinco primeros libros de la crónica general de España*, book 1, ch. 4, fol. xxiiir.

16. Pedro de Alcocer, *Hystoria, o descripcion dela Imperial cibdad d Toledo, con todas la cosas acontecidas en ella, desde su principio, y fundacion, adonde se tocan, y refieren muchas antiguedades, y cosas notables de la hystoria general de España, agora nueuamente impressa* [Toledo: J. Ferrer, 1554] (Toledo, 1973), book 1, chap. 3, iiir–iiiiv.

17. Giuseppe Marcocci, *The Globe on Paper: Writing Histories of the World in Renaissance Europe and the Americas*, trans. Richard Bates (Oxford, 2020), 35; Don Cameron Allen, *The Legend of Noah: Renaissance Rationalism in Art, Science, and Letters* (Urbana, 1949), 113–137.

18. Jean Bodin, *Method for the Easy Comprehension of History*, trans. Beatrice Reynolds (New York, 1969), 334–338.

19. On the life and works of Jean Bodin, see Marie-Dominique Couzinet, *Jean Bodin* (Rome, 2001).

20. Bodin, *Method*, 334. On Jean Bodin and history writing, see Ann Blair, "Learning in the Life of a Sixteenth-Century French Nobleman: The Case of Nicolas de Livre, Friend of Jean Bodin," in *Historians and Ideologues: Essays in Honor of Donald R. Kelley*, ed. Anthony Grafton and John Hearsey McMillan Salmon (Rochester, NY, 2001), 3–39, 21. On the reception of Jean Bodin in the Iberian Peninsula, see Martim de Albuquerque, *Jean Bodin na Península Ibérica: Ensaio de história das ideias políticas e de direito público* (Paris, 1978).

21. Although I use Beatrice Reynolds's translation of the *Methodus*, I changed the named from Megasthenes to Metasthenes. In the original Latin, Bodin cites Metasthenes, a writer invented by Annius of Viterbo. See Jean Bodin, *Methodus ad facilem historiarum cognitionem, ab ipso recognita, et multo quam antea locupleitor. Cum indice rerum memorabilium copiosissimo* (Paris: Martinus Iuvenis, 1572), 539.

22. Alessandro Galimberti, "Gli Ebrei e storiografia giudaica nella *Methodus* di Jean Bodin," in *Storichi antichi e storichi moderni nella Methodus di Jean Bodin*, ed. Guiseppe Zecchini and Alessandro Galimberti (Milan, 2012), 10–25.

23. Bodin, *Method*, 338.

24. Bodin, *Method*, 338.

25. Grafton, *What Was History?*, 176.

26. Bodin, *Method*, 343.

27. Paolo Desideri, "Popoli antichi e moderni nella *Methodus* di Jean Bodin," in Zecchini and Galimberti, *Storichi antichi e storichi moderni nella Methodus di Jean Bodin*, 85–108.

28. Marian J. Tooley, "Bodin and the Medieval Theory of Climate," *Speculum* 28, no. 1 (January 1953): 64–83, 73–74. Also see Ann Blair, "Bodin's Philosophy of Nature," in *The Theater of Nature: Jean Bodin and Renaissance Science* (Princeton, NJ, 1991), 121.

29. Angus Vine, "Etymologies, Names and the Search for Origins: Deriving the Past in Early Modern England," *Seventeenth Century* 21, no. 1 (2006): 1–21, 1. Also see Vine, *In Defiance of Time* (Oxford, 2010).

30. Claudes-Gilbert Dubois, *Mythe e langage au seizième siècle* (Bordeaux, 1970), 16 and 20–21; Jean Céard, "De Babel ála Pentecôte: Lá transformation du mythe de la confusion des langues au XVIe siècle," *Bibliothèque d'humanisme* 42, no. 3 (1980): 577–94; Marie Luce Demonet, "Renaissance étymologiques," in *Lexique 14: L'étymologie de l'antiquité à la renaissance*, ed. Claude Buridant (Lille, 1998), 57–67.

31. Juan Luis Vives, *Los comentarios de Juan Luis Vives a 'La ciudad de Dios' de San Agustín*, trans. Rafael Cabrera Petit (Valencia, 2000), chaps. 11, 4. See Augustine of Hippo, *The City of God*, trans. Marcus Dods (New York, 1950), 536–37.

32. Conrad Gessner, *Mithridates*, trans. Bernard Colombat and Manfred Peters (Geneva, 2009), fol. 2v.

33. George J. Metcalf, *On Language Diversity and Relationship from Bibliander to Adelung*, ed. T. Van Hal and R. Van Rooy (Amsterdam, 2013), 85–104, esp. 87.

34. Michael T. Ward, "Bernardo de Aldrete and Celso Cittadini: Shared Sophistication in Renaissance Linguistic Investigation," *Hispanic Review* 61 (1993): 65–85; Lucia Binotti, "Historicizing Language, Imagining People: Aldrete and Linguistics Politics," in *Cultural Capital, Language and National Identity in Imperial Spain* (Woodbridge, 2012), chap. 7, 149–72, esp. 154–55.

35. Metcalf, *On Language Diversity and Relationship from Bibliander to Adelung*, 25.

36. On the life and works of Juan de Valdés, see Daniel Crews, *Twilight of the Renaissance: The Life of Juan de Valdés* (Toronto, 2008). On the *Diálogo*, see Ignacio Navarrete, "Juan de Valdés, Diego Hurtado de Mendoza, and the Imperial Style in Spanish Poetry," *Renaissance and Reformation* 28 (2004): 3–25.

37. Juan de Valdés, *Diálogo de la lengua*, ed. José Enrique Laplana (Barcelona, 2010), 59–60. On Juan de Valdés, see Juan M. Lope Blanch, introduction to *Diálogo de la lengua*, ed. Juan M. Lope Blanch (Madrid, 1969), 7–30; Rita Hamilton, "Juan de Valdés and Some Renaissance Theories on Language," in *Bulletin of Hispanic Studies* 30, no. 119 (1953): 125–135.

38. See Gregorio Mayans y Siscar, *Orígenes de la lengua española, compuestos por varios autores*, 2 vols. (Madrid: Juan de Zúñiga, 1737), vol. 1, 179–80.

39. Manuel Taboada Cid, "Lingüística hispánica renacentista: Lenguas y dialectos en las gramáticas españolas de los siglos XVI, y XVII (1492–1630)," *Verba: Anuario Galego de filoloxía* 16 (1989): 77–95.

40. Juan de Valdés, *Diálogo de lengua*, ed. José Enrique Laplana (Barcelona, 2010), 130–35. All subsequent citations are to this edition.

41. Antonio de Nebrija, *Gramática de la lengua castellana*, facsimile ed. (Madrid, 1991), prologue.

42. Valdés, *Diálogo de lengua*, 130–135.

43. Valdés, *Diálogo de lengua*, 139.

44. Werner Bahner, *La lingüística española del Siglo de Oro: Aportaciones a la conciencia lingüística en la España de los siglos XVI y XVII* (Madrid, 1966), 63–72.

45. Valdés, *Diálogo de lengua*, 154.

46. Navarrete, "Juan de Valdés, Diego Hurtado de Mendoza, and the Imperial Style in Spanish Poetry," 7–10.

47. See Robert L. Surles, "Juan de Valdés' *El Diálogo de la Lengua*: The Erasmian Humanism of a Spanish Expatriate," *College Language Association Journal* 35 (1991): 224–235; Juan Bautista Avalle-Arce, "La estructura del *Diálogo de la lengua*," in *Dintorno de una época dorada* (Madrid, 1978), 57–72; Ana Vian Herrero, "La mimesis conversacional en el *Diálogo de la lengua* de Juan de Valdés," *Criticón* 40 (1987): 45–79.

48. See, for instance, Carol Kidwell, *Pietro Bembo: Lover, Linguist, Cardinal* (Montreal, 2004), 217–226.

49. Valdés, *Diálogo de lengua*, 125–126.

50. Garibay, *Los XL libros d'el compendio historial*, book 4, chap. 2, 85–87.

51. Garibay, *Los XL libros d'el compendio historial*, book 4, chap. 2, 85–87.

52. Garibay, *Los XL libros d'el compendio historial*, book 4, chap. 4, 91.

53. Garibay, *Los XL libros d'el compendio historial*, book 4, chap. 4, 91.

54. Mayans, *Orígenes de la lengua*, 49.

55. Mayans, *Orígenes de la lengua*, 49. Also see William Stenhouse, *Reading Inscriptions and Writing Ancient History: Historical Scholarship in the Late Renaissance* (London, 2005).

56. Juan Madariaga Orbea and Frederick H. Fornoff, *Anthology of Apologists and Detractors of the Basque Language* (Reno, NV, 2006), 201.

57. Isidore of Seville, *The Etymologies of Isidore of Seville*, trans. Stephen A. Barney, W. J. Lewis, J. A. Beach, and Oliver Berghof (Cambridge, 2010), book 9, chap. 109.

58. Ambrosio de Morales, *Las antigüedades de las ciudades de España: Que van nombradas en la Coronica con la aueriguacion de sus sitios y nombres antiguos* [1575], 2 vols. (Madrid: Benito Cano, 1792), vol. 1, 112–113.

59. Francisco Javier Perea Siller, "Pablo de Céspedes (1548–1608): Argumentación lingüística y legitimación histórica," in *Sociedad Española de Historiografía Lingüística: Congreso (3° 2002. Vigo)*, ed. Miguel Ángel Esparza Torres, B. Fernández Salgado, and H. J. Niederehe (Hamburg, 2002), 641–52. On the life of Pablo de Céspedes, see Francisco M. Tubino, *Pablo de Céspedes: Obra premiada por voto unánime de la Academia de Nobles Artes de San Fernando en el certamen de 1866* (Madrid, 1868).

60. "Documento XIII," in Jesús Rubio Lapaz, *Pablo de Céspedes y su círculo: Humanismo y contrarreforma en la cultura andaluza del renacimiento al barroco* (Granada, 1993), 312–315. Also see Mercedes García-Arenal and Fernando Rodríguez Mediano, "Les antiquités hébraiques dans l'historiographie espagnole à l'époque moderne," *Dix-septième siècle* 66 (2015): 79–91, esp. 81–83.

61. On the *relaciones topográficas*, see Alfredo Alvar Ezquerra, *Relaciones topográficas de Felipe II: Madrid*, 3 vols. (Madrid, 1993), vol. 3.

62. Dominique Reyre, "Topónimos hebreos y memoria de la España judía en el Siglo de Oro," *Criticón* 65 (1995): 31–53, esp. 33; Francisco Javier Perea Siller, "Benito Arias Montano y la identificación de Sefarad, exégesis poligráfica de Abdías 20," *Helmántica* 51 (2000): 199–218.

63. Reyre, "Topónimos hebreos," 33.

64. Genesis 10:26. See Emilia Fernández Tejero and Natalio Fernández Marcos, "Scriptural Interpretation in Renaissance Spain," in *Hebrew Bible / Old Testament: The History of Its Interpretation*, vol. 2: *From the Renaissance to the Enlightenment*, ed. Magne Saebø. (Göttingen, 2008), 231–253.

65. Andrés Poza, *De la antigua lengua, poblaciones, y comarcas de las Españas, en que de paso se tocan algunas cosas de la Cantabria* (Bilbao: Mathías Mares, 1587), fol. 22r.

66. Poza, *De la antigua lengua*, fol. 22r.

67. Poza, *De la antigua lengua*, fol. 30r.

68. Poza, *De la antigua lengua*, 30v.

69. Poza, *De la antigua lengua*, 9v.

70. On the Basque in the New World, see Fernando Serrano, *Vascos y extremeños en el Nuevo Mundo durante el siglo XVII: Un conflicto por el poder* (Mérida, 1993); Serrano, *Euskal Herria y el Nuevo Mundo: La contribución de los vascos a la formación de las Américas*, ed. Ronald Escobedo Mansilla, Ana de Zaballa Beascoechea, and Oscar Alvarez Gila (Vitoria, 1996).

71. For a brief biography of Baltasar Echave Orio, see Madariaga Orbea and Fornoff, *Anthology of Apologists and Detractors of the Basque Language*, 216–217.

72. Baltazar de Echave, *Discursos de la lengua cántabra-bascongada compuestos por Balthasar de Echave, natural de la Villa de Çumaya en la provincia de Guipuzcoa, y vezino de Mexico* (Mexico: Henrrico Martínez, 1607).

73. Hernando de Ojeda, "Fray Hernando de Ojeda de la orden de Santo Domingo a su amigo, Balthasar de Echave en loor de esta obra," in Echave, *Discursos de la lengua cántabra-bascongada*.

74. Echave, prologue to *Discursos de la lengua cántabra-bascongada*.

75. Echave, *Discursos de la lengua cántabra-bascongada*, 83r–84v.

76. Olivia Constable, *To Live Like a Moor: Christian Perceptions of Muslim Identity in Medieval and Early Modern Spain* (Philadelphia, 2018), 1–14.

77. On the life and works of Martí de Viciana, see Sebastiá García Martínez, "Estudio preliminar," in Rafael Martí de Viciana, *Crónica de la ínclita y coronada ciudad de Valencia* (Valencia, 1983), vol. 1, 24–222.

78. Rafael Martí de Viciana, *Libro de las alabanças de las lenguas hebrea, griega, latina, castellana, y valenciana. Copiado por Marin de Viziana y consagrado al Ilustre Senado de la Inclyta y coronada ciudad de Valencia* (Valencia: J. Navarro, 1574).

79. Viciana, *Libro de la alabanças*, A5v.

80. Viciana, *Libro de las alabanças*, A6r.

81. Francisco Núñez Muley, *A Memorandum for the President of the Royal Audiencia and Chancery Court of the City and Kingdom of Granada*, trans. Vincent Barletta (Chicago, 2007), 92.

82. On the life and career of Morales, see Katherine Elliott van Liere, "The Missionary and the Moorslayer: James the Apostle in Spanish Historiography from Isidore of Seville to Ambrosio de Morales," *Viator* 37 (2006): 519–43; Richard Kagan, *Clio and the Crown* (Baltimore, 2009), 106–114; Sebastián Sánchez Madrid, *Arqueología y humanismo: Ambrosio de Morales* (Córdoba, 2002); and Enrique Redel, *Ambrosio de Morales: Estudio biográfico* (Córdoba, 1909).

83. Ambrosio de Morales, *Apologia de Ambrosio de Morales, con una información al Consejo del Rey Nuestro Señor, hecha por su orden y mandamiento en defensa de los Anales de Geronymo Çurita* (Zaragoza: en el Colegio de S. Vicente Ferrer por Iuan de Lanaja y Quartanet, 1610), 1r–25v, 6r.

84. Morales, *Las antigüedades de las ciudades*, vol. 1, 2.

85. Morales, *Las antigüedades de las ciudades*, vol. 1, lxxvii.

86. Morales, *Las antigüedades de las ciudades*, vol. 1, 2–3.

87. Morales, *Las antigüedades de las ciudades*, vol. 1, 8.

88. Morales, *Las antigüedades de las ciudades*, vol. 1, 3.

89. Morales, *Las antigüedades de las ciudades*, vol. 1, 80–82.

90. Morales, *Las antigüedades de las ciudades*, vol. 1, 81–82.

91. Morales, *Las antigüedades de las ciudades*, vol. 2, 8.

92. Morales, *Las antigüedades de las ciudades*, vol. 2, 15.

93. Morales, *Las antigüedades de las ciudades*, vol. 2, 27.

94. Morales, *Las antigüedades de las ciudades*, vol. 2, 28.

95. Morales, *Las antigüedades de las ciudades*, vol. 1, 202.

96. Diego de Guadix, *Recopiliación de algunos nombres arábigos que los árabes pusieron a algunas ciudades: Y otras muchas cosas* [1593], ed. Elena Bajo Pérez and Felipe Maíllo Salgado (Gijón, 2005), 149–150.

97. Mercedes García-Arenal and Fernando Rodríguez Mediano, "Jerónimo Román de la Higuera and the Lead Books of Sacromonte," in *The Conversos and Moriscos in Late Medieval Spain and Beyond* (Leiden, 2009), 243–68.

98. *Gramática de la lengua vulgar de España, impresa en Lovaina por Bartolomé Gravio en 1559* (Zaragoza, 1892), 18.

99. *Gramática de la lengua vulgar de España*, 19–21.

100. On the life of Diego de Guadix, see Elena Bajo Pérez and Felipe Maíllo Salgado, introduction to Guadix, *Recopilación de algunos nombres arábigos*, 115–31; Darío Cabanelas, "Tres arabistas franciscanos de los siglos XVII Y XVIII," in *Homenaje a la profesora Elena Pezzi*, ed. Antonio Escobedo Rodríguez (Granada, 1992), 21–36.

101. Guadix, *Recopilación de algunos nombres arábigos*, 150–151.

102. Guadix, *Recopilación de algunos nombres arábigos*, 149.

103. Guadix, *Recopilación de algunos nombres arábigos*, 174.

104. Guadix, *Recopilación de algunos nombres arábigos*, 461.

105. Guadix, *Recopilación de algunos nombres arábigos*, 883.

106. Juan de Mariana, *Historia general de España compuesta enmendada y añadida, por el padre Juan de Mariana; con el sumario y tabla* [Lat. 1592, Sp. 1601] (Madrid: Joaquín de Ibarra, 1780), vol. 1, book 1, chap. 7, 13–14.

107. Mariana, *Historia general de España*, vol. 1, book 1, chap. 7, 13–14.

108. Jerónimo Román y Zamora, *Republicas del mundo dividida en tres partes*, book 2: *De las Republicas de las Indias Occidentales* (Salamanca: En casa de Iuan Fernández, 1595), chap. 9, 161v–162r.

109. Bernardo de Aldrete, *Del origen y principio de la lengua castellana, ò Romance que oy se vsa en España* (Rome: Carlo Vulliet, 1606), book 3, chap. 3, 284.

110. Aldrete, *Del origen*, book 3, chap. 13, 356–57.

111. Francisco Cervantes de Salazar, *Crónica de la Nueva España*, ed. Manuel Magallón y Cabrera and Agustín Millares Carlo (Madrid, 1914), book 1, chap. 1, 16.

112. Aldrete, *Del origen*, 284.

113. Kathryn A. Woolard, "Is the Past a Foreign Country? Time, Language Origins, and the Nation in Early Modern Spain," *Journal of Linguistic Anthropology* 14, no. 1 (2004): 57–80, 66. Also see Kathryn A. Woolard, "Bernardo de Aldrete, Humanist and Laminario," *Al-Qantara* 24, no. 2 (2003): 449–476; and Woolard, "Bernardo de Aldrete and the Morisco Problem: A Study in Early Modern Spanish Language Ideology," *Comparative Study of Society and History* 44, no. 3 (2002): 446–480.

114. Aldrete, *Del origen*, 177.

115. George Metcalf distinguishes between the comparative method of the nineteenth century and the efforts of early modern scholars: "As far as the 16th and 17th centuries were concerned, a language might show historical progression, but its laws changed at random." Metcalf, *On Language Diversity and Relationship from Bibliander to Adelung*, 52; Giuliano Bonfante, "Ideas on the Kinship of European Languages from 1200 to 1800," *Cahiers d'histoire mondiale* 1 (1953–54): 679–699.

116. García-Arenal and Rodríguez Mediano, *The Orient in Spain*, chap. 17.

117. Prologue to *Diccionario de Autoridades* (Madrid: En la Imprenta de la Real de la Academia Española, 1726), vol. 1, v.

118. Prologue to *Diccionario de Autoridades*, vol. 1, xlviii. Also see Maria Águeda Moreno Moreno, "Modelos y métodos de la lexicografía etimológica: La 'maldición' de Babel hasta el tesoro (1611) de S. de Covarrubias," *Revista argentina de historiografía linguistica* 5, no. 1 (2013): 1–25.

119. See María José Martínez Alcalde, *Las ideas lingüísticas de Gregorio Mayans* (Valencia, 1992), 145–209; Enrique Jiménez Ríos, "El cultivo de la etimología en España durante el siglo XVIII," *Moenia* 9 (2003): 253–274.

120. Mayans y Siscar, *Orígenes de la lengua española*, 62.

121. Metcalf, *Linguistic Diversity and Relationship from Bibliander to Adelung*, 38–39.

122. Mayans y Siscar, *Orígenes de la lengua española*, 59–71. Regarding the words derived from Arabic, see Gregorio Mayans y Siscar, *Epistolario*, vol. 6, ed. Antonio Mestre Sanchis Valencia, 1977).

123. Gregorio Mayans y Siscar, *Conversación sobre el Diario de los Literatos en España* (1737), in *Obras completas*, vol. 3, ed. Antonio Mestre Sanchis (Valencia, 1984), 427.

CHAPTER THREE: Language and the Ancient History of the Americas

1. Alonso de Zorita, *Relación de la Nueva España*, ed. Ethelia Ruiz Medrano and José Mariano Leyva, 2 vols. (México, 2011), vol. 1, pt. 1, chap. 1, 133–135.

2. I have consulted both the manuscript of Zorita's *History of New Spain* and the modern edition. I cite the marginal summaries that Zorita included to guide the reader but are not

always in the modern edition. See Alonso de Zorita, *Historia de la Nueva España*, mss. IBIS RB II/59, BRP, fols. 1r–2v. Also see Genesis 2:18–20; and Toribio Benavente Motolinía, *Memoriales, o libro de las cosas de la Nueva España y los naturales a ella*, ed. Edmundo O'Gorman (México, 1971), pt. 1, chap. 1, 19.

3. *Pomar y Zurita*, ed. Joaquín García Icazbalceta (México, 1891), chap. 9, 92. Motolinía also remarks on the etymology of Michoacan. "Por lo cual la etimología e interpretación de su nombre le conviene muy bien y es conforme a su propiedad, como son todos los nombres de esta lengua." Motolinía, *Memoriales*, pt. 1, chap. 67, 280.

4. Georges Baudot, *Utopia and History in Mexico: The First Chroniclers of Mexican Civilization* trans. Bernard R. Ortiz de Montellano and Thelma Ortiz de Montellano (Niwot, CO, 1995), 491–526. In 1556 the Spanish Crown sought to more tightly regulate the publication and sale of books about the Americas by subordinating their censorship and licensing to the Council of Indies—the administrative body in charge of advising the king on all matters related to the governance of the American viceroyalties. After 1556, with a few notable exceptions, the Crown's policy brought about a shift in the dissemination of works concerning the Americas, many of which began to circulate mainly in manuscript form, among other reasons, to protect state secrets. See Stephen C. Mohler, "Publishing in Colonial Spanish America: An Overview," *Inter-American Review of Bibliography* 28, no. 3 (1978): 259–273, 262.

5. For the genesis of both projects under the cosmographer-chronicler Juan López de Velasco, see María M. Portuondo, *Secret Science: Spanish Cosmography and the New World* (Chicago, 2009), 212–223.

6. For a general review of lexicography and grammar in the American viceroyalties in the colonial period, see Otto Zwartjes, "Missionary Traditions in South America," and "Missionary Traditions in Mesoamerica," in *The Cambridge World History of Lexicography*, ed. John Considine (Cambridge, 2019), 555–596; Alan Durston, "Indigenous Languages and the Historiography on Latin America," *Storia della Storiografia* 67, no. 1 (2015): 51–65.

7. Anthony Pagden, "Dispossessing the Barbarian: Rights and Property in Spanish America," in *Spanish Imperialism and the Political Imagination: Studies in European and Spanish-American Social and Political Theory 1513–1830* (New Haven, CT, 1998), 13–36.

8. Rolena Adorno, *The Polemics of Possession in Spanish American Narrative* (New Haven, CT, 2008), 1–20.

9. Carlos Sempat Assadourian, "La despoblación indígena en Perú y Nueva España durante el siglo XVI y la formación de la economía colonial," in *Historia Mexicana* 38, no. 3 (January–March 1989): 419–453.

10. Francisco Cervantes de Salazar, *Crónica de la Nueva España*, ed. Manuel Magallón y Cabrera and Agustín Millares Carlo (Madrid, 1914), book 4, chap. 17, 301–302.

11. Cervantes de Salazar, *Crónica de la Nueva España* [1914], book 4, chap. 17, 301–302.

12. Although the *Chronicle* was never published, the seventeenth-century historian Antonio de Herrera y Tordesillas (1549–1625) used it extensively for his *General History of Castilians on the Islands and Mainland of the Ocean Sea* (c. 1615). I have also consulted the manuscript available in the BNE. See Fransisco Cervantes de Salazar, *Crónica de la Nueva España, su descripción, la calidad y temple de ella, la propiedad y naturaleza de los indios*, BNE, Ms. 2011. The first of the five books of *Chronicle* deal with the pre-Hispanic past, the remainder present a history of the conquest of Mexico. On the life and works of Cervantes de Salazar, see Diane M. Bono, *Cultural Diffusion of Spanish Humanism in New Spain: Francisco Cervantes de Salazar's "Diálogo de la Dignidad del Hombre"* (New York, 1991); Enrique González González, "A Humanist in the New World: Francisco Cervantes de Salazar (c.1518–75)," in *Neo-Latin and the Humanities: Essays in Honour of Charles E. Fantazzi*, ed. Luc Deitz, Timothy Kircher, and Jonathan Reid (Toronto, 2014), 235–258.

13. On Cervantes de Salazar's influences and friends, see Elena Pellús Pérez, "Entre el Renacimiento y el Nuevo Mundo: Vida y obras de Hernán Peréz de Oliva (1494?–1531)" (Madrid, 2015); William Atkinson, "Hernán Pérez de Oliva: A Biographical and Critical Study," *Revue hispanique*, no. 71 (1927): 309–483; Dolores Clavero, "Algunas cosas de Hernán Cortés y México: Una lectura humanista de la Segunda carta de relación de Hernán Cortés," *Revista canadiense de estudios hispánicos* 20, no. 2 (Winter 1996): 213–224.

14. Cervantes de Salazar, *Crónica de la Nueva España* [1914], book 1, chap. 17, 33.

15. Cervantes de Salazar, *Crónica de la Nueva España* [1914], book 1, chap. 17, 33–34.

16. Cervantes de Salazar, *Crónica de la Nueva España* [1914], book 3, chap. 20, 176.

17. Paolo Rossi, *The Dark Abyss of Time: The History of the Earth and the History of Nation from Hooke to Vico*, trans. Lydia G. Cochrane (Chicago, 1984), 238.

18. Alonso de Molina, *Aqui comiença un vocabulario en lengua castellana y mexicana, compuesto por el muy reverendo padre fray Alonso de: Guardian del convento de Sant Antonio de Tetzcuco dla orden de los frayles Menores* (Mexico City: Juan Pablos, 1555), prologue. This work was amplified and reprinted in 1571. See Alonso de Molina, *Vocabulario en lengua castellana y mexicana* (Mexico: En casa de Antonio de Spinosa, 1571).

19. Molina, *Aqui comiença un vocabulario*, prologue. For an interpretation of Molina's prologues, see Miguel Ángel Esparza Torres, "Los prólogos de Alonso Molina (c. 1514–1585): Destrucción de una ideología," *Revista de estudos ibéricos*, no. 2 (2005): 69–91; Esparza Torres, "Los prólogos de Alonso de Molina al *Vocabulario* (1555) y al *Arte* (1571)," *Filología y lingüística: Estudios ofrecidos a Antonio Quilis* 2 (2006): 1701–1718.

20. See, for instance, Yanna Yannakakis, "Making Law Intelligible: Networks of Translation in Mid-Colonial Oaxaca," in *Indigenous Intellectuals: Knowledge, Power, and Colonial Culture in Mexico and the Andes*, ed. Yanna Yannakakis and Gabriela Ramos (Durham, NC, 2014), 79–99.

21. Domingo de Santo Tomás, *Gramática o Arte de la lengua general de los indios de los reinos del Perú. Nuevamente compuesta por el maestro Fray Domingo morador en los dichos reynos* (Valladolid: Francisco Fernandez de Cordoua, 1560), prologue.

22. "Instrucion y memoria, de las relaciones que se han de hazer, para la descripcion de las Indias, que su Magestad manda hazer," in *Ordenanzas de la Hacienda Real (en Indias)*, BNE, Ms. 3035, 42r–53r. Also see Sylvia Vylar, "La trajectoire des curiosités espagnoles sur les Indes: Trois siècles 'd'interrogatorios' et 'relaciones,'" *Mélanges de la casa Velásquez* 6 (1970): 247–308.

23. On Juan de Ovando's life, see Stafford Poole, *Juan de Ovando: Governing the Spanish Empire in the Reign of Phillip II* (Norman, OK, 2004).

24. Alfredo Alvar Ezquerra, introduction to *Relaciones topográficas de Felipe II, Madrid*, ed. Alfredo Alvar Ezquerra, transcribed by María Elena García Guerra and María de los Angeles Vicioso Rodríguez, 4 vols. (Madrid, 1993), vol. 3, 29–45, 38.

25. Juan Pimentel, "The Iberian Vision: Science and Empire in the Framework of a Universal Monarchy, 1500–1800," *Osiris*, 2nd ser., 15 (2001): 17–30.

26. See Barbara Mundy, preface to *The Mapping of New Spain: Indigenous Cartography and the Maps of the "Relaciones Geográficas"* (Chicago, 1996), 61–133.

27. Francisco Javier Campos y Fernández de Sevilla, "Las *Relaciones topográficas de Felipe II*: Perspectivas de unas fuentes históricas monumentales sobre Castilla la Nueva en el siglo XVI," in *La ciencia en el Monasterio del Escorial actas del Simposium, 1/4-IX-1993*, ed. Francisco Javier Campos y Fernández de Sevilla, 2 vols. (San Lorenzo del Escorial, 1994), vol. 1, 381–430.

28. Cited in Campos y Fernández de Sevilla, "Las *Relaciones topográficas de Felipe II*," 389–390.

29. *Relaciones topográficas de Felipe II, Madrid*, vol. 4, 204.

30. See Joan-Pau Rubiés, "Instructions for Travellers: Teaching the Eye to See," *History and Anthropologie* 9, nos. 2/3 (2010): 139–90.

31. *Cuestionarios para la formación de las relaciones geográficas de Indias, siglos XVI/XIX*, ed. Francisco de Solano, Pilar Ponce, and Antonio Abellán (Madrid, 1988), 81.

32. Solano, Ponce, and Abellán, *Cuestionarios*, 82.

33. Solano, Ponce, and Abellán, *Cuestionarios*, 82.

34. Solano, Ponce, and Abellán, *Cuestionarios*, 82.

35. Patricia Seed, *Ceremonies of Possession in Europe's Conquest of the New World, 1492–1640* (New York, 1995).

36. See Gene Rhea Tucker, "Place-Names, Conquest, and Empire: Spanish and Amerindian Conceptions of Place in the New World" (PhD diss., University of Texas at Arlington, 2011).

37. *Relaciones geográficas de México siglo XVI*, ed. Rene de Acuña, 10 vols. (México, 1986), vol. 7, 32. Acuña notes that the glyph that represented the town was a curved mountain. He claims that the toponym might be shortened since the etymology means "place of Culhua."

38. Acuña, *Relaciones geográficas de México*, vol. 7, 37. Also see Francis Karttunen, "Itztapalli: Paving Stone / Flagstone," in *An Analytical Dictionary of Nahuatl* (Norman, 1992), 108.

39. Acuña, *Relaciones geográficas de México*, vol. 7, 113. Acuña notes that it "signifies lugar del juego de pelota."

40. Answers to the questionnaires in New Spain often were accompanied by maps, especially in towns where mendicant orders had established missions. Many were drawn by Indigenous illustrators and, as Barbara Mundy has shown, reveal local traditions of map making and of conceptualizing space. Mundy, *The Mapping of New Spain*, 61–134.

41. Marcos Jiménez de la Espada, *Relaciones geográficas de Indias, Peru*, 4 vols. (Madrid: Tip. de M. G. Hernández, 1881).

42. Jiménez de la Espada, *Relaciones geográficas de Indias, Peru*, vol. 1, 79–82. On the differences between the questionnaires of the viceroyalties of Peru and Mexico, see Eva Stoll, "La voz índigena en las relaciones geográficas del siglo XVI," *Cuadernos de la Alfal*, May 2019, 216–230.

43. Portuondo, *Secret Science*, 220–223.

44. Alvar Ezquerra, introduction to *Relaciones topográficas de Felipe II, Madrid*, vol. 3, 29. Also see Howard Cline, "The Relaciones Geográficas of the Spanish Indies, 1577–1648," in *Handbook of Middle American Indians*, 16 vols. (Austin, TX, 1972), vol. 12, 183–242, 188.

45. Ezquerra, *Relaciones topográficas de Felipe II, Madrid*, vol. 1, 43.

46. Ezquerra, *Relaciones topográficas de Felipe II, Madrid*, vol. 1, 103.

47. *Relaciones topográficas de los pueblos de España hechas por orden del Sr. Felipe II: Copiadas de las originales que existen en la Real Biblioteca de El Escorial y se pasaron a la Academia en virtud de orden de S.M. para sacar la copia*, RAH, Ms. 9-3954-60, fols. 150–159.

48. Ezquerra, *Relaciones topográficas de Felipe II, Madrid*, vol. 2, 707.

49. *Relaciones topográficas de los pueblos de España hechas por orden del Sr. Felipe II*, fols. 13v–18v.

50. Ezquerra, *Relaciones topográficas de Felipe II, Madrid*, vol. 2, 859–860.

51. See Ezquerra, introduction to *Relaciones topógraficas de Felipe II, Madrid*, vol. 3, 35–36; and José Miguel Morán Turina, *La memoria de las piedras: Anticuarios, arqueólogos y coleccionistas de antigüedades en la España de los Austrias* (Madrid, 2010), 201–221.

52. Abraham Ortelius, *Synonymia geographica sive populorum, regionum, insularum, urbium* [. . .] (Antwerp: Christophe Plantin, 1578).

53. Richard Kagan, *Urban Images of the Hispanic World 1493–1793* (New Haven, CT, 2000), 27–29.

54. Max Deardorff, "Republics, Their Customs and the Law of the King: Convivencia and Self-Determination in the Crown of Castile and Its American Territories 1400–1700," *Rechtsgeschichte / Legal History* 26 (2018): 162–199.

55. Solano, Ponce, and Abellán, *Cuestionarios*, 97–105.

56. Bernabé Cobo, *Fundación de Lima, escripta por el padre Bernabé Cobo de la Compañia de Jesús*, BCS, ms. 58-04-11, book 1, chap.1, fols. 7r–v.

57. Cobo, *Fundación de Lima*, book 1, chap. 4, fol. 23r.

58. Cobo, *Fundación de Lima*, book 1, chap. 4, fol. 33r–v.

59. Cobo, *Fundación de Lima*, book 1, chap. 4, fol. 34v.

60. Kathy Eden, *Friends Hold All in Common: Tradition, Intellectual Property and the Adages of Erasmus* (New Haven, CT, 2001), 4.

61. Ari Wesseling, "Dutch Proverbs and Expressions in Erasmus' Adages, Colloquies, and Letters," *Renaissance Quarterly* 55, no. 1 (Spring 2002): 81–147.

62. Desiderius Erasmus, *Collected Works of Erasmus: Adages Ii1 to Iv100*, trans. Margaret Mann Phillips, annotated by R. A. B. Mynors (Toronto, 1982), 14.

63. Erasmus, *Collected Works*, 16–17. George Huxley studied the ways Aristotle used proverbs as historical evidence of previous societies in his writings. For instance, in the *Politics*, he used a Naxian proverb to explain the rise of tyrants. See George Huxley, "Aristotle as Antiquary," *Greek, Roman and Byzantine Studies* 14, no. 3 (Fall 1973): 271–286.

64. Erasmus, *Collected Works*, 17.

65. Esteban de Garibay y Zamalloa, "Refranes Vacongados, Recogidos y ordenados por Esteban de Garibay y Zamalloa," in *Memorial histórico español: Colecció de documentos, opusculos y antigüedades que publica la Real Academia de la Historia* (Madrid, 1854), vol. 7, 630–659.

66. Juan de Valdés, *Diálogo de lengua*, ed. José Enrique Laplana (Barcelona, 2010), 125–126.

67. On the life of Hernán Núñez, see Helen Nader, "The Greek Commander' Hernán Núñez de Toledo, Spanish Humanist and Civic Leader," *Renaissance Quarterly* 31 (1978): 463–485.

68. Other contemporary collections include the Frenchman Charles Bovelles's *Proverbiorum vulgarium libri tres* (1531) and, from the Italian collection, *Opera quale contiene le dieci tavole de proverbi, sententie, detti, et modi di parlare che hoggidi da tutt'homo nel comun parlare d'Italia si usano* (Torino, 1535).

69. On the collection, see Abraham Madroñal, "*Los refranes o proverbios en romance* (1555), de Hernán Núñez Pinciano," *Rlit* 64, no. 127 (2002): 5–39; and introduction to Hernán Núñez, *Refranes o proverbios en romance*, ed. Louis Combet, Julia Sevilla Muñoz, Germán Conde Tarrío, and Josep Guia I Marín, 2 vols. (Madrid, 2001), vol. 1, ix–xii. On the prohibition and expurgation of Erasmus's sayings in the Catholic world and in Spain, see Antonio Serrano Cueto, "Los *Adagia* de Erasmo en el *Index Expurgatorius* de Amberes (1571): El alcance de la censura dirigida por Arias Montano," *Calamus Renascens* 1 (2000): 363–383.

70. Hernán Núñez, *Refranes o proverbios en romance* [. . .] (Salamanca: Juan de Canova, 1555).

71. "Clérico, frayle o judío, no lo tengas por amigo." "Del mar, la sal, de la muger, mucho mal." Núñez, *Refranes o proverbios en romance* [1555], 25r, 30r.

72. Núñez, *Refranes o proverbios en romance* [1555], 50v.

73. Hernán Núñez, *Refranes o proverbios en romance*, ed. Louis Combet, Julia Sevilla Muñoz, Germán Conde Tarrío, and Josep Guia I Marín, 2 vols. (Madrid, 2001), vol. 1, ix–xii.

74. On the *Philosophía vulgar* and Juan Mal Lara, see M. Bernal Rodríguez, *Cultura popular y humanismo: Estudio de la philosophía vulgar de Juan de Mal Lara* (Madrid, 1982);

Américo Castro, "Juan de Mal Lara y su filosofía vulgar," in *Hacia Cervantes* (Madrid, 1967), 167–209; Inmaculada Osuna Rodríguez, "Juan de Mal Lara, humanista y traductor," in *La philosophía vulgar de Juan Mal Lara: Facsimile edition*, ed. José J. Labrador Herraiz and Ralph A. DiFranco (México, 2012), 29–45.

75. Juan Mal Lara, *La Philosophia vulgar de Juan Mal Lara vecino de Sevilla* (Seville: En la calle de Sierpe. En casa de Hernando Diaz, 1568), 7.

76. Mal Lara, *Philosophia vulgar*, preambulos.

77. Mal Lara, *Philosophia vulgar*, preambulos.

78. Mal Lara, *Philosophia vulgar*, preambulos, 12. Also see Ángel Iglesias Ovejero, "La figura etimológica en la paremiología clásica," in *Estado actual de los estudios sobre el Siglo de Oro: actas del II Congreso Internacional de Hispanistas del Siglo de Oro*, ed. Manuel García Martín (Salamanca, 1993), vol. 2, 519–528.

79. See Howard F. Cline, "A Note on Torquemada's Native Sources and Historiographical Methods," *Americas* 25, no. 4 (April 1969): 372–386.

80. Juan Estalleras, "The College of Tlatelolco and the Problem of Higher Education for Indians in 16th-Century Mexico," *History of Education Quarterly* 2, no. 4 (December 1962): 234–243, 235; W. Michael Mathes, "Humanism in Sixteenth and Seventeenth Century Libraries of New Spain," *Catholic Historical Review* 82, no. 3 (July 1996): 412–435.

81. José Mallea-Olaetxe, "The Private Basque World of Juan Zumarraga, First Bishop of Mexico," *Revista de historia de América*, no. 114 (July–December 1992): 41–60; Joaquín García Icazbalceta, *Biografía de d. Fr. Juan de Zumárraga, primer obispo y arzobispo de Mejíco* (Madrid, 1929).

82. See Iraís Hernández Suárez, *El horizonte de enunciacion novohispano en Fray Andrés de Olmos* (México, 2008); Georges Baudot, "Fray Andrés Olmos y la penetración del luteranismo en México: Nuevos datos y documentos," *Nueva revista de filología híspanica* 40, no. 1 (1992): 223–232.

83. Judith M. Maxwell and Craig A. Hanson, *Of the Manners of Speaking That the Old Ones Had: The Metaphors of Andrés de Olmos in the TULAL Manuscript* (Salt Lake City, 1992), 5; Also see Nancy Farriss, *Tongues of Fire: Language and Evangelization in Colonial Mexico* (Oxford, 2018), chap. 6.

84. Jerónimo Mendieta cited in Ángel María Garibay, *Historia de la literatura náhuatl* [1953–1954] (México, 2000), 527–530.

85. The Franciscan chroniclers Mendieta and Juan Torquemada also used Olmos as a source in their own writings. See Maxwell and Hanson, *Of the Manners of Speaking That the Old Ones Had*, 7.

86. Maxwell and Hanson, *Of the Manners of Speaking That the Old Ones Had*, 1–52.

87. Victoria Ríos Castaño, "Translation Purposes and Target Audiences in Sahagún's *Libro de la Rhetorica* (c. 1577)," in *Missionary Linguistics V / Lingüistica Misionera V: Translation Theories and Practices*, ed. Otto Zwartjes, Klaus Zimmermann, and Martina Scharder-Kniffi (Amsterdam, 2012), 53–84, 57; Camilla Townsend, *Annals of Native America: How the Nahuas of Colonial Mexico Kept Their History Alive* (Oxford, 2016), 1–16, 17–55.

88. Zwartjes, "Missionary Traditions in South America," 559.

89. Jeanette Fravrot Peterson, "Rhetoric as Acculturation: The Anomalous Book 6," in *The Florentine Codex: An Encyclopedia of the Nahua World in Sixteenth-Century Mexico*, ed. Jeanette Fravrot Peterson and Kevin Terraciano (Austin, 2019), 167–183.

90. For a review on how scholars classify and count these speeches, see Thelma Suvillan, "The Rhetorical Orations, or Huehuetlatolli, Collected by Sahagún," in *Sixteenth-Century Mexico: The Work of Sahagún*, ed. Munro S. Edmonson (Albuquerque, 1974), 79–109.

91. Cited in Á. M. Garibay, *Historia de la literatura nahuátl*, 527–530.

92. On the stages of alphabetization of the Mexican language, see Elena Días Rubio and Jesús Bustamante García, "La alfabetización de le lengua Nahuatl," *Historiographia lingüistica* 11, nos. 1/2 (January 1984): 189–211.

93. Á. M. Garibay, *Historia de la Literatura Nahuátl*, 527–530; Kagan, *Clio and the Crown*, 184.

94. Bernardino de Sahagún, *Historia general de las cosas de Nueva España*, ed. Ángel María Garibay (México, 2016), book 1, prologue. On the censorship of Sahagún's work in 1577, see José Toribio Medina, *Historia de la imprenta en los antiguos dominios españoles de América y Oceanía*, ed. Guillermo Feliu Cruz (Santiago de Chile, 1958), vol. 1, 6–7.

95. Michael Mathes, *Santa Cruz de Tlatelolco: La primera biblioteca académica de las Americas* (México, 1982).

96. Sahagún, *Historia general*, book 1, prologue, 65.

97. On Calepinus and his uses, see Ann Blair, *Too Much to Know: Managing Scholarly Information before the Modern Age* (New Haven, CT, 2012), 48.

98. Peterson, "Rhetoric as Acculturation," 167–183.

99. David Boruchoff, "The Conflict of Natural History and Moral Philosophy in *De Antiquitatibus Novae Hispaniae* of Francisco Hernández," *Revista canadiense de estudios hispánicos* 17, no. 2 (1993): 241–258.

100. Peterson, "Rhetoric as Acculturation," 171.

101. Ríos Castaño, "Translation Purposes and Target Audiences in Sahagún's Libro de la Rhetorica (c. 1577)," 54.

102. Miguel León Portilla, *La filosofía Nahuátl estudiada en sus fuentes* (México, 1959), 7–26.

103. Sahagún, *Historia general*, book 6, chap. 41, 390. For this section I have also relied on the English translation of the Nahuatl text of Charles E. Dibble and Arthur Anderson. See *Florentine Codex: General History of the Things of New Spain*, trans. Charles E. Dibble and Arthur Anderson, 13 vols. (Santa Fe, NM, 1969), vol. 7, book 6, 224.

104. Sahagún, *Historia general*, book 6, chap. 41, 658. Also see Dibble and Anderson, *Florentine Codex*, vol. 7, book 6, 219.

105. Giuliano Gliozzi, *Adamo e il Nuovo Mondo: La nascita dell'antropologia come ideologia coloniale: Dalle genealogie bibliche alle teorie razziali (1500–1700)* (Florence, 1977), 16–17. Anthony Pagden, *The Fall of Natural Man: The American Indian and the Origins of Comparative Ethnology* (Cambridge, 1982), 10–56.

106. López de Gómara, *Historia general de las Indias*, 294r–294v.

107. See Gonzalo Fernandez de Oviedo, *Historia general y natural de las Indias, islas y tierra-firme del mar océano*, ed. D. José Amador de los Ríos, 4 vols. (Madrid, 1851–1855), vol. 1, book 2, chap. 3, 14–18.

108. Jorge Cañizares Esguerra, *How to Write the History of the New World* (Stanford, CA, 2001), 64–69.

109. Bruce Mannheim, "Lexicography of Colonial Quechua," in *Wörterbücher: Ein internationales Handbuch zur Lexikographie*, ed. F. J. Hausmann, O. Reichmann, H. E. Wiegand and L. Zgusta (Berlin, 1991), vol. 3, 2676–2684.

110. Wiebke Arndt, "Alonso de Zorita: Un funcionario colonial de la corona Española," in Zorita, *Relación de la Nueva España*, vol. 1, 17–58. Also see Benjamin Keene, *Life and Labor in Ancient Mexico: The Brief and Summary Relation of the Lords of New Spain by Alonso de Zorita* (New Brunswick, NJ, 1994); and Ralph H. Vigil, *Alonso de Zorita: Royal Judge and Christian Humanist, 1512–1585* (Norman, OK, 1987).

111. Benjamin Keene, "Reflections on Alonso de Zorita and His Brief and Summary Relation of the Lords of New Spain," in *Essays in the Intellectual History of Colonial Latin America* (Boulder, CO, 1998), 130–139.

112. Vigil, *Alonso de Zorita*, 241–264.

113. Zorita, *Relación de la Nueva España*, vol. 1, part 1, chap. 3, 145.

114. See Rolena Adorno, "Obedezco pero No Cumplo: Surviving Censorship in Early Modern Spain," *Textual Cultures* 13, no. 1 (Spring 2020): 29–74.

115. Jerónimo Román y Zamora, *Republicas del mundo divididas en XXVII libros*, Libro Primero de la Republica de los Indios Occidentales (Medina del Campo: Francisco del Canto, 1595), chap. 11, 138v.

116. Zorita, *Relación de la Nueva España*, vol.1, part 1, chap. 2, 139.

117. Zorita, *Relación de la Nueva España*, vol.1, part 1, chap. 2, 139. Pablo Nazareo was the Indigenous rector of the College of Santa Cruz de Tlatelolco. See Andrew Laird, "The Teaching of Latin to the Native Nobility in Mexico in the Mid-1500s: Contexts, Methods, and Results," in *Learning Latin and Greek from Antiquity to the Present*, ed. Elizabeth P. Archibald, William Brockliss, and Jonathan Gnoza (Cambridge, 2015), 118–135, 133.

118. Zorita, *Relación de la Nueva España*, vol.1, 111.

119. Diego Durán, *Historia de las Indias de Nueva España e Islas de la Tierra Firme* [1579], ed. José F. Ramirez, 2 vols. (México, 1867), vol. 1, chap. 1.

120. Jerónimo de Mendieta, *Historia eclesiástica Indiana*, ed. Joaquín García Icazbalceta, 2 vols. (México, 1870), vol. 1, book 2, chap. 32.

121. Zorita, *Relación de la Nueva España*, vol. 1, pt. 1, chap. 6, 184.

122. Icazbalceta, *Pomar y Zurita*, chap. 9, 94.

123. Zorita, *Relación de la Nueva España*, vol. 1, pt. 1, chap. 2, 142–143.

124. Zorita, *Relación de la Nueva España*, vol. 1, pt. 1, chap. 27, 307–309.

125. Zorita, *Relación de la Nueva España*, vol. 1, pt. 1, chap. 27, 307–309. I based this transcription on the manuscript, see Zorita, *Historia de la Nueva España*, mss. IBIS RB II/59, BRP, fol. 164v.

126. Exodus 15.

127. Zorita, *Relación de la Nueva España*, vol. 1, pt. 1, chap. 27, 307–309.

128. Questions about whether *mayeques* were a significant class, because the term rarely appears in the surviving corpus of Nahuatl documents of New Spain, have also been central in contemporary appraisals of Zorita's work. Most importantly, modern scholars have argued about the nature of the *calpolli* and whether it was in fact based on kinship ties. Benjamin Keen explains that Zorita's definitions probably originated from Francisco de las Navas, OFM, whose investigations took place in the Puebla-Tlaxcala region and not in the Valley of Mexico, where the seat of pre-Hispanic, and then Spanish, power stood. In regions like Texcoco, in the Valley of Mexico, the *calpolli* was dominated by the state. At the eve of the Spanish conquest, the *calpolli* had become an "administrative unit or ward, providing tribute, labor, and military service to the state." Even if it was ruled by hereditary elite families, the *calpolli*'s "original kinship basis" was either sharply eroded or no longer extant in these regions. Keene, "Reflections on Alonso de Zorita and His Brief and Summary Relation of the Lords of New Spain," 136–137.

129. On the accuracy of Zorita's explanations, see James Lockhart, *The Nahuas after the Conquest: A Social and Cultural History of the Indians of Central Mexico, Sixteenth through Eighteenth Centuries* (Stanford, CA, 1992), 97, 112.

130. Icazbalceta, *Pomar y Zurita*, 77.

131. Benjamin Keen, editor's introduction to Alonso de Zorita, *Life and Labor in Ancient Mexico: The Brief and Summary Relation of the Lords of New Spain* (Norman, OK, 1994), 1–78.

132. José Luis Egío, "From Castilian to Nahuatl, or from Nahuatl to Castilian? Reflections and Doubts about Legal Translation in the Writings of Judge Alonso de Zortia (1512–1585?)," *Rechtsgeschichte / Legal History* 24 (2016): 122–153.

133. Icazbalceta, *Pomar y Zurita*, 146.

134. Bartolomé de las Casas, *Apologética historia sumaria*, ed. Edmundo O'Gorman (México, 1967), book 3, chap. 45, 237–241.

135. "Las Ordenanzas de 1571 del Real y Supremo Consejo de las Indias: Texto facsimilar de la edición de 1585; Notas de Antonio Muro Orejón," in *Anuario de estudios americanos*, January 1957, 363–423, 373.

136. Icazbalceta, *Pomar y Zurita*, 184.

137. Egío, "From Castilian to Nahuatl, or from Nahuatl to Castilian?," 137.

138. Zorita explained how in the Psalms, the Egyptians are called a "barbarous people" because they were "unfaithful," even though "they created almost all sciences," as Plato and Aristotle claimed about mathematics and astronomy. The poet Martial referred to those ignorant of Latin and of Roman customs as barbarians. So, in this usage, barbarian meant foreign, not lacking civility. The Romans and the Greeks called the people of different languages barbarians. Numerous other authors, like Cardinal Adrian in his work on Latin and Gisbertius Longlius in his annotations to Plautus, confirmed this usage in ancient times. Saint Paul, in Corinthians, described how if he preached in a language foreign to the people that he was seeking to convert, he would be taken as a barbarian. Icazbalceta, *Pomar y Zurita*, 146–148.

139. Cabello Valboa was also a contemporary of Pedro Sarmiento de Gamboa. Sarmiento de Gamboa's *History of the Incas* (1572) sought to prove, in seventy chapters, that the Inca kings had been tyrants and were thus the illegitimate rulers of the Andes. The viceroy Francisco de Toledo commissioned Sarmiento de Gamboa to accompany him to Huarochirí, Jauja, Huamanga, and Cuzco in the 1570s to collect information about the Incas. On Sarmiento de Gamboa's theory of origin and how it followed Annius's genealogies, see Soledad González Díaz, "Genealogía de un origen: Túbal el falsario y la Átlantida en la *Historia de los Incas* de Pedro Sarmiento de Gamboa," in *Revista de Indias* 72, no. 225 (2012): 497–526.

140. Miguel Cabello Valboa, *Miscelánea Antártica*, al pío y curioso lector. Cabello de Valboa's work survives in two manuscript copies. One is located in the library of the University of Texas at Austin, and a subsequent copy is preserved in the New York Public Library. The critical edition is based on the Texas manuscript. For this section I have worked with both the critical edition and the University of Texas manuscript *Miscelanea Anthartica, donde se describe, el origen, de nuestros Indios Occidentales, deduzido desde Adan, y la Erection y principia del imperio de los Reyes Ingas de el Piru. Vidas y guerras que tu vieron: cosas notables q hicieron, computados los años de sus nascimientos y muertes, y de lo q por el Universso y va subcediendo; durantes sus edades y tiempos. Dirigido a Don Ferdo. de Torres y Portugal, conde del Villar, vissorrey gouer. y capn. general de estos reynos y prouincias del Piru: electo año -1584. Por Miguel Cabello Valboa clerigo presbitero del Arçobispado de los Reyes en el Piru. Natural de la Villa de Archidona en Andaluzia, Año -1586*, JGI 1946; Miguel Cabello Valboa, *Miscelánea Antártica*, ed. Isaías Lerner (Seville, 2011).

141. On the concept of *varietas* and its uses by Cabello Valboa, see Sonia V. Rose, "*Varietas Indiana*: Le cas de la *Miscelánea Antártica* de Miguel Cabello Valboa," *Bulletin de l'Institut français d'études andines* 30, no. 3 (2001): 413–425.

142. Cabello Valboa, *Miscelánea Antártica* [2011], al pío y curioso lector.

143. Cabello Valboa, *Miscelánea Antártica* [2011], part 1, chap. 6, 67–68. On the interpretation of this passage, see Isaías Lerner, "Teorías de Indios: Los orígenes de los pueblos del continente americano y la Biblia Políglota de Amberes (1568–1573)," *Colonial Latin American Review* 19, no. 2 (2010): 231–245; and María Portuondo, *The Spanish Disquiet: The Biblical Natural Philosophy of Benito Arias Montano* (Chicago, 2019), 178–179.

144. Cabello Valboa, *Miscelánea Antártica* [2011], part 2, chap. 6, 145–150.
145. Domingo de Santo Tomás, *Lexicón o Vocabulario de la lengua general del Perú llamada quichua* (Valladolid: Francisco Fernandez de Cordoua, 1560), "cazco," 84r.
146. Cabello Valboa, *Miscelánea Antártica* [2011], part 2, chap. 5, 139–144.
147. Cabello Valboa, *Miscelánea Antártica* [2011], part 3, chap. 3, 269–275.
148. Cabello Valboa, *Miscelánea Antártica* [2011], part 3, chap. 3, 269–275.
149. Jeremy Mumford, *Vertical Empire: The General Resettlement of Indians in the Colonial Andes* (Durham, NC, 2012), 34.
150. Cabello Valboa, *Miscelánea Antártica* [2011], part 3, chap 3, 269–275.
151. See Mirko Tavoni, "The 15th-Century Controversy on the Language Spoken by the Ancient Romans: An Inquiry into Italian Humanist Concepts of 'Latin,' 'Grammar,' and 'Vernacular,'" *Historiographia Linguistica* 9, no. 3 (January 1982): 237–264. Also see Frederic Clark, "*Antiquitas* and the *Medium Aevum*: The Ancient/Medieval Divide and Italian Humanism," in *Remembering the Middle Ages in Early Modern Italy*, ed. Lorenzo Pericolo and Jessica N. Richardson (Turnhout, 2015), 19–41.
152. Toon Van Hal and John P. Considine, "Introduction: Classifying and Comparing Languages in Post-Renaissance Europe," *Language & History* 53 (2010): 63–69.
153. Sahagún, *Historia general*, book 10, 562–568.
154. Sahagún, *Historia general*, book 10, 562–568.
155. *Huehuehtlahtolli: Testimonios de la antigüa palabra* [1600], facsimile ed., trans. Librado Silva Galeana, introductory study by Miguel León-Portilla (México, 1988), prologue dedicated to the Licenciado Antonio Maldonado.
156. *Huehuehtlahtolli*, prologue.
157. *Huehuehtlahtolli*, prologue.
158. *Huehuehtlahtolli*, prologue.
159. *Huehuehtlahtolli*, A5.
160. Antonio de León Pinelo, *Epitome de la biblioteca oriental i occidental, nautica i geografica* (Madrid: Juan Gonzalez, 1629), xxxviii.
161. Bernardo Jose de Aldrete, *Del origen y principio de la lengua castellana o romance que oi se usa en España* (Rome: Carlo Wllietto, 1606), book 1, chap. 22, 145–148.
162. Aldrete, *Origen*, book 1, chap. 22, 145–148.
163. Aldrete, *Origen*, book 1, chap. 22, 145–148.

CHAPTER FOUR: Language and the Secrets of Nature

1. On the history of the alliance among Texcoco, Tenochtitlan, and Tlacopan, see Camilla Townsend, *Fifth Sun: A History of the Aztecs* (Oxford, 2019), 45–46. On the creation of Nezahualcoyotl's image as a lawmaker and poet in postconquest traditions, see Jongson Lee, *The Allure of Nezahualcoyotl: Pre-Hispanic History, Religion, and Nahua Poetics* (Albuquerque, 2008), 1–16. Throughout this chapter I have made use of the *Gran diccionario náhuatl* of the Universidad Nacional Autónoma de México (https://gdn.iib.unam.mx). I have also consulted the printed versions of the following dictionaries: Francis Karttunen, *An Analytical Dictionary of Nahuatl* (Norman, OK, 1992); Joe R. Campbell, *A Morphological Dictionary of Classical Nahuatl: A Morpheme Index to the Vocabulario en lengua mexicana y castellana of Fray Alonso de Molina* (Madison, WI, 1985); and Alonso de Molina, *Arte de la lengua mexicana y castellana compuesta por el muy R.P. Fray Alonso De Molina, de la orden de señot Sant Francisco, de nueuo en esta segunda impression corregida, emendada y añádida, mas copiosa y clara que la Primera* (México: En casa de Pedro Balli, 1576).
2. Francisco Hernández, *Antigüedades de la Nueva España*, ed. and trans. Ascensión H. de León Portilla (Madrid, 1986), book 2, chap. 13, 132–133. I have also consulted the facsimile

of the manuscript; see Francisco Hernández, *De antiquitatibus Novae Hispaniae* (México, 1926).

3. Hernández, *Antigüedades* [1986], chap. 13, 133.

4. Hernández, *Antigüedades* [1986], chap. 14, 134.

5. Hernández, *Antigüedades* [1986], chap. 13, 133.

6. Francisco Hernández, *Obras completas de Francisco Hernández*, 8 vols., ed. Germán Somolinos D'Ardois (México, 1959–1985), vol. 3: *Historia Natural de la Nueva España*, chap. 213.

7. Gianna Pomata and Nancy G. Siraisi, introduction to *Historia: Empiricism and Erudition in Early Modern Europe*, ed. Gianna Pomata and Nancy G. Siraisi (Cambridge, MA, 2005), 7.

8. "The Instructions of Philip II to Francisco Hernandez (January 11, 1570)," in *The Mexican Treasury: The Writings of Dr. Francisco Hernández*, ed. Simon Varey, trans. Rafael Chabrán, Cynthia L. Chamberlin, and Simon Varey (Stanford, CA, 2000), 46. Also see Ernesto Schafer, "Los protomedicatos en Indias," *Anuario de Estudios Americanos*, no. 1 (1946): 1029–46.

9. Jesús Bustamante García, "La empresa naturalista de Felipe II y la primera expedición científica en suelo americano: La creación del modelo expedicionario renacentista," in *Felipe II, 1527–1598: Europa y la Monarquía Católica*, ed. José Martínez Millán (Madrid, 1998), 39–59.

10. For a history of the missionary orders in New Spain, see Georges Baudot, *Utopia and History in Mexico: The First Chroniclers of Mexican Civilization (1520–1569)*, trans. Bernard R. Ortiz de Montellano and Thelma Ortiz de Montellano (Niwot, CO, 1995).

11. James Lockhart, *The Nahuas after the Conquest: A Social and Cultural History of the Indians of Central Mexico, Sixteenth through Eighteenth Centuries* (Stanford, CA, 1992), 14–58 and 427–451. On the adoption of the Roman alphabet by Indigenous scholars and scribes, see Camilla Townsend, *Annals of Native America: How the Nahuas of Colonial Mexico Kept Their History Alive* (Oxford, 2016), 15–54; and Nancy Farriss, *Tongues of Fire: Language and Evangelization in Colonial Mexico* (Oxford, 2018), 83–110.

12. On population decline in Mexico City, see Noble David Cook, *Born to Die: Disease and New World Conquest, 1492–1650* (New York, 1998). Hernández himself wrote about the illness that spread in different parts of New Spain and that the natives called the *cocoliztli*. See "On the Illness in New Spain in the Year 1576, Called the *Cocoliztli* by the Indians," in Hernández, *The Mexican Treasury*, 83–84.

13. Alvarez López Enrique, "El Dr. Francisco Hernández y sus comentarios a Plinio," *Revista de Indias* 3, no. 2 (1942): 251–290.

14. José Luis Benítez Miura, "Cartas inéditas dirigidas a Felipe II por el protomédico de Indias doctor Francisco Hernández desde la ciudad de Méjico, 1571–1575," *Anuario de estudios americanos* (Seville) 7 (January 1, 1950): 367–409, 400.

15. In his will, Hernández mentions how his works were also translated into Nahuatl. These translations do not survive, as they most likely perished in the Escorial fire of 1671. See Agustín Barreiros, "El testamento del Doctor Francisco Hernández," *Boletin de la Real Academia de la Historia* 94 (1929): 475–497.

16. Geoffrey Parker, *The World Is Not Enough: The Imperial Vision of Philip II of Spain*, Charles Edmondson Historical Lectures (Waco, TX, 2001), 10–13.

17. Carla Rahn Phillips, "Twenty Million People United by an Ocean: Spain and the Atlantic World beyond the Renaissance," *Renaissance Quarterly* 62 (2009): 1–60, esp. 33.

18. "Quien, pues, creyera que estava un tan grande mundo como este, distante tantos millares de leguas que por el oçeano se caiman al nuestro, gentes nuevas, nuevos ritos y religion y habitaçion debaxo de la linea equinoccial, contra el pareçer de tantos y tan graves authores? quien tanta diversidad de cielo y suelo? y en el plátas y animals tan admirables por no dezir monstruosos, y tanta copia de muy fertiles minerales? Y que se havia de caminar a

el, tan osada y diestramente en confiança de tan mudables vientos y tan fragiles vasos?" Francisco Hernández, *Historia natural libros vii y viii*, BNE, Ms. 2864, fol. 7r.

19. "Discurre aqui de pasada por las coberturas de los animales; porque, segun se entiende de díversos lugares de Aristóteles y deste mismo author, . . . como las salamanque-sas, lagartos, tortugas y serpientes; o de espinas como el erizo y puerco espin; o de vello, como las cabra; o de vello y espinas como el hoitztlacuatzin." Hernández, *Historia natural libros vii y viii*, BNE, Ms. 2864, fol. 3r ("hoitztlacuatzin" is inserted).

20. Two versions of this book survive. In the more polished version, the copyist forgot to insert the name *hoitztlacuatzin*, which is later added. Francisco Hernández, *Historia natural libros vii, x, xi y xii*, BNE, Ms. 2780, fol. 3v.

21. See Juan Pimentel, "The Iberian Vision: Science and Empire in the Framework of a Universal Monarchy, 1500–1800," *Osiris* 15 (2001): 17–30.

22. Pliny, *Natural History*, vol. 1, trans. Horace Rackham, Loeb Classical Library 330 (Cambridge, MA, 1938), 1.3.1.

23. "(Conquistadora del Mundo) de esto fue pronostico la cabeça humana, que segun cuenta Livio y Dionisio Alicarnaseo queriendo Tarquino Superbo levantar el capitolio los que abriã los cimientos del templo le hallarõ. La cual dixerõ los adivinos significar q Roma havia se der cabeça del mundo, como despues lo fue, conquistadas por ella (segun el presente Plinio dize) las mas de las naciones." Francisco Hernández, *Historia natural libros i, ii, iii, iv, y v*, BNE, Ms. 2869, fols. 79r–79v; and Hernández, *Historia natural libros i, ii, iii*, BNE, Ms. 2862, fols. 242r–242v.

24. David A. Boruchoff, "New Spain, New England, and the New Jerusalem: The 'Translation' of Empire, Faith, and Learning (Translatio Imperii, Fidei Ac Scientiae) in the Colonial Missionary Project," *Early American Literature* 43, no. 1 (2008): 5–34; and J. H. Elliott, "A Wider World," in *Spain, Europe & the Wider World, 1500–1800* (New Haven, CT, 2009), 36.

25. On the relationship between the Pliny's *Natural History* and Rome's imperial outlook, see Mary Beagon, *The Elder Pliny on the Human Animal: Natural History*, book 7 (New York, 2005), 1–58; Beagon, *Roman Nature: The Thought of Pliny the Elder* (Oxford, 1992); Sorcha Carey, *Pliny's Catalogue of Culture: Art and Empire in the Natural History* (Oxford, 2003); Trevor Morgan Murphy, *Pliny the Elder's Natural History the Empire in the Encyclopedia* (Oxford, 2004); Aude Doody, "Pliny's Natural History: Enkuklios Paideia and the Ancient Encyclopedia," *Journal of the History of Ideas* 70, no. 1 (2009): 1–21.

26. *Medicinas, drogas y alimentos vegetales del Nuevo Mundo: Textos e imágenes españolas que los introdujeron en Europa*, ed. José M. López Piñero, José Luis Fresquet Febrer, María Luz López Terrada, and José Pardo Tomás (Madrid, 1992).

27. The dogs that aid in the conquest appear in Gonzalo Fernández de Oviedo, *Historia general y natural de las Indias, islas y tierra-firme del mar océano*, ed. D. José Amador de los Ríos, 4 vols. (Madrid: Imprenta de la Real Academia de la Historia, 1851–1855), vol. 1, book 2, chap. 13, published first in 1535 and widely read all over Europe. Also see Ricardo Piqueras, "Los perros de la guerra o el 'canibalismo canino' en la conquista," *Boletín americanista* 56 (2006): 186–202.

28. "Y aun me conto Bernal Perez de Vargas hombre de gran bõdad y virtud y varia erudicion, mayormente en lẽguas y mathematicas, que en el descubrimiento que se hizo de Cybola al norte de nueva hespaña y casi al noreste de la Florida, topo el exército muchas recuas de perros, de los cuales cada uno lleva carga de cuatro çelemines de Hespaña de tlaolli, y que nosotros llamamos maiz." Hernández, *Historia natural libros vii y viii*, BNE, Ms. 2864, fol. 233r.

29. "A la sazon que esto escrivia me conto un soldado de peru que, en la provincia de Homagua, hai unas aves tan grandes que arrebatan los indios en las uñas y se los van

comiendo y despedaçando por el aire. Y descubren con ellas el oro que hai en grande abundançia por aquella tierra, y encima hazen nidos y sacan sus crias, y añadio el modo que tienen los de la tierra para defenderse dellas. Dezia tambien haver visto algunas de sus plumas cada una de las quales eran tan gruessa como un mediano braço. Si estos fuessen griphos o bueytres o otra nueva casta de aves yo no sabia determinarlo, ni obligo a nadie a que crea dello mas de lo que quisiere: aunque no me pareçio cosa indigna de advertirse en este lugar, entre otras monstruosidades que sabemos llevar cada dia de esas tierras." Hernández, *Historia natural libros vii y viii*, BNE, Ms. 2864, fol. 15v–16r. The previous draft of this book contains an interesting anecdote about what could have possibly been the gryph's claw, although it has been crossed out: "Yo he visto aqui en Toledo en poder de un ciudadano . . . una uña maçica de tamaño de un . . . grande de toro . . . que dize ser de gripho. A mi no me consta de su fiera pero eso oso affirmarse, que si no es gripho, no sera a lo menos animal inferior al que nos describen los antiguos por el en grandeza y fiereza, juzgando como dizen por la uña al leon." Hernández, *Historia natural libros vii, x, xi y xii*, BNE, Ms. 2870, fols. 17r–17v. Xavier Loya believes that perhaps Hernández was reproducing the first reports of the Andean condor. See Xavier Loya, "Natural Medicine and Herbal Medicine in Sixteenth-Century America," in *Science in Latin America: A History*, ed. Juan José Saldaña (Austin, TX, 2006), 36–39. On private collections of natural and artificial wonders in Seville and their American objects, see José Miguel Morán Turina and Fernando Checa Cremades, *El coleccionismo en España: De la cámara de maravillas a la galería de pinturas* (Madrid, 1985), 94–95.

30. Hernández, *Obras completas*, vol. 3, 380.

31. Hernández, *Obras completas*, vol. 4, 304.

32. "Y entre tanto persuadamosnos que la vida inventa cada dia y halla cosas nuevas las cuales a veces son de tal suerte que ello no recibe agravio, antes gloria la antiguedad." Francisco Hernández, *Historia natural libros ix, x, y xi*, BNE, Ms. 2865, fol. 217r.

33. Germán Somolinos D'Ardois, "Plinio, España y la época de Hernández," in *Historia natural de Cayo Plinio Segundo, Trasladada y anotada por el Doctor Francisco Hernández, Obras completas* (México, 1966), vol. 4, ix–xxiii.

34. María del Carmen Serrano, in her study of Álvar Gómez de Castro, reproduces a letter of 1552 in which the humanist describes to Ambrosio de Morales these meetings: "Pues a los óptimos varones y al mismo tiempo doctísimos médicos compañeros míos en el estudio de Plinio, les he prometido el oro y el moro, y según pienso no nos engaña la esperanza. Sobre nuestros trabajos te escribiré en otra ocasión." See María del Carmen Vaquero Serrano, *El maestro Álvar Gómez: Miografía y prosa inédita* (Toledo, 1993), 129; Vaquero Serrano, *En el entorno del maestro Álvar Gómez, Pedro del Campo, María de Mendoza, y los Guevara* (Toledo, 1996), 89–101.

35. Arantxa Domingo Malvadi, *Bibliofilia humanista en tiempos de Felipe II: La biblioteca de Juan Páez de Castro* (Salamanca, 2011), 525.

36. "Passo los que creen derrogar estos trabajos, estar Plinio puesto en lĕgua Ytaliana, a los q les (si gustan desgañarse) ruego depongan aquella voluntad, y si de ser desengañados, traigã a la memoria que ya que este en Italiano, no lo esta en hespañol, y si ynterpretado, no illustrado, mayormente q la de Landino no es traslazion sino confusion." Hernández, *Historia natural libros i, ii, iii, iv, y v*, BNE, Ms. 2869, fols. 9r–9v; Hernández, *Historia natural libros i, ii, iii*, BNE, Ms. 2862, fols. 6r–6v. Also see Pliny the Elder, *Historia naturale di latino in volgare tradotta per Christophoro Landino* (Venetia: Gabriel Jolito di Ferrarii, 1543); and Rowan Cerys Tomlinson, "'Plusieurs choses qu'il n'avoit veuës': Antoine Du Pinet's Translation of Pliny the Elder (1562)," *Translation & Literature* 21, no. 2 (2012): 145–161.

37. "Biense vee estar este lugar depravado. No quise castigarle, mas pondre las palabras de Theophrasto de donde Plinio saco las suyas quales quiera aquellas fueron, de donde aqui en

quiera podra restituirle y emendarle, que son el quinto libro de las causas de la plantas. Cap. 13." Francisco Hernández, *Historia natural libros xvii, xviii, xix, y xx*, BNE, Ms. 2867, fol. 74v.

38. Anthony Grafton, *Bring Out Your Dead: The Past as Revelation* (Cambridge, MA, 2001), introduction, 5–15; Brian W. Ogilvie, *The Science of Describing: Natural History in Renaissance Europe* (Chicago, 2006), 82–138.

39. Ogilvie, *The Science of Describing*, 116.

40. Guillaume Rondelet, *Libri de piscibus marinis, in quibus verae piscium effigies expressae sunt*, 2 vols. (Lyon: apud Matthiam Bonhomme, 1554), vol. 1, book 16, 453–458.

41. "Llaman los indios manati una speçie destos, distincta de las dos que pinta Rondolethio, lo cual se arguye, aliende de la similitud de sus forma, de criar sus hijos a los pechos, de la piedra que ambos crian en la cabeça y de otras cosas en que se hallan ser semejantes. Ignoraron esto los modernos, como la figura del pesce reverso, el cual por su estraña hechura damos pintado, segun le vimos pegado a un tiburon y en el estomago de otro pasando a estas Indias de n. Hespaña." Hernández, *Historia natural libros ix, x, y xi*, BNE, Ms. 2865, fol. 33r.

42. "Hallase en el oceano otra especie de bezerro que, aunque realmente lo es por la grande conveniencia que tiene con este que hemos hablado es en algo diferente. Mucho se engaña Bellonio tiniendo por bezerro de los antiguos, aquel cuyo retracto se ve en su libro de aquatilibus debuxado segun podra ver el que le cotejare con las antiguas descripciones." Hernández, *Historia natural libros ix, x, y xi*, BNE, Ms. 2865, fol. 33v. See Pierre Belon, *Petri Bellonii Cenomani de aquatilibus* (Paris: Apud C. Stephanum, 1553), book 1, chap. 4, 19–21.

43. Hernández, *Obras completas*, vol. 3, book 5, chap. 51, 401.

44. Ogilvie, *The Science of Describing*, 82–138.

45. "En las Indias Occidentales me dizen haverlas grandissimas, pero destas daremos relacion con el favor divino en la historia natural que, por mandado del muy invictisimo Philippo segundo, señor nuestro, tengo de ir a aquellas partes a scrivir, donde si nro Señor fuere servido se tractara de todo lo que Plinio en esta suya del Viejo Mundo scrive." Hernández, *Historia natural libros vii y viii*, BNE, Ms. 2864, fol. 153v.

46. Ogilvie, *The Science of Describing*, 94.

47. "(Murcielago) Tomo este animal el nombre en Griego y en Latin del tiempo en que sale a buscar de comer, llamandose vespertilo . . . y en español y otras lenguas, de la forma de raton y corta vista que tiene por que se dize murcielago, que es tanto como raton, o mur ciego, y rat-penat de otros, que raton alado (por alas)." Hernández, *Historia natural libros ix, x, y xi*, BNE, Ms. 2865, fol. 174r.

48. "(Phiseter) Palabra griega por la cual se significa a esta nacion el pescado dicho de los latinos flator y de los italianos capidolio y de los Hespañoles bufeo. La causa destos nombres es la muchedumbre de agua, que arroja por la fistula o canal, que en el es mas ancha y de mayor amplitud, que en todos los otros pescados que el vulgo llama ballenas." Hernández, *Historia natural libros ix, x, y xi*, BNE, Ms. 2865, fols. 9r–9v.

49. Gessner divided each of his entries on animals into eight sections, labeled A to H. Section A was devoted to compiling all of an animal's names in ancient and modern languages. Section H was devoted to philological questions. See Laurent Pinon, "Conrad Gessner and Historical Depth of Renaissance Natural History," in *Historia: Empiricism and Erudition in Early Modern Europe*, ed. Gianna Pomata and Nancy G. Siraisi (Cambridge, 2005), 248–251.

50. William B. Ashworth, "Emblematic Natural History of the Renaissance," in *Cultures of Natural History*, ed. J. A. Secord, N. Jardine, and E. C. Spary (Cambridge, 1996), 17–37, 18.

51. Mary J. Carruthers, "Etymology: The Energy of a Word Unleashed," in *The Craft of Thought: Meditation, Rhetoric, and the Making of Images, 400–1200* (Cambridge, 1998), 155–160.

52. Paolo Rossi, *Logic and the Art of Memory: The Quest for a Universal Language*, trans. Stephen Clucas (Chicago, 2000), 8–11.

53. Earline Jennifer Ashworth, "Traditional Logic," in *The Cambridge History of Renaissance Philosophy*, ed. Charles B. Schmitt and Quentin Skinner (Cambridge, 1998), vol. 1, 143–172.

54. Juan Luis Vives, "Tratado del Alma," in *Obras completas de Juan Luis Vives*, trans. Lorenzo Riber, 2 vols. (Madrid, 1947), vol. 2, book 2, chap. 6, 1205–1205.

55. Cited in Francisco Javier Perea Siller, "Los límites de la arbitrariedad linguistica en Vives, Huarte de San Juan y el Brocense," in *Estudios linguisticos y literarios in memoriam Eugenio Coseriu (1921–2002)*, ed. María Luisa Calero Vaquera and Fernando Riveras Cárdenas (Córdoba, 2004), 327–345, 332.

56. Isidore of Seville, *The Etymologies of Isidore of Seville*, trans. Stephen A. Barney, W. J. Lewis, J. A. Beach, and Oliver Berghof (Cambridge, 2010), book 9, chap. 1, 190.

57. Juan Luis Vives, *Los comentarios de Juan Luis Vives a "La ciudad de Dios" de San Agustín*, trans. Rafael Cabrera Petit (Valencia, 2000), chaps. 11, 6.

58. "Y porq en esta parte deseara alguno saber, qual fue la primera lengua ē que hablaron los hombres, dire mi pareçer brevemente allegandome al de Sancto Agustin en el capitulo XI del libro XVI. de la ciudad de Dios, do quiere que la lengua hebraica fuese aquella en que primero se hablase, porque antes de la confusión de Babilonia todos usavan de un lenguaje, como aparece en el capítulo XII del Genesis, y este seria el mas excellente y quedaria en los buenos y que no cõspiraron en el desvariado edificio de la torre, ansi como se confundio en los malos por razon de castigo. Uno de los justos fue Heber, de que quien la lengua hebraica tomo el nombre, y ansi pareçe que su lengua fuesse la primera y se salvasse en el sin corrupcion alguna." Hernández, *Historia natural libros vii y viii*, BNE, Ms. 2864, fols. 8r–8v; Hernández, *Historia natural libros vii, x, xi y xii*, BNE, Ms. 2870, fols. 9r–9v. Also see Augustine of Hippo, *The City of God*, trans. Marcus Dods, DD (New York, 1950), 536–37.

59. See Vivien Law, *The History of Linguistics in Europe from Plato to 1600* (Cambridge, 2003), 104; Clifford Ando, "Augustine on Language," *Revue de études Augustiniennes* 40 (1994): 45–78.

60. Marie Luce Demonet, "Renaissance étymologiques," in *Lexique 14: L'étymologie de l'antiquité à la renaissance*, ed. Claude Buridant (Lille, 1998), 57–67, 58–59.

61. John M. Lenhart, *Language Studies in the Franciscan Order: A Historical Sketch*, Franciscan Studies 5 (New York, 1926); See Varro, *De lingua latina*, trans. Roland G. Kent, Loeb Classical Library, 333–334, 2 vols. (Cambridge, MA; London, 1977); Daniel J. Taylor, "Varro and the Origins of Latin Linguistic Theory," in *L'hèritage des grammairiens latins de l'Aniquitè aux Lumièrs*, ed. Irène Rosier (Louvain, 1988), 37–48.

62. "En el capitulo catorce del libro diez y seis de Rhodigino podras ver algo a proposito de la fuerça que tienen las palabras pronunciadas para hazer mudança y effecto en las cosas." Hernández, *Historia natural libros vii y viii*, BNE, Ms. 2864, fol. 163r. See Lodovicus Caelius Rhodiginus, *Lodouici Caelij Rhodigini Lectionum Antiquarum Libri XXX* (Basel: Per Ambrosium et Aurelium Frobenios fratres, 1566), 417–420.

63. Demonet, "Renaissance étymologiques," 57.

64. Demonet, "Renaissance étymologiques," 57–58; Allison Coudert, "Some Theories of a Natural Language from the Renaissance to the Seventeenth Century," in *Magia naturalis und die Entstehung der modernen Naturwissenschaften: Symposion der Leibniz-Gesellschaft, Hannover, 14. und 15. November 1975*, ed. Albert Heinekamp Redaktion and Dieter Mettler (Wiesbaden, 1978); John Considine, *Dictionaries in Early Modern Europe: Lexicography and the Making of Heritage* (Cambridge, 2008), 19–55.

65. For a full account of Gessner's efforts, see George J. Metcalf, "Gessner's Views on the Germanic Languages," in *On Language Diversity and Relationship from Bibliander to Adelung*, ed. T. Van Hal and R. Van Rooy (Amsterdam, 2013), 77–84, 81.

66. For the challenges of communicating about specific animals and plants before the development and implementation of Linneaus's binomial system, see Dániel Margoscy, "'Refer to Folio and Number': Encyclopedias, the Exchange of Curiosities, and Practices of Identification before Linnaeus," *Journal of the History of Ideas* 71, no. 1 (January 2010): 63–89.

67. "Llamose ansi por razon de su grandiossisima ligereza: porque tigre en lengua de Armenia quiere dezir saeta, y por la misma causa el rio Tigris uno de los quatro que la scriptura testifica salir del paraiso terrenal que son, Indomilo, Euphrates, y Tigris, dizen haverse llamado de esta manera: aunque a otros agrada mas, ser la occasion de su nombre la velocidad y vehemencia de esta furia, a quien asemeja. De los demas nombres que en otras lenguas tienen no dire alguno porque pienso (mediante dios) poner al fin de este libro una tabla que contenga por la orden del alphabeto los nombres de las plantas animales y minerales las mas de las lenguas y ansi cumplir con las mas de las naciones, por no embaraçar el texto de mis comentarios con cosas semejantes, que parecen derrogar la authoridad de la obra." Hernández, *Historia natural libros vii y viii*, BNE, Ms. 2864, fols. 168r–168v.

68. "Muchos hay destos en Hespaña, mayormente en esta corte del rey Philippo Segundo, nro. señor, y por eso no trahire lugares de autores que pertenezcan a su descripcion, pues de sus medicinas y nombres como tambien haremos en los demas animales. . . . Lease tambien la descripcion de aquel animal o monstruo, semejante a puerco espin mas que a otro animal alguno que el año de mill y quinientos cincuenta se truxo por diversas partes de hombres que ganavan de comer a mostrarle, en Conrado Gesnerio, donde tracta del puerco espin, porque aunque le descrive Cardano en su libro de Subtilitate esta vedado. Yo no le decrivo ansi porque le vido mucha gente entonçes como por que le descriven estos authores que he allegado y me cumple porcuar en obra tan larga de ser breve." Hernández, *Historia natural libros vii y viii*, BNE, Ms. 2864, fols. 217v–218r.

69. Hernández, *Obras completas*, vol. 3, 299; Karttunen, *An Analytical Dictionary of Nahuatl*, 258.

70. See Conrad Gessner, *Conradi Gesneri Historiae animalium*, 5 vols. (Zürich: C. Froschouerum, 1551), 156.

71. "Las quales razones son tan flacas que yo no le aconsejara que por ellas juzgara tan cosa mayormente en bestias que nunca vido . . . lo del nombre tiene poca fuerça, como estos se topen a cada paso en diversas lenguas significando diversisismas cosas, mayormente que pudo ser el imponedor algun paniaguado de este pareçer." Hernández, *Historia natural libros vii y viii*, BNE, Ms. 2864, fols. 215r–215v.

72. Katharine Park, "Observation in the Margins, 500–1500," *Histories of Scientific Observation*, ed. Lorraine Daston and Elizabeth Lunbeck (Chicago, 2011), 15–44, esp. 18.

73. Park, "Observation in the Margins," esp. 18 and 35.

74. Juan Mal Lara, preambles to *La philosophia vulgar de Ioan Mal Lara vecino de Sevilla* (Seville: En la calle de Sierpe. En casa de Hernando Diaz, 1568), preamble 8, sig. b1v–b2r.

75. Gianna Pomata, "Observation Rising: Birth of an Epistemic Genre, 1500–1650," in Daston and Lunbeck, *Histories of Scientific Observation*, 45–80, esp. 49.

76. Ogilvie, *The Science of Describing*, 15.

77. José Luis Checa Cremades, "Impresos de *Historia natural* en la Biblioteca del Monasterio de El Escorial: Hacia una nueva síntesis explicativa," *Boletín de la ANABAD* 39, nos. 3/4 (1989): 549–564.

78. "(Persas) Vinieron con el rei Nabuchadnasar segundo de este nombre persas y chaldeos, segun testifican Josepho y Strabon, y aun, como dizen algunos tambien hebreos: Y ansi quieren haver dado nombres a no pocas ciudad de Hespaña, que duran hasta el dia de hoy, como Açeca y Escalona, Yepes, Maqueda y otros semejantes, los quales nombres fueron

primero de ciudades de Palestina." Hernández, *Historia natural libros i, ii, iii*, BNE, Ms. 2862, fol. 249v.

79. Adam Beaver, "Nebuchadnezzar's Jewish Legions: Sephardic Legends' Journey from Biblical Polemic to Humanist History," *After Conversion: Iberia and the Emergence of Modernity*, ed. Mercedes García-Arenal (Leiden, 2016), 21–65.

80. "Fundaron estos mismos la ciudad de Vertobriga; hallose a par de Frexenal (segun me dixo Arias Montano, varon excellente en theologia, lenguas y varia erudicion, el qual assiste al presente a la impression de la biblia quadrilingue, que dizen del cardinal en Flandes por mandado del invictissimo phillipo.2. Sro. No. sin grande comodo de la religion christiana) una piedra donde esta escripto Republica vertodeicensis y aunque difiere mucho la graduacion de Ptolomeo, de la que oy responde a Frexenal, no se debe dar poco credito a las inscripciones, mayormente estando en depravacion, si no queremos dezir error de las longitudes y latitudes de aqueste sumo matematico." Hernández, *Historia natural libros i, ii, iii, iv, y v*, BNE, Ms. 2869, fol. 99r.

"Hallose, tengo por relación de Arias Montano, varon principal en letras y lenguas, conocido por la impression de la Biblia Regia, en Frexenal, una piedra donde se lee respublica vertodeicensis, y aunque difiere mucho la graduacion de Ptolomeo de la q oy tiene Frexenal no se deve dar poco credito a las inscripciones, mayormente en tanta depravacion de numeros, como son los que causa la antiguedad de las cosas." Hernández, *Historia natural libros i, ii, iii*, BNE, Ms. 2862, fol. 256r.

In addition to this reference, another tentative link between Montano and Hernández might be found in the books that the natural historian drew from to compose his commentaries. The correspondence between Juan de Ovando and Montano of the late 1560s, before Hernández set out for the New World, reveals numerous instances in which Ovando requests from the Plantin Press books like Leo Africanus's *Description of Africa*, the *Hyeroglyphica* of Piero Valeriano, and the works of Olaus Maugnus, which all appear frequently cited in the commentaries. See *La correspondencia de Benito Arias Montano con el Presidente de Indias Juan de Ovando: Cartas de Benito Arias Montano conservadas en el Instituto de Valencia de Don Juan*, ed. and trans. Baldomero Macías Rosendo (Huelva, 2008).

81. Ambrosio de Morales, *Las antigüedades de las ciudades de España: que van nombradas en la Coronica con la aueriguacion de sus sitios y nombres antiguos* [1575], 2 vols. (Madrid: Benito Cano, 1792), vol. 1, 256.

82. "Tomo el nombre de Afer del linage de Abraham si no se llamo ansi (segun otros quieren) por estar agena de frio, como sitiada entre los dos tropicos y que no pasa del tercero clima. O (segun Leon Africano) de Iphrico, rey de Arabia Felix, que dizen los eruditos de los moros haver venido alli con exercito ahuyentado al rey de Asyria, o de faraca verbo Arabigo que significa dividir, a causa de estar divisa de Europa, como diremos, por el mar mediterraneo y de Egipto y de Asia [por el] Nilo." Francisco Hernández, *Historia natural libros iv, v y xi*, BNE, Ms. 2863, fol. 72r.

83. For an explanation of Plato's *Cratylus* and his etymological theories, see David Sedley, *Plato's "Cratylus"* (Cambridge, 2003).

84. The bibliography on Hernández's expedition and the precedents to the expedition is enormous. For a representative selection, see Hernández, *The Mexican Treasury*.

85. See Francisco Dominguez, "Carta del geógrafo Francisco Dominguez á Felipe II desde Méjico el 30 de diciembre de 1581 sobre que S.M. mande al virrey D. Martinez Enrique remita," in *Colección de documentos inéditos para la historia de España*, 112 vols. (Madrid, 1842–1896), vol. 1, 379–84.

86. Marcy Norton, *Sacred Gifts, Profane Pleasures: A History of Tobacco and Chocolate in the Atlantic World* (Ithaca, NY, 2008), 109–15.

87. Charles Boxer, *Two Pioneers of Tropical Medicine: Garcia d' Orta and Nicolás Monardes* (London, 1963), 19–29.

88. Anthony Grafton, *New Worlds, Ancient Texts: The Power of Tradition and the Shock of Discovery* (Cambridge, MA, 1992), 177.

89. *Medicinas, drogas y alimentos vegetales del Nuevo Mundo: Textos e imágenes españolas que los introdujeron en Europa*, ed. José M. López Piñero, José Luis Fresquet Febrer, María Luz López Terrada, and José Pardo Tomás (Madrid, 1992), 149.

90. Nancy G. Siraisi, *Medieval and Early Renaissance Medicine: An Introduction to Knowledge and Practice* (Chicago, 1990), 115–152. On Nahua understandings of disease and the body, see Alfredo López Austen, *The Human Body and Ideology Concepts of the Ancient Nahuas*, trans. Thelma Ortiz de Montellano and Bernard Ortiz de Montellano (Salt Lake City, 1988), 161–180.

91. José Pardo Tomás, *Oviedo, Monardes, Hernández, el tesoro natural de américa: Colonialismo y ciencia en el siglo XVI* (Madrid, 2002), 94. Also see Antonio Barrera Osorio, *Experiencing Nature: The Spanish American Empire and the Early Scientific Revolution* (Austin, 2006), 102–127.

92. Barrera Osorio, *Experiencing Nature*, 122.

93. On the emergence of the concept of local knowledge in early modern Europe, see Alix Cooper, *Inventing the Indigenous: Local Knowledge and Natural History in Early Modern Europe* (Cambridge, 2007), 14–19.

94. Ogilvie, *The Science of Describing*, 94; Rafael Chabrán and Simon Varey, "'An Epistle to Arias Montano': An English Translation of a Poem by Francisco Hernández," *Huntington Library Quarterly* 55, no. 4 (Autumn 1992): 620–634.

95. Germán Somolinos D'Ardois, "Plinio, España y la época de Hernández," in Hernández, *Obras completas*, vol. 4, ix–xxiii.

96. A surviving receipt demonstrates that Hernández borrowed from Cervantes de Salazar a work by the German botanist Hieronymus Bock (d. 1554). Germán Somolinos D'Ardois, "Vida y obra de Francisco Hernández," in Hernández, *Obras completas*, vol. 1, 164–167.

97. J. Jorge Klor de Alva argues that confessional practices were central in shaping Sahagún's ethnographic approach and led to forms of questioning that sometimes were casual and other times coerced, and which relied on both written and oral questionnaires. See J. Jorge Klor de Alva, "Sahagún and the Birth of Modern Ethnography: Representing, Confessing, and Inscribing the Native Other," in *The Work of Bernardino de Sahagún: Pioneer Ethnographer of Sixteenth-Century Aztec Mexico*, ed. J. Jorge Klor de Alva, H. B. Nicholson, and Eloise Quiñones Keber (Austin, TX, 1988), 31–52, esp. 42.

98. José Luis Benítez Miura, "Cartas inéditas dirigidas a Felipe II por el protomédico de Indias doctor Francisco Hernández desde la ciudad de Méjico, 1571–1575," *Anuario de estudios americanos* (Seville) 7 (January 1, 1950): 367–409, esp. 400.

99. On the practices of "learned empiricism," see Pomata and Siriasi, introduction to *Historia: Empiricism and Erudition in Early Modern Europe*, 1–38, esp. 23.

100. On the importance of the genre of the *relación*, Renaissance ideas about translation, and Hernández's work alongside Indigenous physicians, see Jaime Marroquín Arredondo, "The Method of Francisco Hernández: Early Modern Science and the Translation of Mesoamerica's Natural History," in *Translating Nature: Cross Cultural Histories of Early Modern Science*, ed. Ralph Bauer and Jaime Marroquín Arredondo (Philadelphia, 2019), 45–69.

101. In his treaty on medicine, Agustín Farfán praises Hernández several times. See Agustín Farfán, *Tractado brebe de medicinas y de todas las enfermedades* (México: en casa de Pedro Ocharte, 1592), 206r–209r.

102. In the introduction to his synthesis of Hernández's work, Recchi describes the importance of botanical names and briefly ancient theories about the imposition of names—namely, the positions that had become associated with Plato and Aristotle. Nardo Recchi, *De materia medica Novae Hispaniae Philippi Secundi Hispaniarum ac Indiarum regis invictissimi iussu*, Codex Lat 5, JCB, fol. 7v.

103. Hernández, *Obras completas*, vol. 3, book 16, chap. 74, 123.

104. Hernández, *Obras completas*, vol. 3, book 12, chap. 41, 39.

105. Hernández, *Obras completas*, vol. 3, book 17, chap. 50, 40–41.

106. Hernández, *Obras completas*, vol. 3, book 1, chap. 1, 295–296.

107. Hernández, *Obras completas*, vol. 3, book 3, chap. 12, 371–372.

108. Hernández, *Obras completas*, vol. 3, book 5, chap. 2, 390–391.

109. "Esto se ve mas claro en las Indias Occidentales, donde en pocas leguas de distancia se varia muchas vezes el lenguaje: Cosa por cierto admirable, pues cada uno dellos segun yo creo podria ser muy perfecto y copioso si fuese dilatado de Indios sabios, que tuviessen cuenta con conoçer e imponer a su modo nombre a las cosas: porque no me curo de la variedad de las letras y pronunçiaçion, pues es tan natural y blanda a los naturales cuanto estraña a los estrangeros." Hernández, *Historia natural libros vii y viii*, BNE, Ms. 2864, fol. 8r; and Hernández, *Historia natural libros vii, x, xi y xii*, BNE, Ms. 2870, fol. 9r.

110. Francisco Hernández, *Antigüedades de la Nueva España*, trans. Joaquín García Pimentel in Hernández, in *Obras completas*, vol. 6 (1984), 90–91.

111. Boruchoff, "The Conflict of Natural History and Moral Philosophy in *De Antiquitatibus Novae Hispaniae* of Francisco Hernández," 241–58.

112. "Ornithogale. O leche de gallina, es vulgar a Hespaña [crossed out: donde yo la he dado a muchos a conocer], y no le dan otro nombre que el anitiguo. de lo qual me maravillo mucho en esta y en otras innumerables yerbas a nosotros sin nombre, como en las indias donde la gente estan inculta y barbara, de tan grande numero de yerbas que hay de algunas de las quales se sabe virtud y de otras no, casi no se halla ninguna que con nombre particular no sea dellos nombrada y conocida, por ventura nasce de lengua antiquissima casta y muy natural, y no ensuciada jamas del comercio con otras naciones." Francisco Hernández, *Historia natural libros xxi, xxii, xxiii, xiv, y xv*, BNE, Ms. 2868, fol. 34r.

113. Bernardino de Sahagún, *Historia general de las cosas de Nueva España*, ed. Alfredo López Agustín and Josefina García Quintana (México, 2000), 339–340.

114. Hernández, *Antigüedades de la Nueva España* [1984], 59–60.

115. See Carla Mulford, "Huehuetlahtolli' Early American Studies, and the Problem of History," *Early American Literature* 30, no. 2 (1995): 146–151.

116. Hernández, *Antigüedades de la Nueva España* [1984], 134.

117. The *titici*, or Indigenous doctors, were, in Hernández's view, mere empirical doctors who did not understand humoral theory or the causes of disease. See Hernández, *Antigüedades de la Nueva España* [1984], 100–101.

118. See Marcy Norton, "The Quetzal Takes Flight: Microhistory, Mesoamerican Knowledge and Early Modern Natural History," in *Translating Nature*, 119–147.

119. Beagon, *The Elder Pliny on the Human Animal*, book 7, 1–58.

120. Hernández, *Antigüedades de la Nueva España* [1984], 105.

121. Hernández, *Antigüedades de la Nueva España* [1984], 117–119. Also see Gliozzi, "Gli Ebrei nel Nuovo Mondo," in *Adamo e il Nuovo Mondo: La nascita dell'antropologia come ideologia coloniale: Dalle genealogie bibliche alle teorie razziali (1500–1700)* (Florence, 1977), 49–110.

122. Juan Gil, "Escribir en Latín: Ventajas y e inconvenientes," in *Tradición clásica y universidad*, ed. Francisco L. Lisi Bereterbide (Madrid, 2010), 259–301.

123. Hernández, *Antigüedades de la Nueva España* [1984], 91.

124. See José María López Piñero and José Pardo Tomás, *Nuevos materiales y noticias sobre la "Historia de las plantas de Nueva España" de Francisco Hernández* (Valencia, 1994); Jesús Bustamante García, "Francisco Hernández, Plinio del Nuevo Mundo: Tradición clásica, teoría nominal y sistema terminológico indígena en una obra Renacentista," in *Entre dos mundos: Fronteras culturales y agentes mediadores*, ed. B. Ares Queija and S. Gruzinski (Seville, 1997), 243–68.

125. Raquel Álvarez Peláez, introduction to *De materia medica Novae Hispaniae: Manuscrito de Recchi*, trans. and ed. Florentino Fernández González and Raquel Álvarez Peláez, 2 vols. (Madrid, 1998), vol. 1, 36–43.

126. See Domingo Ledezma, "Una legitimación imaginativa del Nuevo Mundo: La *Historia naturae, maxime peregrinae* de Juan Eusebio Nieremberg," in *El saber de los Jesuitas, historias naturales y el Nuevo Mundo*, ed. Luis Millones and Domingo Ledezma (Frankfurt, 2005), 53–83.

127. "Sobre la impresión, a costa de su majestad, de ciertos libros que el doctor Francisco Hernández hizo en Nueva España sobre animales, aves, serpientes, plantas y sobre la historia y descripción de las tierras occidentales y una traducción al romance de la Historia natural de Plinio. R. [Annotated in the margin:] Mirese más en lo que vendria a costar la impressión, y se podria sacar della, que hauiendo de costar mucho los libros creo que pocos los comprarian, y si seria mejor que se hiziese vn exemplar de mano con sus figuras de pintura y lo de mas que contiene, de que se fuessen sacando en volumenes pequeños y manuales las materias sumariamente de manera que ni la impression destos viniese a ser tan costosa, ni despues dexasse de correr, y venderse con que paresce que se conseguiría el fin de la vtilidad publica." Consulta del Consejo de Indias 03-20 Madrid, 1578," AGI, INDIFERENTE, 739, N.60.

128. On the life of Antonio Nardo Recchi, see Giuseppe Gabrieli, *Contributi alla storia della Accademia dei Lincei*, vol. 1 of *Storia dell'accademia dei Lincei studi*, 2 vols. (Rome: Accademia nazionale dei Lincei, 1989). Recchi's manuscript survives in the John Carter Brown Library. See Nardo Recchi, *De materia medica Novae Hispaniae Philippi Secundi Hispaniarum ac Indiarum regis invictissimi iussu*, Codex Lat 5, JCB. Of special interest in the discussion of nomenclature and early modern language theory are fols. 7v–11v.

129. Francisco Hernández, *Rerum medicarum Novae Hispaniae thesaurus* (Rome: ex typographeio Vitalis Mascardi, 1651), 6. In these sections I follow the spelling of plant names in the Cesi edition.

130. Hernández, *Rerum medicarum Novae Hispaniae thesaurus*, 6.

131. Hernández, *Rerum medicarum Novae Hispaniae thesaurus*, 7–8.

132. Hernández, *Rerum medicarum Novae Hispaniae thesaurus*, 8.

133. Bustamante García, "Francisco Hernández, Plinio del Nuevo Mundo," 243–268.

134. Álvarez Peláez, introduction to *De materia medica Novae Hispaniae*, 58–70.

135. Ogilvie, *The Science of Describing*, 215.

136. See Phillip Sloan, "John Locke, John Ray and the Problem of the Natural System," *Journal of the History of Biology* 5, no. 1 (Spring 1972): 1–53.

137. López Piñero and Pardo Tomás, *Nuevos materiales y noticias*, 265. The appendix of this book is a reproduction of the *Index* of Montpellier.

138. Bustamante García, "Francisco Hernández, Plinio del Nuevo Mundo," 256.

139. Campbell, *A Morphological Dictionary of Classical Nahuatl*, 186.

140. See Francisco Guerra, "Aztec Science and Technology," *History of Science* 8 (1969): 32–52; and Francisco del Paso y Troncoso, "Primer estudio: La botánica entre los nahuas y otros estudios," *Anales—Museo Nacional de Antropología de México* 3 (1886): 140–235.

141. Bustamante García, "Francisco Hernández, Plinio del Nuevo Mundo," 243–68.

142. Miguel Ángel Esparza Torres, "Nebrija y los modelos de los misioneros linguistas del Náhuatl," in *Missionary Linguistics III = Lingüística Misionera III: Morphology and Syntax; Selected Papers from the Third and Fourth International Conferences on Missionary Linguistics, Hong Kong / Macau, 12–15 March 2005, Valladolid, 8–11 March 2006,* ed. Otto Zwartjes, Gregory James, and Emilio Ridruejo (Amsterdam, 2007), 3–40. R. Joe Campbell and Mary L. Clayton study of Molina's dictionary shows that "Molina had Nebrija's dictionary at hand, but he neither felt obliged to include words in his dictionary because Nebrija used them, nor did he limit himself to those items contained in Nebrija." See R. Joe Campbell and Mary L. Clayton, "Bernardino de Sahagún's Contributions to the Lexicon of Classical Nahuatl," in Klor de Alva, Nicholson, and Quiñones Keber, *The Work of Bernardino de Sahagún,* 295–314, esp. 297.

143. See Johannes Reuchlin, *Recommendation Whether to Confiscate, Destroy and Burn All Jewish Books,* ed. Peter Wortsman (New York, 2000).

144. Demonet, "Renaissance étymologiques," 57–59.

145. Bustamante García, "Francisco Hernández, Plinio del Nuevo Mundo," 243–68.

146. Casimiro Gómez Ortega, "Ad lectorem prefatio," in Francisco Hernández, *Opera: Cum edita, tum inedita, ad autographi fidem et integritatem expressa, impensa et jussu regio,* ed. Casimiro Gómez Ortega, 3 vols. (Madrid: ex typographia Ibarrae Heredum, 1790), vol. 1, xiv–xv.

147. Antonio Lafuente and Nuria Valverde, "Linnean Botany and Spanish Imperial Biopolitics," in *Colonial Botany: Science, Commerce, and Politics in the Early Modern World,* ed. Londa Schiebinger and Claudia Swan (Philadelphia, 2007), 134–147.

148. Cited in Lafuente and Valverde, "Linnean Botany and Spanish Imperial Biopolitics," 138.

CHAPTER FIVE: The Rudiments of All Languages

1. "Cacique: vocablo arabigo tomado del Ebreo, que al gran principe llaman עקין çakin y סבא çagui." "El licenciado Valverde tractado de etymologias de voces castellanas en otras lenguas, Castellana, Hebrea, Griega, Arabe," in Papeles Varios, BNE, Ms. 9934, fol. 155v. The treatise has two versions: the first composed c. 1579, and the second c. 1586. For the dating of this treatise, I follow Francisco Javier Perea Siller's reconstruction of the life and works of Bartolomé Valverde. See Francisco Javier Perea Siller, "El tratado de etymologias de voces Castellanas (c. 1579/c. 1586): Hebraísmo y etimologías," in *La historiografía lingüística como paradigma de investigación,* ed. Antonio Salvador Plans, Carmén Galán Rodríguez, José Carlos Martín Camacho, María Isabel Rodríguez Ponce, Francisco Jiménez Calderón, Elena Fernández de Molina Ortés, and Ana Sánchez (Madrid, 2016), 609–629. I have also examined the Escorial copy of this manuscript, BME L-I-2, fols. 123r–134v. The entry on the *cacique* is on 127v.

2. Diego de Guadix would also etymologize *cacique* as an Arabic word. See Diego de Guadix, Elena Bajo Pérez, and Felipe Maíllo Salgado, *Recopiliación de algunos aombres arábigos que los árabes pusieron a algunas ciudades, Y a otras muchas cosas* [1593] (Gijón, 2005), 461. The etymologist Francisco del Rosal (c. 1537–c. 1613), who compiled the first unpublished etymological dictionary of the Castilian language, also included *cacique* in this work and attributed to it Hebrew origins. See Francisco del Rosal, *Origen y etimología de todos los vocablos originales de la lengua castellana, ms., c. 1610,* ed. Enrique Gómez Aguado (Madrid, 1992), 128. Esther Hernández explains that *cacique* is a Taíno word first documented in Hispaniola in Columbus's diary (1492). The Spanish spread its usage throughout the Americas. The *Glosario etimólogico taíno-español histórico y etnográfico* (1941) derives the word from the Arawak verb *kassikóan,* which means "to inhabit or have a home." See Esther

Hernández, *Vocabulario en lengua castellana y mexicana de Fray Alonso de Molina: Estudio de los indigenismos léxicos y registros de las voces españoles internas* (Madrid, 1996), 74–76.

3. "Hay tambien en todas las lenguas muchos vocablos recebidos de comunicacion de las naciones vecinas, ó de las contrataciones, como lo vemos en las tierras mercantiles donde no hay puridad de una lengua sola, y en el lenguaje de los soldados, y en los que vienen de Indias hablando sus propias lenguas mezclan muchos vocablos peregrinos de las tierras donde han estado, y asi se van perdiendo unos y trocando otros." "El licenciado Valverde tractado de etymologias de voces castellanas en otras lenguas, Castellana, Hebrea, Griega, Arabe," in Papeles Varios, BNE, Ms. 9934, fol. 135r.

4. "Pues no hallandose como digo la etimologia del vocablo Castellano en la Arabiga ni la Latina, lo seguro y asertado es acudir a la fuente y madre de todas, la legua Ebrea, mayormente habiendo dexado es España los judios infinitos vocablos, como provaré despues, y como el otro porfiando Julio Cesar sobre un vocablo, le dijo: Civitatem Caesar dare nobis potes verba dare non potes. Asi por el contrario aunque los Reyes Catolicos echaron los judios de España, loc vocablos que ellos habian introducido no pudieron." "El licenciado Valverde tractado de etymologias de voces castellanas en otras lenguas, Castellana, Hebrea, Griega, Arabe," in Papeles Varios, BNE, Ms. 9934, fol. 134v.

5. "No importa mucho declararse una misma cosa con diferentes nombres ó silaba . . . asi como las purgas y xaraves siendo unos mesmos a veces no parecen diferentes, pore star disfrazados e encubiertos con algunos colores ó olores; pero el Medico que sabe la fuerza y virtud de cada uno, juzga con verdad ser unos mesmos, por que aquel disfrace nada le impide ni estorva; asi tambien el que sabe bien de etimologias tiene ojo á la fuerza y propiedad del vocablo, y no se turba ni engaña, por ver que se añada, quite, trueque letra o letras." "El licenciado Valverde tractado de etymologias de voces castellanas en otras lenguas, Castellana, Hebrea, Griega, Arabe," in Papeles Varios, BNE, Ms. 9934, fols. 135v–136r.

6. Perea Siller, "El tratado de etymologias de voces Castellanas (c. 1579/c. 1586)," 624–626.

7. Perea Siller, "El tratado de etymologias de voces Castellanas (c. 1579/c. 1586)," 624–626.

8. *Biblia Sacra Hebraice, Chaldaice, Graece, et Latine: Philippi II. Reg. Cathol. pietate, et studio ad Sacrosanctae Ecclesiae usum*, ed. Benito Arias Montano, 8 vols. (Antwerp: Plantin, 1569–1573). See Giuliano Gliozzi, *Adamo e il Nuovo Mondo: La nascita dell'antropologia come ideologia coloniale: dalle genealogie bibliche alle teorie razziali (1500–1700)* (Florence, 1977), 152–184; Isaías Lerner, "Teorías de Indios: Los orígenes de los pueblos del continente americano y la Biblia Políglota de Amberes (1568–1573)," *Colonial Latin American Review* 19, no. 2 (2010): 236–237. For an inventory of some of the copies of the Antwerp Polyglot that have been discovered in Mexican libraries, see Theodor Dunkelgrün, "The Multiplicity of Scripture: The Confluence of Textual Traditions in the Making of the Antwerp Polyglot Bible (1568–1573)" (PhD diss., University of Chicago, 2012), 491–492; Also see César Manrique Figueroa, "From Antwerp to Veracruz: Looking for Books from the Southern Netherlands in Mexican Colonial Libraries," *De Gulden Passer* 87, no. 2 (2009): 1–16.

9. Franciscus Raphelengius (1539–1597) published the nine antiquarian treatises of the Antwerp Polyglot in a standalone volume entitled the *Antiquitatum iudaicarum libri IX* in 1593. See Benito Arias Montano, *Antigüedades hebráicas: Antiquitatum iudaicarum libri IX; Tratados exegéticos de la Biblia Regia*, ed. Luis Gómez Canseco and Sergio Fernández López, Bibliotecha Montaniana 25 (Huelva, 2013). For a history of the printing of the *Apparatus* and its contents (vols. 6–8), see Dunkelgrün, "The Multiplicity of Scripture," 341n129.

10. On Benito Arias Montano's life, see Ben Rekers's *Benito Arias Montano* (London, 1972). Also see C. Doetsch, *Benito Arias Montano: Extractos de su vida* (Madrid, 1920). For a biography of Montano and how he developed his intellectual interests and influences, see Dunkelgrün, "The Multiplicity of Scripture," 125–180.

11. F. Javier Campos y Fernández de Sevilla, *Arias Montano en la Biblioteca Real y en el Gabinete de Estampas del Escorial* (Escorial, 2010).

12. Rekers, *Benito Arias Montano*, 130.

13. Baldomero Macías Rosendo, "*De arcano sermone* en el marco de la Biblia Políglota de Amberes," in Benito Arias Montano, *Libro de José o sobre el lenguaje arcano*, ed. and trans. Luis Gómez Canseco, Fernando Navarro Antolín, and Baldomero Macías Rosendo (Huelva, 2006), 31.

14. Marcel Batallion, *Erasmo y España: Estudios sobre la historia espiritual del siglo XVI*, trans. Antonio Alatorre (México, 1966), 740–42; Alastair Hamilton, *The Family of Love* (Cambridge, 1981).

15. Montano's library also included a great number of Erasmus's works. See José Luis Gonzalo Sánchez-Molero, "La biblioteca de Arias Montano en el Escorial," in *Benito Arias Montano y los humanistas de su tiempo*, vol. 1, ed. José María Maestre Mastre, Eustaquio Sánchez Salor, Manuel Antonio Díaz Gito, Luis Charlo Brea, and Pedro Juan Galán Sánchez (Mérida, 2006), 91–110.

16. On Erasmus's attitude regarding Hebrew, see Erika Rummel, "The Textual and Hermeneutic Work of Desiderius Erasmus of Rotterdam," *Hebrew Bible / Old Testament: The History of Its Interpretation*, vol. 2: *From the Renaissance to the Enlightenment*, ed. Magne Saebø (Göttingen, 2008), 215–230.

17. Emilia Fernández Tejero and Natalio Fernández Marcos, "Scriptural Interpretation in Renaissance Spain," in Saebø, *Hebrew Bible / Old Testament*, 231–253, 232; Natalio Fernández Marcos, "Biblismo y erasmismo en la España del siglo XVI," in *Biblia y humanismo: Textos, talantes y controversias del siglo XVI español* (Madrid, 1997), 15–26.

18. Cipriano de la Huerga, *Obras completas*, 10 vols. (León, 1990–2005).

19. Fernández Tejero and Fernández Marcos, "Scriptural Interpretation in Renaissance Spain," 231–253. Also see Natalio Fernández Marcos, "Cipriano de la Huerga," in *Biblia y humanismo*, 47–84.

20. See *Apologia: L'autodifesa di Pico di fronte al Tribunale dell'Inquisizione*, ed. Paolo Fornaciari (Florence, 2010); *Oration on the Dignity of Man: A New Translation and Commentary*, ed. Francesco Borghesi, Michael Papio, and Massimo Riva (Cambridge, 2012); Brian P. Copenhaver, "Number, Shape, and Meaning in Pico's Christian Cabala: The Upright *Tsade*, the Closed *Mem*, and the Gaping Jaws of *Azazel*," in *Natural Particulars: Nature and Disciplines in Renaissance Europe*, ed. Anthony Grafton and Nancy Siraisi (Cambridge, MA, 1999), 25–76. Also see Yaacob Dweck, *The Scandal of Kabbalah: Leon Modena, Jewish Mysticism, Early Modern Venice* (Princeton, NJ, 2011), 8–10.

21. See Crofton Black, "Eucherius of Lyon, Giovanni Pico della Mirandola and Sixtus of Siena: Early Christian Exegesis," in *Giovanni Pico e la cabbala*, ed. Fabrizio Lelli (Florence, 2014), 231–258.

22. Fernández Tejero and Fernández Marcos, "Scriptural Interpretation in Renaissance Spain," 238–239.

23. Colin P. Thompson, *Fray Luis de León and the Golden Age of Spain* (Cambridge, 1988).

24. Luis de León, *De los nombres de Cristo*, in *Obras completas castellanas de Fray Luis de León*, ed. Felix Garcia (Madrid, 1957), vol. 1, 417. This work was published four times. See, for instance, Luis de León, *De los Nombres de Christo: En tres libros* (En Salamanca: en casa de Iuan Fernandez, 1595), book 1, 9r–9v.

25. For the similarities and differences between Montano's and Fray Luis de León's approaches, see Natalio Fernández Marcos, "*De los nombres de Cristo* de Fray Luis de León y *De arcano sermone* de Arias Montano," *Sefarad* 48 (1988): 245–270.

26. "Prólogo de Benito Arias Montano Hispalense a la edición Regia de la Biblia Sacra Cuadrilingüe," in *Prefacios de Benito Arias Montano a la Biblia Regia de Felipe II*, ed. and trans. María Asunción Sanchez Manzano (Salamanca, 2006), 20–21; Benito Arias Montano, "De divina scripturae dignitate, linguarum usu & Catholici Regis Consilio, Praefatio," in *Biblia Sacra Hebraice, Chaldaice, Graece, et Latine: Philippi II. Reg. Cathol. pietate, et studio ad Sacrosanctae Ecclesiae usum*, ed. Benito Arias Montano, 8 vols. (Antwerp: Christophe Plantin, 1569–1573), vol. 1, 13.

27. Arias Montano, *Prefacios*, 14–15; Antwerp Polyglot, vol. 1, 9.

28. Benito Arias Montano, *Liber generationis et regenerationis Adam, sive, De historia generis humani: Operis magni pars prima, id est, Anima* (Antwerp: Jan Moretus, 1593). The first volume of the second part was published in 1601. See Benito Arias Montano, *Naturae historia, prima in magni operis Corpore pars* (Antwerp: Jan Moretus, 1601). For a study Arias Montano's natural philosophical studies and its relationship to his biblical scholarship, see María Portuondo, *The Spanish Disquiet: The Biblical Natural Philosophy of Benito Arias Montano* (Chicago, 2019), chaps. 7 and 8.

29. See Gliozzi, *Adamo e il Nuovo Mondo*, 16–17; Fernández Tejero and Fernández Marcos, "Scriptural Interpretation in Renaissance Spain," 231–253. On Montano's perception of the New World, see Fernando Navarro Antolín, Luis Gómez Canseco, and Baldomero Macías Rosendo, "Fronteras del humanismo: Arias Montano y el Nuevo Mundo," in *Orbis incognitus: Avisos y legajos del nuevo mundo*, 2 vols., ed. Fernando Navarro Antolín and Luis Navarro García (Huelva, 2007), vol. 1, 101–136.

30. Benito Arias Montano, "Hebraicorum Bibliorum Veteris Testamenti Latina interpretatione," in *Prefacios*, 90–91; Antwerp Polyglot, vol. 7, praefatio, 3.

31. Arias Montano, *Prefacios*, 36–37; Antwerp Polyglot, vol. 1, 23.

32. Vicente Becarés Botas, "Las ideas lingüísticas y el método de Arias Montano," in *El humanismo extremeño II: Estudios presentados a las segundas Jornadas organizadas por la Real Academia de Extremadura en Frenegal de la Sierra en 1997*, ed. Marqués de la Encomienda, Carmelo Solís Rodríguez, Francisco Tejada Vizuete, Manuel Terrón Albarrán y Antonio Viudas Camarasa, and Santiago Castelo (Trujillo, 1998), 25–39.

33. Macías Rosendo, "*De arcano sermone* en el marco de la Biblia Políglota de Amberes," 24.

34. Macías Rosendo, "*De arcano sermone* en el marco de la Biblia Políglota de Amberes," 24.

35. Natalio Fernández Marcos and Emilia Fernández Tejero, "Pagnino, Servet y Arias Montano Avatares de una traducción latina de la Biblia Hebrea," *Sefarad* 63, no. 2 (2003): 283–329.

36. Macías Rosendo, "*De arcano sermone* en el marco de la Biblia Políglota de Amberes," 33–34.

37. Dunkelgrün, "The Multiplicity of Scripture," 232.

38. Macías Rosendo, "*De arcano sermone* en el marco de la Biblia Políglota de Amberes," 24–26.

39. Macías Rosendo, "*De arcano sermone* en el marco de la Biblia Políglota de Amberes," 28.

40. Sylvaine Hänsel, *Benito Arias Montano (1527–1598): Humanismo y arte en España* (Huelva, 1999), 39; Macías Rosendo, "*De arcano sermone* en el marco de la Biblia Políglota de Amberes," 28–29.

41. For an account of Montano's final years in Seville and his engagement with the botanist Francisco Hernández, see Guy Lazure, "Mecenazgo y clientelismo en los años sevillanos de Benito Arias Montano: Genealogía social e intelectual de un humanista," in Maestre Mastre et al., *Benito Arias Montano y los humanistas de su tiempo*, 111–24.

42. Benito Arias Montano, "Benedicti Ariae Montani Hispalensis in Librum Chaleb sive de terrae promissae partitione. Praefatio," in *Prefacios*, 188–89; Antwerp Polyglot, vol. 8, praefatio, 1.

43. Zur Shalev, "Benjamin of Tudela, Spanish Explorer," *Mediterranean Historical Review* 25, no. 1 (June 2010): 17–33.

44. Benito Arias Montano, "Benedicti Ariae Montani Hispalensis in Chaldaicarum Paraphraseon libros et interpretationes," in *Prefacios*, 65–67; Antwerp Polyglot, vol. 2, 2.

45. See Serena Franceschi, "Las reminiscencias garcilasianas en la Paráfrasis del Cantar de los Cantares de Benito Arias Montano," *Revista de filología románica*, no. 15 (1998): 205–214. The full text can be found in *Tras las huellas de humanistas Extremeños: Arias Montano-Pedro de Valencia*, ed. Abdón Moreno García (Badajoz, 1996), 21–71.

46. Arias Montano, "Prólogo de Benito Arias Montano," in *Prefacios*, 41.

47. Benito Arias Montano, "Benedicti Ariae Montani Hispalensis in latinam ex hebraica veritate veteris testamenti interpretationem," in *Prefacios*, 88–89; Antwerp Polyglot, vol. 7, *Hebraicorum bibliorum Veteris Testamenti Latina interpretationi, opera olim Xantis Pagnini Lucensis* [. . .], 2.

48. Benito Arias Montano, "Benedicti Ariae Montani in librum de Hebraicis idiotismis. Praefatio," in *Prefacios*, 102–103, Antwerp Polyglot, vol. 7, *Communes et familiares hebraicae linguae idiotismi*, 1.

49. Benito Arias Montano, "Benedicti Ariae Montani in librum de actione. Praefatio," in *Prefacios*, 124–125; Antwerp Polyglot, vol. 8, *Liber Ieremiae, siue De actione*, 1.

50. For Quintilian *actio* can be divided into two elements: voice and movement. He further explained how "there is no Proof . . . which is so secure that it does not lose its force unless it is assisted by the assurance of the speaker." For Cicero *actio* was a "sort of language," a "kind of eloquence." Quintilian, *The Orator's Education*, vol. 5: *Books 11–12*, ed. and trans. Donald A. Russell, Loeb Classical Library 494 (Cambridge, MA, 2002), 11.3.

51. Arias Montano, "Benedicti Ariae Montani in librum de actione. Praefatio," *in Prefacios*, 139; Antwerp Polyglot, vol. 8, *Liber Ieremiae, siue De actione*, 5.

52. *Collected Works of Erasmus: Spiritualia and Pastoralia; Exomologesis and Ecclesiastes*, ed. Frederick J. McGinness, Alexander Dalzell, and Frederick J. McGinness, trans. Michael John Heath and James L. P. Butrica (Toronto, 2015), vol. 67. Also see John O'Malley, "Erasmus and the History of Sacred Rhetoric: The *Ecclesiastes* of 1535," *ERSY* 5 (1985): 1–29; and Kirk Essary, "Fiery Heart and Fiery Tongue: Emotion in Erasmus' *Ecclesiastes*," *Erasmus Studies* 36 (2016): 5–34.

53. Kathy Eden, *Hermeneutics and the Rhetorical Tradition: Chapters in the Ancient Legacy and its Humanist Reception* (New Haven, CT, 1997), 73–74.

54. Shalev, "Benjamin of Tudela," 20.

55. Benito Arias Montano, "Benedicti Ariae Montani Hispalensis in Librum Thubal-Cain, sive de mensuris. Praefatio," in *Prefacios*, 150–151; Antwerp Polyglot, vol. 8, *Thubal Cain, siue De mensuris sacris liber*, 5.

56. Shalev, "Benjamin of Tudela," 20.

57. Arias Montano, "Benedicti Ariae Montani Hispalensis in Librum Thubal-Cain, sive de mensuris. Praefatio," *in Prefacios*, 152–153; Antwerp Polyglot, vol. 8, *Thubal Cain, siue De mensuris sacris liber*, 5.

58. Benito Arias Montano, "Benedicti Ariae Montani Hispalensis in Librum Phaleg . . . Praefatio," in *Prefacios*, 170–171; Antwerp Polyglot, vol. 8, *Phaleg, siue De gentium sedibus primis, orbisque Terrae situ liber*, 5.

59. Arias Montano, "Benedicti Ariae Montani Hispalensis in Librum Chaleb [. . .] Praefatio," *in Prefacios*, 190–191; Antwerp Polyglot, vol. 8, *Chaleb siue De Terrae promissae partitione*, 1.

60. Benito Arias Montano, "Ejemplar o arquetipo de las construcciones sagradas," ed. and trans. José Solís de los Santos, in Benito Arias Montano, *Antigüedades hebráicas: Antiquitatum iudaicarum libri IX. Tratados exegéticos de la Biblia Regia*, ed. Luis Gómez Canseco and Sergio Fernández López (Huelva: Universidad de Huelva, 2013), 554; Antwerp Polyglot, vol. 8, *Exemplar, sive de sacris fabricis liber*, "Noah, sive de arcae fabrica et forma," book 1, 3.

61. Arias Montano, "Ejemplar o arquetipo de las construcciones sagradas," 554.

62. "Benito Arias Montano Saluda al Lector Estudioso de las Sagradas Letras," in Arias Montano, *Libro de José o sobre el lenguaje Arcano*, 91; Antwerp Polyglot, vol. 8, *Liber Ioseph sive De arcano sermone*.

63. For the dissemination of the works of Hor Apollo, see Don Cameron Allen, *Mysteriously Meant: The Rediscovery of Pagan Symbolism and Allegorical Representation in the Renaissance* (Baltimore, 1970), 112–133.

64. Allen, *Mysteriously Meant*, 112–133, 115. Also see Daniel Stolzenberg, *Egyptian Oedipus: Athanasius Kircher and the Secrets of Antiquity* (Chicago, 2013), 129–150.

65. Pierio Valeriano, *Hieroglyphica, sive De Sacris Aegyptiorum literis commentarii* (Basel: M. Isengrin, 1556). Also see Julia Haig Gaisser, *Pierio Valeriano on the Ill Fortune of Learned Men: A Renaissance Humanist and His World* (Ann Arbor, 1999).

66. Arias Montano, *Libro de José o sobre el lenguaje Arcano*, 91.

67. Arias Montano, *Libro de José o sobre el lenguaje Arcano*, 251.

68. Arias Montano, *Libro de José o sobre el lenguaje Arcano*, 221; Antwerp Polyglot, vol. 8, *Liber Ioseph sive De arcano sermone*, chap. 61, 52–53.

69. "Benito Arias Montano Saluda al Lector Estudioso de las Sagradas Letras," in Arias Montano, *Libro de José o sobre el lenguaje Arcano*, 91; Antwerp Polyglot, vol. 8, *Liber Ioseph sive De arcano sermone*.

70. For a detailed interpretation of the translation, see Shalev, "Benjamin of Tudela," 17–33.

71. Also see Zur Shalev, *Sacred Words and Worlds: Geography, Religion, and Scholarship, 1550–1700* (Leiden, 2012).

72. Benito Arias Montano to Juan de Ovando, Antwerp, October 6, 1571, in Marcos Jiménez de la Espada, *El codigo Ovandino* (Madrid, 1891), 34.

73. Benito Arias Montano, "Benito Arias Montano saluda al ilustrísimo señor Juan de Ovando, Presidente de los Reales Consejos de Indias y de Hacienda," in *La correspondencia de Benito Arias Montano con el presidente de Indias Juan de Ovando*, trans. Baldomero Macías Rosendo (Huelva, 2008), 357.

74. See Fernando Navarro Antolín, Luis María Gómez Canseco, and Baldomero Macías Rosendo, "Fronteras del humanismo: Arias Montano y el Nuevo Mundo," in *Orbis incognitvs. Avisos y legajos del Nuevo Mundo: Homenaje al profesor Luis Navarro García*, ed. Fernando Navarro Antolín (Huelva, 2007), vol. 1, 101–136.

75. Benito Arias Montano, *Historia de la naturaleza: Primera parte del Cuerpo de la Obra magna*, ed. Fernando Navarro Antolín, trans. Andrés Oyola Fabián, Regla Fernández Garrido, Guillermo Galán Vioque and Solís de los Santos (Huelva, 2002), 355.

76. Portuondo, *The Spanish Disquiet*, 178–179.

77. Benito Arias Montano, "Phaleg o libro sobre los primeros asentamientos de las tribus y su lugar en el orbe de la tierra," ed. and trans. Fernando Navarro Antolín, in *Antigüedades hebráicas*, 366.

78. Arias Montano, "Phaleg o libro sobre los primeros asentamientos de las tribus y su lugar en el orbe de la tierra," 366.

79. Arias Montano, "Phaleg o libro sobre los primeros asentamientos de las tribus y su lugar en el orbe de la tierra," 366.

80. Abraham Ortelius, *Synonymia geographica sive populorum, regionum, insularum, urbium* [. . .] (Antwerp: Christopher Plantin, 1578), 235.

81. The first two books of the *Moral and Natural History of the Indies* initially appeared in Latin and accompanied Acosta's meditation about conversion entitled *De procuranda indorum salute* (1588). On Acosta's ideas about accommodation in Christian teaching, see Sabine MacCormack, *Religion in the Andes: Vision and Imagination in Early Colonial Peru* (Princeton, NJ, 1991), 249–280. On Acosta's ethnography and its reception see Anthony Pagden, *The Fall of Natural Man: The American Indian and the Origins of Comparative Ethnology* (Cambridge, 1987), 146–197.

82. José de Acosta, *Historia moral y natural de las Indias* (Seville: Impresso en la casa de Juan de Leon, 1590), proemio al lector. I have also consulted the modern critical English edition, which builds upon the annotations of Edmundo O'Gorman. See José de Acosta, *Natural and Moral History of the Indies*, trans. Frances López Morillas (Durham, NC, 2002).

83. Fermín del Pino Díaz, "*La historia natural y moral de las Indias* como género: Orden y génesis literaria de la obra de Acosta," *Historica* 24, no. 2 (2000): 295–326. For a biography of Acosta, see Simón Valcárcel Martínez, "El Pàdre José de Acosta," *Thesaurus* 44, no. 2 (1989): 389–428; For a recent biography of Acosta, see Claudio M. Burgaleta, *Jose de Acosta, S.J. (1540–1600): His Life and Thought* (Chicago, 1999); and Andrés Prieto, *The Theologian and the Empire: A Biography of José de Acosta (1540–1600)* (Leiden, 2024).

84. On the organization of the book and its reliance on the idea of the "Great Chain of Being," see Mackenzie Cooley, *The Perfection of Nature: Animals, Breeding, and Race in the Renaissance* (Chicago, 2022), 177–195, esp. 183.

85. For a detailed discussion of the implications of the Ophir identification and its reception in the rest of Europe, see Gliozzi, *Adamo e il nuovo mondo*, 152–184.

86. Acosta, *Historia moral y natural de las Indias*, book 1, chap. 13, 49–54.

87. Acosta, *Historia moral y natural de las Indias*, book 1, chap. 13, 49–51.

88. Ann Blair, "Mosaic Physics and the Search for a Pious Natural Philosophy in the Late Renaissance," *Isis* 91 (2000): 32–58; Martin Muslow, "Ambiguities of the Prisca Sapientia in Late Renaissance Humanism," trans. Janita Hämäläinen, *Journal of the History of Ideas* 65 (2004): 1–13.

89. Acosta, *Historia moral y natural de las Indias*, book 1, chapter 16, 56–61.

90. Jean Bodin, *Method for the Easy Comprehension of History*, trans. Beatrice Reynolds (New York, 1969), 301.

91. Anthony Grafton, "Renaissance Histories of Art and Nature," in *The Artificial and the Natural: An Evolving Polarity*, ed. Bernadette Bensaude-Vincent and William R. Newman (Cambridge, MA, 2007), 185–210, 196.

92. See *Renaissance Invention: Stradanus's Nova Reperta*, ed. Lia Markey (Evanston, IL, 2020).

93. Acosta, *Historia moral y natural de las Indias*, book 1, chapter 16, 56–61.

94. Luis de León, *Escritos sobre América*, ed. and trans. Andrés Moreno Mengíbar and Juan Martos Fernández (Madrid, 1999), 50–51.

95. Luis de León, *Escritos sobre América*, 76–77.

96. Luis de León, *Escritos sobre América*, 70–71.

97. For controversies surrounding the reading of Obadiah and the use of the toponym Sefarad for Spain, see Beaver, "Nebuchadnezzar's Jewish Legions," 39–51.

98. Acosta, *Historia moral y natural de las Indias*, book 1, chap.15, 54–56. For a study of Acosta's historical method and its sixteenth-century context, see Anthony Grafton, "José de Acosta: Renaissance Historiography and New World Humanity," in *The Renaissance World*, ed. John Jeffries Martin (New York, 2007), 166–188.

99. Acosta, *Historia moral y natural de las Indias*, book 1, chap. 22, 73–78.

100. Acosta, *Historia moral y natural de las Indias*, book 1, chap. 23, 78–82.

101. Acosta, *Historia moral y natural de las Indias*, book 1, chap. 25.

102. Juan de Tovar, "Historia de la Benida de los Yndios a poblar a Mexico de las partes remotas de Occidente los sucesos y pregrinaciones del camino su govierno," JCB, ms. Codex Ind 2, 1r.

103. Acosta, *Historia moral y natural de las Indias*, book 7, chap. 2, 452–455.

104. Acosta, *Historia moral y natural de las Indias*, book 7, chap. 3, 455–458.

105. "Y es de adveritir que aunque dizen que salieron de siete cuevas no es por que avitaran en ellas porque tenian sus casas y sementeras con mucho orden y policia de republica, sus Dioses, ritos y ceremonias por ser gente muy politica como se hecha bien de ver en el modo y traça de los del nuevo Mexico de donde ellos vinieron que son muy conformes en todo. Usase en aquella provincia tener cada linaje su sitio y lugar conocido, el qual señalavan en una cueva, diciendo la cueva de tal y linaje o desendencia, como en españa se dize la casa de los Velascos, Mendoça." Juan de Tovar, "Historia de la Benida de los Yndios a poblar a Mexico de las partes remotas de Occidente los sucesos y pregrinaciones del camino su govierno," JCB, ms. Codex Ind 2, 1v.

106. Acosta, *Historia moral y natural de las Indias*, book 7, chap. 3, 455–458.

107. Acosta, *Historia moral y natural de las Indias*, book 3, chap. 27.

108. Acosta, *Historia moral y natural de las Indias*, book 6, chap. 8, 410–412.

109. "Vi entonces toda esta historia con caracteres y Hieroglificas que yo no entendia y assi fue necesario que los sabios de Mexico Tezcuco y Tulla se viesen conmigo por mandado del mismo Virrey y con ellos yendome diziendo y narrando las cosas en particular hize una historia bien cumplida." Juan de Tovar, "Historia de la Benida de los Yndios a poblar a Mexico de las partes remotas de Occidente los sucesos y pregrinaciones del camino su govierno," JCB, ms Codex Ind 2, "Respuesta del Padre Joan de Tovar."

110. "Y esta es la autoridad que eso tiene que para mi es mucha porque demas de que lo vi en sus mismos libros lo trate antes del cocolistle con todos los ancianos que supe sabian de esto y niguno discrepa como cosa muy notoria entre ellos." Juan de Tovar, "Historia de la Benida de los Yndios a poblar a Mexico de las partes remotas de Occidente los sucesos y pregrinaciones del camino su govierno," JCB, ms Codex Ind 2, "Respuesta del Padre Joan de Tovar."

111. "Digo como queda referido que tenian figuras y Hieroglificos con que pintavan las cosas en esta forma que las cosas que tenian figuras las ponian con sus propias ymagines y para las cosas que no avya ymagen propia tenian otros caracteres significativos de aquello y con estas cosas figuravan cuanto querian. . . . Pero es de advertir que aunque tenian diversas figuras y caracteres con que escrivian las cosas no eran tan suficientemente como nra escritura que sin discrepar por las mismas palabras refierese cada uno lo que esta escrito solo concordavan en los conceptos pero para tener memoria entera de las palabras y traça de los parlamentos que hacian los oradores y los muchos cantares que todos sabian sin discrepar palabra . . . aunque los figuravan con sus caracteres-pero para conservarlos con las mismas palabras que los dixeron, sus oradores y poetas avya cada dia exercicio de ello en los collegios de los moços principales." Juan de Tovar, "Historia de la Benida de los Yndios a poblar a Mexico de las partes remotas de Occidente los sucesos y pregrinaciones del camino su govierno," JCB, ms Codex Ind 2, "Respuesta del Padre Joan de Tovar."

112. Acosta, *Historia moral y natural de las Indias*, book 6, chap. 5, 408–411.

113. A manuscript of Arias Montano's poems reveals that the Spanish Hebraist also belonged, at least during his university years, to the Guevara's circle. In this poetic work, Montano dedicates a number of verses to Diego de Guevara, alongside the poems in honor of

his teacher Cipriano de la Huerga and of the Sevillian chronicler Pedro Mexía (1497–1551), the doctor Francisco de Arce (1493–1580), and Ambrosio de Morales. As Antonio Cerrano Cueto has shown, during his studies at Alcalá de Henares (c. 1548–1552), Montano participated in the poetic competitions that the university celebrated to commemorate special events, even obtaining the title of *poeta laureatus* in 1552, while Morales served as chair of the contest. Diego de Guevara had also been the recipient of this distinction. Antonio Serrano Cueto, "Aportación a la biografía de Diego de Guevara: Cuatro epístolas de Ambrosio de Morales, un epigrama de Arias Montano," *Revista de estudios latinos (RELat)* 5 (2005): 257–274. Also see BNE, Ms. 155. A. Holgado Redondo, "Hacia un corpus de la poesía latina de Benito Arias Montano," *Revista de estudios Extremeños* 2 (1987): 93.

114. Felipe de Guevara, *Comentarios de la pintura que escribió Don Felipe de Guevara; se publican por la primera vez con un discurso preliminar y algunas notas de Don Antonio Ponz* (Madrid: por D. Gerónimo Ortega, Hijos de Ibarra y comp., 1788), 234–36. See Horapollo, *The Hieroglyphics of Horapollo*, trans. George Boas (Princeton, NJ, 1993).

115. Erik Iversen, *The Myth of Egypt and Its Hieroglyphs in the European Tradition* (Princeton, NJ, 1993), 57–87, 64.

116. Brian A. Curran, "'De sacrum litterarum aegyptiorum interpretatione' Reticence and Hubris in Hieroglyphic Studies of the Renaissance: Pierio Valeriano and Annius of Viterbo," *Memoirs of the American Academy in Rome* 43/44 (1998/1999): 139–182; *The Egyptian Renaissance: The Afterlife of Ancient Egypt in Early Modern Italy* (Chicago, 2007).

117. Jorge Cañizares Esguerra, *How to Write the History of the New World: Histories, Epistemologies, and Identities in the Eighteenth-Century Atlantic World* (Stanford, CA, 2002), 70–71. For the history of how Western grammatology understood Maya writing for more than five hundred years using categories like "hieroglyph" and writing, see Byron Ellsworth Hamann, "How Maya Hieroglyphs Got Their Name: Egypt, Mexico, and China in Western Grammatology since the Fifteenth Century," *Proceedings of the American Philosophical Society* 152, no. 1 (March 2008): 1–68. For a study of the structural and historical relations among Maya writing and iconography, see *Word and Image in Maya Culture: Explorations in Language, Writing, and Representation*, ed. William F. Hanks and Don S. Rice (Salt Lake City, 1989). For a comparative study of the linguistic theories of Acosta and Fray Luis de León, see Karl A. Kottman, "Fray Luis de León and the Universality of Hebrew: An Aspect of 16th and 17th Century Language Theory," *Journal of the History of Philosophy* 13, no. 3 (July 1975): 297–310.

118. Benito Arias Montano, *Libro de la generación y regeneración del hombre, o Historia del género humano: Primera parte de la Obra magna, esto es, Alma*, ed. Fernando Navarro Antolín and trans. Luis María Gómez Canseco, Baldomero Macías Rosendo, Miguel Angel Vinagre Lobo, Domingo F. Sanz, introduction by Luis Gómez Canseco (Huelva, 1999), book 3, chap. 4, 235.

119. Benito Arias Montano, *Tratado sobre la fe que había que revelarse / Tractatus tertius de fide, quae revelanda erat; Adán o de la lengua, intérprete del pensamiento humano, y de los rudimentos comunes a todas las lenguas / Adam, sive de humani sensus interprete lengua*, ed. and trans. Fernando Navarro Antolín and Luis Gómez Canseco (Huelva, 2009), 80.

120. "Un tratado de grámatica general y comparada de Benito Arias Montano: *Adam, sive de humani sensus interprete lingua communibusque linguarum rudimentis*," in Maestre Mastre et al., *Benito Arias Montano y los humanistas de su tiempo*, 163–224.

121. See George J. Metcalf, "Theodor Bibliander (1505–1564) and the Language of Japhet's Progeny," in *On Language Diversity and Relationship from Bibliander to Adelung*, ed. T. Van Hal and R. Van Rooy (Amsterdam, 2013), 57–75.

122. Theodor Bibliander, "Argumentum libri," in *De ratione communi omnium linguarum et literarum commentarius*, trans. and ed. Hagit Amirav and Hans-Martin Kim (Geneva, 2011), 20–25.
123. Umberto Eco, *The Search for the Perfect Language*, trans. James Fentress (Malden, MA, 1997), 74.
124. Eco, *The Search for the Perfect Language*, 80.
125. Luis de León, "Prólogo, Cantar de los Cantares," in *Obras Completas Castellanas de Fray Luis de León*, ed. Felix Garcia (Madrid, 1957), 74.
126. Dominique Reyre, "Cuando Covarrubias arrimaba el hebreo a su castellano," in *Criticón*, no. 67 (1997): 5–20.
127. Reyre, "Cuando Covarrubias arrimaba el Hebreo a su Castellano," 19–20.
128. Juan M. Lope Blanch, "Las fuentes americanas del Tesoro de Covarrubias," *Actas del Sexto Congreso de la Asociación Internacional de Hispanistas celebrado en Toronto del 22 al 26 de agosto de 1977*, digital ed. (Toronto: Department of Spanish and Portuguese, University of Toronto, 1980), 467–472. Also see José Durand, "Perú y Ophir en Garcilaso el Inca, el Jesuita Pineda y Gregorio Garcia," in *Histórica* 3, no. 2 (December 1979): 35–55.
129. Sebastian de Covarrubias, *Tesoro de la lengua castellana o española compuesto por el licenciado don Sebastian de Covarrubias Orozco, capellan de su magestad mastrescuela y canonigo de la santa iglesia de Cuenca, y consultor del Santo Oficio de la Inquicision* (Madrid: Luis Sanchez, 1611), 586v–586r.
130. Covarrubias, *Tesoro de la lengua castellana o española*, 168r.
131. Covarrubias, *Tesoro de la lengua castellana o española*,168r.
132. Covarrubias, *Tesoro de la lengua castellana o española*, 170v.
133. Covarrubias, *Tesoro de la lengua castellana o española*, 461r.
134. Covarrubias, *Tesoro de la lengua castellana o española*, 557r. Also see Francisco López de Gomara, *La historia general de las indias y todo lo acaesido en ellas dende que se ganaron hastaagora y la conquista de Mexico y de la Nueva España* (Antwerp: Martin Nuncio, 1554), chap. 67. Also see Lope Blanch, "Las fuentes americanas del Tesoro de Covarrubias," 468.

EPILOGUE

1. Antonio León Pinelo, *Epitome de la biblioteca oriental y occidental, nautica i geografica de don Antonio de León Pinelo* (Madrid: por Juan Gonzalez, 1629), discurso apologetico.
2. León Pinelo, *Epitome*, tabla de las lenguas, *4r.
3. Lorenzo Hervás y Panduro, *Catálogo de las lenguas de las naciones conocidas, y numeración, division, y clases de éstas, según la diversidad de sus idiomas y dialectos*, 6 vols. (Madrid: Imprenta de la Administración del Real Arbitrio de Beneficiencia, 1800). For a study of Hervás y Panduro's approach, see Manuel Breva Claramonte, "Data Collection and Data Analysis in Lorenzo Hervás: Laying the Ground for Modern Linguistic Typology," in *History of Linguistics in Spain / Historia de la lingüística en España*, ed. E. F. K Koerner and Hans Josef Niederehe (Philadelphia, 2001), vol. 2, 265–280.
4. "Señor Dn Zenón Alonso. Mi estimadísimo Amigo y Señor. Ha llegado la ocasión de hacer ver la anterioridad de mis proyectos en este ramo de amena literatura, en que desfallecí por falta de auxilios y protección. Desde mi llegada a este Reyno puse en execución mis designios de formar la colección de libros impresos y manuscritos principalmente en los idiomas de nras. Americas, y formar las listas de las palabras mas comunes en defecto a los vocabularios completos. Mi fin se dirijia a depositar estos tesoros en alguna Academia de las bellas letras, rezelando quan precipitadamente caminaban estos idiomas a la región del olvido con la extinción de estas barbaras naciones; y viendo al mismo tiempo desde lejos que

debia renacer el gusto por estas preciosas antigüedades, pero tal vez con el desconsuelo imponderable no de hallarlas, ni de saber si existieron." Letter from José Celestino Mutis, AGI, INDIFERENTE, 1342A, N.1, Doc.5a, fols. 16r–17v.

5. See Justo Pastor Benítez, *La vida solitaria del doctor Gaspar Rodríguez de Francia* (Asunción, 1984).

6. Peter Burke, *Languages and Communities in Early Modern Europe* (Cambridge, 2004), 160–172; Benedict Anderson, *Imagined Communities: Reflections on the Origins and Spread of Nationalism* (London, 1983), 47–66; and José del Valle and Luis Gabriel-Stheeman, "Nacionalismo, hispanismo, y cultura monoglósica," in *La batalla del idioma: La intelectualidad hispánica ante la lengua* (Frankfurt, 2004), 15–34.

7. See Bárbara Cifuentes, "The Politics of Lexicography in the Mexican Academy in the Late Nineteenth Century," and Elvira Narvaja de Arnoux, "Grammar and the State in the Southern Cone in the Nineteenth Century," in *A Political History of Spanish: The Making of a Language*, ed. José del Valle (Cambridge, 2013), 167–78, 152–166.

8. José del Valle, "Lingüística histórica e historia cultural: Notas sobre la polémica entre Rufino José Cuervo y Juan Valera," in *La batalla del idioma: La intelectualidad hispánica ante la lengua*, ed. José del Valle and Luis Gabriel-Stheeman (Frankfurt, 2004), 93–108.

9. Rufino José Cuervo, *Diccionario de construcción y régimen de la lengua castellana* (París, 1886), xxi–xxii.

10. Rufino José Cuervo, *Apuntaciones críticas sobre el lenguaje bogotano* (Bogotá, 1867–1872), vi–vii.

Selected Bibliography

This bibliography does not cite all the sources that appear in the notes but focuses on the primary works cited throughout the book. The notes provide full citations of secondary works.

MANUSCRIPT SOURCES

Archivo General de Indias, Seville

Consulta del Consejo de Indias 03–20 Madrid, 1578." INDIFERENTE, 739, N.60.
"Expediente causado con motivo de ciertas noticias pedidas por la Emperatriz de Rusia, Catalina II, sobre lenguas indígenas de las provincias españolas de Ultramar, para la realización de un Diccionario Universal." 1787–1792. INDIFERENTE, 1342A, N.1.

Benson Latin American Collection, University of Texas, Austin

"Ameca, 1579." In *Relaciones Geográficas of Mexico and Guatemala, 1577–1585.* Nettie Lee Benson Latin American Collection, University of Texas Libraries, University of Texas at Austin. Box 1, folder XXIII-10. 14 folios.
Cabello Valboa, Miguel. *Miscelanea Anthartica, donde se describe, el origen, de nuestros Indios Occidentales, deduzido desde Adan, y la Erection y principia del imperio de los Reyes Ingas de el Piru. Vidas y guerras que tu vieron: cosas notables q hicieron, computados los años de sus nascimientos y muertes, y de lo q por el Universso y va subcediendo; durantes sus edades y tiempos. Dirigido a Don Ferdo. de Torres y Portugal, conde del Villar, vissorrey gouer. y capn. general de estos reynos y prouincias del Piru: electo año -1584. Por Miguel Cabello Valboa clerigo presbitero del Arçobispado de los Reyes en el Piru. Natural de la Villa de Archidona en Andaluzia, Año -1586.* JGI 1946.

Biblioteca Capitular y Colombina, Seville

Cobo, Bernabé. *Fundación de Lima, escripta por el padre Bernabé Cobo de la Compañia de Jesús* [año de 1639]. Ms. 58-04-11.
Cueva, Juan de la. *Obras de Juan de la Cueva* [1603–1608]. Ms. 56-03-04, 56-03-05.
Cueva, Juan de la. *Segundo Coro febeo de romances historiales* [seventeenth century]. Ms. 56-03-06.

Biblioteca Medicea Laurenziana, Florence

Sahagún, Bernardino de. *Historia general de las cosas de Nueva España*. Med.Palat.218-220.
https://florentinecodex.getty.edu.

Biblioteca Nacional de España, Madrid

Cervantes de Salazar, Fransico. *Crónica de la Nueva España, su descripción, la calidad y temple de ella, la propiedad y naturaleza de los indios*. Ms. 2011.
Hernández, Francisco. *Historia natural libros i, ii, iii, iv, y v*. Ms. 2869.
Hernández, Francisco. *Historia natural libros i, ii, iii*. Ms. 2862.
Hernández, Francisco. *Historia natural libros iv, v y xi*. Ms. 2863.
Hernández, Francisco. *Historia natural libros ix, x, y xi*. Ms. 2865.
Hernández, Francisco. *Historia natural libros vii, x, xi y xii*. Ms. 2870.
Hernández, Francisco. *Historia natural libros vii y viii*. Ms. 2864.
Hernández, Francisco. *Historia natural libros xiii, xiv, xv*. Ms. 2871.
Hernández, Francisco. *Historia natural libros xii, xiii, xiv, xv, xvi, y xvi*. Ms. 2866.
Hernández, Francisco. *Historia natural libros xvii, xviii, xix, y xx*. Ms. 2867.
Hernández, Francisco. *Historia natural libros xxi, xxii, xxiii, xiv, y xv*. Ms. 2868.
Hernández, Francisco. *Obras del bienaventurado Sanct Dionisio Areopagita, traducidas por el Doctor Francisco Hernandez, medico e historiador de Philippo segundo y su protomedico general en todas las Yndias Occidentales*. Ms. 10813.
Lull, Raymond. *Blanquerna*. Translated by José Andreu. Ms. 5611.
Ordenanzas de la Hacienda Real (en Indias). Ms. 3035, 42r–53r.
Páez de Castro, Juan. "Notas y adiciones a la dedicatoria de una historia anónima del Emperador Carlos V" [c. 1550]. Ms. 23083/4.
Páez de Castro, Juan. "Método para escribir la Historia, por el doctor Juan Páez de Castro, cronista del emperador Carlos V" [1701–1800]. Ms. 18637/1.
Páez de Castro, Juan. "Método para escribir la Historia, por el doctor Juan Páez de Castro, cronista del emperador Carlos V." Ms. 5578, fols. 77r–131v.
Papeles Varios. Ms. 9934.

Bibliothèque nationale de France, Paris

Codex Ixtlilxochitl. Ms. Mexicain 65-71. https://gallica.bnf.fr/ark:/12148/btv1b84701752.

Biblioteca Real del Palacio, Madrid

Oviedo, Gonzalo Fernández de. *Tercera parte de la Historia natural y general de las Indias, Islas y Tierra Firme del Mar Océano, vista y examinada por el Consejo Real por mandado del emperador nuestro señor, la qual hasta el presente no se ha dado al público*. PR II/3042.
Zorita, Alonso de. *Historia de la Nueva España* (1585). PR RB II/59.

Huntington Library, Pasadena, CA

Oviedo, Gonzalo Fernández de. *Historia general y natural de las Indias* [1539–1548]. Ms. HM 177. 2 vols.

John Carter Brown Library, Providence, RI

Recchi, Nardo. *De materia medica Novae Hispaniae Philippi Secundi Hispaniarum ac Indiarum regis invictissimi iussu.* Codex Lat 5.
Tovar, Juan de. *Historia de la benida de los Yndios a poblar a Mexico de las partes remotas de occidente los sucesos y pregrinaciones del camino su govierno.* Codex Ind 2.

Real Academia de la Historia

Relaciones topográficas de los pueblos de España hechas por orden del Sr. Felipe II: Copiadas de las originales que existen en la Real Biblioteca de El Escorial y se pasaron a la Academia en virtud de orden de S.M. para sacar la copia [año de 1773]. Ms. 9/3954-60.

Real Biblioteca de San Lorenzo de El Escorial, El Escorial

&.IV.22, fols. 127–129v., 133, and 137v.
b.III.1, fols. 173r–197v.
I-III-31, fols. 177r–205v.
K-III-8, fols. 331r–390v.
L-I-2, fols. 123r–134v.

Special Collections, Princeton University Library, Princeton, NJ

Paéz de Castro, Juan. Princeton Ms. 174.

ANNOTATED BOOKS

Real Biblioteca de San Lorenzo de El Escorial, El Escorial

Cieza de León, Pedro. *Primera parte de la cronica del Peru.* Antwerp: Martin Nuncio, 1554. Esc. 20.VI.17.
López de Gómara, Francisco. *La historia general de las Indias y todo lo acaecido en ellas dende que se ganaron hasta ahora y de la conquista de Mexico y de la nueva España.* Antwerp: Martin Nuncio, 1554. Esc. 60.IV.29.

Special Collections, Princeton University Library, Princeton, NJ

Núñez, Hernán. *Refranes o proverbios en romance.* Salamanca: Juan de Canova, 1555. EXOV 3166.675.21.

PRINTED SOURCES BEFORE 1800

Agustín, Antonio. *Dialogos de medallas, inscripciones y otras antiguedades.* Tarragona: Felipe Mey, 1587.
Alcocer, Pedro de. *Hystoria, o descripcion dela Imperial cibdad d Toledo, con todas la cosas acontecidas en ella, desde su principio, y fundacion, adonde se tocan, y refieren muchas*

antiguedades, y cosas notables de la hystoria general de España, agora nueuamente impressa. Toledo: Juan Ferrer, 1554.

Aldrete, Bernardo Jose de. *Del origen y principio de la lengua castellana o romance que oi se usa en España*. Rome: Carlo Wllietto, 1606.

Arias Montano, Benito. *Benedicti Ariae Montani Hispalensis Commentaria in duodecim prophetas*. Antwerp: Christophe Plantin, 1571.

Arias Montano, Benito. *Itinerarium Benjamini Tudelensis [. . .] ex Hebraico Latinum factum a B. Aria Montano interprete*. Antwerp: Christophe Plantin, 1575.

Arias Montano, Benito. *Liber generationis et regenerationis Adam, sive, De historia generis humani: Operis magni pars prima, id est, Anima*. Antwerp: Jan Moretus, 1593.

Arias Montano, Benito. *Antiquitatum judaicarum libri IX*. Antwerp: Franciscus Raphelengius, 1593.

Argote de Molina, Gonzalo. *El conde Lucanor compuesto por el excelentissimo principe don Juan Manuel, hijo del Infante don Manuel, y nieto del sancto rey don Fernando*. Sevilla: en casa de Hernando Diaz, 1575.

Argote de Molina, Gonzalo. *Nobleza de Andaluzia*. En Sevilla: Fernando Diaz, 1588.

Biblia Sacra Hebraice, Chaldaice, Graece, et Latine: Philippi II. Reg. Cathol. pietate, et studio ad Sacrosanctae Ecclesiae usum. Edited by Benito Arias Montano. 8 vols. Antwerp: Christophe Plantin, 1569–1573.

Belon, Pierre. *Petri Bellonii Cenomani de aquatilibus*. Paris: Apud C. Stephanum, 1553.

Bodin, Jean. *Methodus ad facilem historiarum cognitionem, ab ipso recognita, et multo quam antea locupleitor. Cum indice rerum memorabilium copiosissimo*. Paris: Martinus Iuvenis, 1572.

Cervantes de Salazar, Francisco. *Obras q[ue]Francisco Ceruantes de Salazar, ha hecho, glosado, y traduzido*. Alcalá de Henares: Juan de Brocar, 1546.

Cittadini, Celso. *Trattato della vera origine, e del processo e nome della nostra lingua: Scritto in vulgar sanese*. Venice: Appresso Gio. Battista Ciotti, 1601.

Covarrubias y Orozco, Sebastián de. *Tesoro de la lengua castellana o española*. Madrid: Luis Sánchez, 1611.

Diccionario de Autoridades. 6 vols. En Madrid: En la Imprenta de la Real de la Academia Española, 1726–1739.

Echave, Baltasar de. *Discursos de la lengua cántabra-bascongada compuestos por Balthasar de Echave, natural de la Villa de Çumaya en la Provincia de Guipuzcoa, y vezino de Mexico*. México: Henrrico Martínez, 1607.

Farfán, Agustín. *Tractado brevede medicina*. México: en casa de Pedro Ocharte, 1592.

Fuentes, Alfonso de. *Quarenta cantos de diversas y peregrinas historias*. Sevilla: Domenico de Robertis, 1550.

Garibay y Zamalloa, Esteban de. *Los XL libros d'el compendio historial de las chronicas y vniuersal historia de todos los reynos de España*. Antwerp: Christophe Plantin, 1571.

Gessner, Conrad. *Conradi Gesneri Historiae animalium*. 5 vols. Zürich: C. Froschouerum, 1551–1587.

Gómez Ortega, Casimiro. "Ad lectorem prefatio." In Francisco Hernández, *Opera: cum edita, tum inedita, ad autographi fidem et integritatem expressa, impensa et jussu regio*, edited by Casimiro Gómez Ortega. 3 vols., vol. 1. Madrid: ex typographia Ibarrae Heredum, 1790.

Guevara, Felipe de. *Comentarios de la pintura que escribió Don Felipe de Guevara; se publican por la primera vez con un discurso preliminar y algunas notas de Don Antonio Ponz*. Madrid: por D. Gerónimo Ortega, Hijos de Ibarra y comp., 1788.

Hernández, Francisco. *Rerum medicarum Novae Hispaniae thesaurus, seu, Plantarum animalium mineralium mexicanorum historia*. Rome: ex typographeio Vitale Mascardi, 1651.

Hernández, Francisco. *Nova plantarum, animalium et mineralium Mexicanorum historia.* Rome: Vitale Mascardi, 1651.

Herrera y Tordesillas, Antonio de. *Historia general de los castellanos en las islas y tierra firme del mar oceano escrita por Antonio de Herrera cronista de castilla y mayor de las Indias.* 9 vols. En Mad[rid]: En la Emplenta Real, 1601–1615.

Instructio[n], y memoria, de las relaciones que se han de hazer, para la descripcion de las Indias, que Su Magestad manda hazer para el buen gouierno y ennoblescimiento dellas [1577]. John Carter Brown Library, Brown University, Providence, RI. BB S7333 1577 1.

Isidore of Seville, *Diui Isidori Hispal. Episcopi Opera* [. . .]: *E vetustis exemplaribus emendata.* Madrid: Typographia Regia, 1599 (apud Ioannem Flandrum, 1597).

Laguna, Andrés de Pedacio Dioscorides Anazarbeo. *Acerca de la materia medicinal, y de los venenose mortiferos, traduzido de lengua Griega, en la vulgar Castellana, & illustrado con claras y substantiales annotationes, y con las figuras de innumeras plantas exquisitas y raras, por Doctor Andres de Laguna Medico de Iulio III. Pont. Maxi.* Salamanca: Mathias Gast, 1570.

Landino. Christophoro. *Historia naturale* [. . .] *di latino in volgare tradotta per Christophoro Landino.* Venice: Gabriel Iolito di Ferrarii, 1543.

León, Luis de. *De los nombres de Christo: En tres libros. Quarta impression en que va añadido el nombre de Cordero, con tres tablas, la vna de los nombres de Christo, otra de la perfecta Casada, la tercera de los lugares de la Scriptura.* Salamanca: en casa de Iuan Fernandez, 1595.

León Pinelo, Antonio. *Epitome de la bibliotheca oriental y occidental, nautica y geográfica de don Antonio de León Pinelo.* Madrid: por Juan Gonzalez, 1629.

Lopes de Castanheda, Fernaõ. *Ho livro primeiro dos dez da historia do descobrimento & conquista da India pelos Portugueses.* Coimbra: Por Iõao da Barreyra, 1554.

López de Velasco, Juan. *Orthographia y pronunciacion castellana.* Burgos, 1582.

Mal Lara, Juan. *La philosophia vulgar de Ioan Mal Lara vezino de Sevilla, primera parte que contiene mil refranes glosados.* Seville: En casa de Hernando Diaz, 1568.

Mariana, Juan de. *Historia general de España compuesta enmendada y añadida, por el padre Juan de Mariana; Con el sumario y tabla.* 2 vols. Madrid: Joaquín de Ibarra, 1780.

Martí de Viciana, Rafael. *Libro de las alabanças de las lenguas hebrea, griega, latina, castellana, y valenciana. Copiado por Marin de Viziana y consagrado al Ilustre Senado de la Inclyta y coronada ciudad de Valencia.* Valencia: J. Navarro, 1574.

Martire d'Anghiera, Pietro. *De orbe novo decades* [. . .] *cura & diligentia.* [. . .] *Antonii Nebrissensis*[. . .]. Alcalá de Henares: Arnaldi Guillelmi de Brocario, 1516.

Mayans y Siscar, Gregorio. *Orígenes de la lengua española, compuestos por varios autores.* Madrid: Juan de Zúñiga, 1737.

Molina, Alonso de. *Aqui comiença un vocabulario en lengua castellana y mexicana, compuesto por el muy reverendo padre fray Alonso de Molina: Guardian del convento de Sant Antonio de Tetzcuco de la orden de los frayles menores.* México: Juan Pablos, 1555.

Molina, Alonso de. *Arte de la lengua mexicana y castellana compuesta por el muy R.P. Fray Alonso De Molina, de la orden de señot Sant Francisco, de nueuo en esta segunda impression corregida, emendada y añádida, mas copiosa y clara que la Primera.* México: En casa de Pedro Balli, 1576.

Monardes, Nicolás. *Historia medicinal de las cosas que se traen de nuestras Indias Occidentales.* Seville: en casa de Alonso Escrivano, 1574.

Morales, Ambrosio de. *La coronica general de España, que continuaua Ambrosio de Morales* [. . .]; *prossiguiendo adelante de los cinco libros, que el Maestro Florian de Ocampo* [. . .] *dexo escritos* [. . .]. Alcalá de Henares: Juan Iñiguez de Léquerica, 1574.

Morales, Ambrosio de. *Las antiguedades de las ciudades de España que van nombradas en la coronica* [. . .]. Alcalá de Henares: En Casa de Juan Iñiguez de Lequerica, 1575.

Morales, Ambrosio de. *Apologia de Ambrosio de Morales, con una información al Consejo del Rey Nuestro Señor, hecha por su orden y mandamiento en defensa de los Anales de Geronymo Çurita.* Zaragoza: en el Colegio de S. Vicente Ferrer por Iuan de Lanaja y Quartanet, 1610.

Morales, Ambrosio de. *Las antigüedades de las ciudades de España: que van nombradas en la Coronica con la aueriguacion de sus sitios y nombres antiguos* [1575]. 2 vols. Madrid: en la Oficina de Don Benito Cano, 1792.

Morales, Ambrosio de. *Opúsculos castellanos de Ambrosio de Morales.* 3 vols. Edited by Francisco Valerio Cifuentes. Madrid: en la Oficina de Don Benito Cano, 1793.

Nebrija, Elio Antonio de. *Dictionarium hispano latinum.* Salamanca: Juan de Porras, 1495.

Nieremberg, Juan Eusebio. *Historia naturae, maxime peregrinae, libris XVI. distincta.* Antwerp: Balthazar Moretus, 1635.

Ocampo, Florián de. *Los cinco primeros libros de la Coronica general de España que recopila el maestro Florian do Campo.* Medina del Campo: Guillermo de Millis, 1553.

Oviedo, Gonzalo Fernández de. *Ouiedo de la natural hystoria de las Indias.* Toledo: Ramon Petras, 1526.

Oviedo, Gonzalo Fernández de. *La historia general de las Indias.* Sevilla: Juan Cromberg, 1535.

Oviedo, Gonzalo Fernández de. *Coronica de las Indias: La hystoria general de las Indias agora nueuamente impressa corregida y emendada.* Salamanca: Juan de Junta, 1547.

Oviedo, Gonzalo Fernández de. *Libro XX dela segunda parte dela general historia delas Indias.* Valladolid: por Francisco Fernandez de Cordoua, Impressor de su Magestad, 1557.

Ortelius, Abraham. *Synonymia geographica sive populorum, regionum, insularum, urbium* [. . .] *&c. variae, pro auctorum traditionibus, saeculorum intervallis, gentiùmque idiomatis & migrationibus, appellationes & nomina* [. . .]. Antwerp: Christophe Plantin, 1578.

Ortelius, Abraham. *Theatre de l'univers: Contenant les cartes de tout le monde. Avec une brieve declaration d'icelles* [translation of *Theatrum orbis terrarum*, 1595]. Antwerp: Plantin Press, 1598.

Perez de Oliva, Fernán. *Las obras del maestro Fernan Perez de Oliua natural de Cordoua* [. . .] *con otras cosas que van añadidas, como se dara razon luego al principio.* Edited by Ambrosio de Morales. Cordoba: Gabriel Ramos Bejarano, 1586.

Poza, Andrés de. *De la antigua lengua, poblaciones, y comarcas de las Españas, en que de paso se tocan algunas cosas de la Cantabria.* Bilbao: Mathias Mares, 1587.

Rhodiginus, Lodovicus Caelius. *Lodouici Caelij Rhodigini Lectionum antiquarum libri xxx.* Basel: Per Ambrosium et Aurelium Frobenios fratres, 1550.

Román y Zamora, Jerónimo. *Republicas del mundo divididas en XXVII libros.* 2 vols. Medina del Campo: Francisco del Canto, 1575.

Román y Zamora, Jerónimo. *Republicas del mundo divididas en tres partes.* Salamanca: En casa de Iuan Fernández, 1595.

Rondelet, Guillaume. *Libri de piscibus marinis, in quibus verae piscium effigies expressae sunt.* 2 vols. Lyon: apud Matthiam Bonhomme, 1554–1555.

Santo Tomás, Domingo de. *Gramática o arte de la lengua general de los Indios de los reinos del Perú. Nuevamente compuesta por el maestro Fray Domingo morador en los dichos reynos.* Valladolid: Francisco Fernandez de Cordoua, 1560.

Santo Tomás, Domingo de. *Lexicón o Vocabulario de la lengua general del Perú llamada quichua.* Valladolid: Francisco Fernandez de Cordoua, 1560.

Sorapán de Rieros, Juan. *Medicina española contenida en proverbios vulgares de nra. lengua, muy provechosa para todo género de estados, para philosophos, y médicos, para teólogos, y juristas para el buen regimiento de la salud, y más larga vida.* Granada: Por Martin Fernandez Zambrano, 1616.

Valeriano, Pierio. *Hieroglyphica, sive, De sacris Aegyptiorum literis commentarii.* Basel: M. Isengrin, 1556.

Viciana, Rafael Martí de. *Libro de las alabanças de las lenguas hebrea, griega, latina, castellana, y valenciana, compiladas por Marin de Viciana y consagradas al Ilustre Senado de la Inclyta y coronada ciudad de Valencia.* Valencia: J. Navarro, 1574.

Zárate, Agustín de. *Historia del descubrimiento y conquista del Peru, con las cosas naturales que señaladamente allí se hallan y los successos que ha avido. La cual escrivia Agustin de Çarate, exerciendo el cargo de Contador de cuentas por su Magestad en aquella provinvia, y en Tierra firme.* Antwerp: En casa de Martin Nuncio, 1555.

Zurita, Jerónimo, *Anales de la Corona de Aragón. Va añadida de nuevo, en esta impresión, en el ultimo tomo una Apología de Ambrosio de Morales, con un parecer del Doctor Juan Páez de Castro, todo en defensa de estos anales.* 6 vols. Zaragosa, Lorenço de Robles, 1610.

SOURCES AFTER 1800

Alcocer, Pedro de. *Hystoria, o descripcion dela Imperial cibdad d Toledo, con todas la cosas acontecidas en ella, desde su principio, y fundacion, adonde se tocan, y refieren muchas antiguedades, y cosas notables de la hystoria general de España, agora nueuamente impressa* [Toledo: J. Ferrer, 1554]. Clásicos Toledanos. Toledo: Instituto Provincial de Investigación y Estudios Toledanos, 1973.

Acosta, José de. *Natural and Moral History of the Indies.* Edited by Jane E. Mangan. Translated by Frances López Morillas. Durham, NC: Duke University Press, 2002.

Anghiera, Pietro Martire. *Décadas del Nuevo Mundo.* Translated and edited by Edmundo O'Gorman. 2 vols. México: J. Porrúa, 1964.

Apologia: L'autodifesa di Pico di fronte al Tribunale dell'Inquisizione. Edited by Paolo Fornaciari. Florence: Edizioni del Galluzzo, 2010.

Arias Montano, Benito. *Libro de la generación y regeneración del hombre, o Historia del género humano: Primera parte de la Obra magna, esto es, Alma.* Edited by Fernando Navarro Antolín. Translated by Luis María Gómez Canseco, Baldomero Macías Rosendo, Miguel Angel Vinagre Lobo, and Domingo F. Sanz. Introduction by Luis Gómez Canseco. Huelva: Universidad de Huelva, 1999.

Arias Montano, Benito. *Historia de la naturaleza: Primera parte del Cuerpo de la Obra magna.* Edited by Fernando Navarro Antolín. Translated by Andrés Oyola Fabián, Regla Fernández Garrido, Guillermo Galán Vioque and Solís de los Santos. Introduction by Luis Gómez Canseco. Bibliotheca Montaniana. Huelva: Servicios de Publicación de la Universidad de Huelva, 2002.

Arias Montano, Benito. *Libro de José o sobre el lenguaje arcano.* Edited and translated by Luis Gómez Canseco, Fernando Navarro Antolín, and Baldomero Macías Rosendo. Huelva: Universidad de Huelva, 2006.

Arias Montano, Benito. *Prefacios de Benito Arias Montano a la Biblia Regia de Felipe II.* Edited and translated by María Asunción Sánchez Manzano. Humanistas españoles 32. Salamanca: Junta de Castilla y León, Consejería de educación y cultura, 2006.

Arias Montano, Benito. *Tratado sobre la fe que había que revelarse / Tractatus tertius de fide, quae revelanda erat: Adán o de la lengua, intérprete del pensamiento humano, y de los*

rudimentos comunes a todas las lenguas / Adam, sive de humani sensus interprete lengua.
Edited and translated by Fernando Navarro Antolín and Luis Gómez Canseco. Huelva:
Universidad de Huelva, 2009.

Arias Montano, Benito. *Antigüedades hebráicas: Antiquitatum iudaicarum libri IX; Tratados
exegéticos de la Biblia Regia.* Edited by Luis Gómez Canseco and Sergio Fernández López.
Bibliotecha Montaniana 25. Huelva: Servicio de Publicaciones Universidad de Huelva,
2013.

Arias Montano, Benito, and Juan Ovando. *La correspondencia de Benito Arias Montano con
el presidente de Indias Juan de Ovando.* Translated and edited by Baldomero Macías
Rosendo. Huelva: Servicio de Publicaciones Universidad de Huelva, 2008.

Augustine of Hippo. *The City of God.* Translated by Marcus Dods, DD. New York: Random
House, 1950.

Barreiros, Agustín. "El testamento del Doctor Francisco Hernández." *Boletin de la Real
Academia de la Historia* 94 (1929): 475–497.

Bibliander, Theodor. *De ratione communi omnium linguarum et literarum commentaries.*
Translated and edited by Hagit Amirav and Hans-Martin Kim. Geneva: Librairie
Droz S.A., 2011.

Caesar, Julius. *The Gallic War.* Translated by H. J. Edwards. Loeb Classical Library 72.
Cambridge, MA: Harvard University Press, 1917, 5–14.

Campbell, Joe R. *A Morphological Dictionary of Classical Nahuatl: A Morpheme Index to the
Vocabulario en lengua mexicana y castellana of Fray Alonso de Molina.* Madison, WI:
Hispanic Seminary of Medieval Studies, 1985.

Cárdenas, Juan de. *Problemas y secretos maravillosos de las Indias* [1591]. México: Bibliófilos
Mexicanos, 1965.

Cervantes de Salazar, Francisco. *Crónica de la Nueva España.* Edited by Manuel Magallón y
Cabrera and Agustín Millares Carlo. Madrid: Hispanic Society of America, 1914.

Cervantes de Salazar, Francisco. *Cartas recibidas de España por Francisco Cervantes de
Salazar (1569–1575).* Edited by Agustín Millares Carlo. México: Antigua Librería Robredo,
1946.

Cicero. *De oratore.* Translated by Harris Rackham and Edward William Sutton. Loeb
Classical Library 348. Cambridge, MA, Harvard University Press, 1942.

Cieza de León, Pedro de. *Obras completas.* 3 vols. Edited by Carmelo Sáenz de Santamaría.
Madrid: Consejo Superior de Investigaciones Científicas, 1984.

Cieza de León, Pedro de. *Crónica del Perú: Segunda parte.* Edited by Francesca Cantù. Lima:
Fondo Editorial, 1986.

Coogan, Michael David, Marc Zvi Brettler, Carol A. Newsom, and Pheme Perkins. *The New
Oxford Annotated Bible with the Apocryphal/Deuterocanonical Books: New Revised
Standard Version.* Augm. 3rd ed. Oxford: Oxford University Press, 2007.

Correas, Gonzalo. *Vocabulario de refranes y frases proverbiales.* Edited by Louis Combet.
Bordeaux: Institut d'etudes iberiques et ibero-americaines de l' Universite de Bordeaux,
1967.

Cuestionarios para la formación de las relaciones geográficas de Indias, siglos XVI/XIX. Edited
by Francisco de Solano, Pilar Ponce, and Antonio Abellán. Madrid: Consejo Superior de
Investigaciones Científicas, 1988.

Cuervo, Rufino José. *Apuntaciones críticas sobre el lenguaje bogotano.* Bogotá: Arnulfo M.
Guarín, 1867.

Cuervo, Rufino José. *Diccionario de construcción y régimen de la lengua castellana.* Paris: A.
Roger y F. Chernoviz Libreros Editores, 1886.

Cueva de la Garoza, Juan de la. *Juan de la Cueva et son "Exemplar poético."* Edited by Frans
  Gustaf Emanuel Walberg. Lund: Imprimiere Håkan Ohlsson, 1904.
Del Rosal, Francisco. *Origen y etimología de todos los vocablos originales de la lengua
  castellana.* Edited by Enrique Gómez Aguado. Madrid: Consejo Superior de Investigacio-
  nes Científicas, 1992.
Domínguez, Francisco. "Carta del geógrafo Francisco Domínguez á Felipe II desde Méjico el
  30 de diciembre de 1581 sobre que S.M. mande al Virrey D. Martinez Enrique Remita." In
  Colección de documentos inéditos para la historia de España, 379–84. Madrid, 1842–1895.
Domingo Malvadi, Arantxa. *Bibliofilia humanista en tiempos de Felipe II: La biblioteca de
  Juan Páez de Castro.* Salamanca: Ediciones Universidad de Salamanca, Área de Publica-
  ciones de la Universidad de León, 2011.
Durán, Diego, and Ángel María Garibay K. *Historia de las Indias de Nueva España e islas de
  la Tierra Firme.* México: Editorial Porrúa, 1967.
Erasmus, Desiderius. *Collected Works of Erasmus: Adages I1 to Iv100.* Translated by Margaret
  Mann Phillips, annotated by R. A. B. Mynors. Toronto: University of Toronto Press, 1982.
Erasmus, Desiderius. *Collected Works of Erasmus.* Vols. 67–68: *Spiritualia and Pastoralia:
  Exomologesis and Ecclesiastes.* Edited by Alexander Dalzell and Frederick J. McGinness.
  Translated by Michael John Heath and James L. P. Butrica. Toronto: University of Toronto
  Press, 2015.
Eustacio, Esteban. "De las cosas necesarias para escribir historia: Memorial inédito del
  Dr. Juan Paéz de Castro al Emperador Carlos V." *La ciudad de Dios* 28 (1892): 604–610.
Eustacio, Esteban. "De las cosas necesarias para escribir historia: Memorial inédito del
  Dr. Juan Paéz de Castro al Emperador Carlos V." *La ciudad de Dios* 29 (1892): 27–37.
Fox Morcillo, Sebastián. *Teoría de la historia y teoría política en el siglo xvi: Sebastián Fox
  Morcillo, De historiae institutione dialogus = Diálogo de la enseñanza de la historia, 1557.*
  Translated and edited by Antonio Cortijo Ocaña. Alcalá: Universidad de Alcalá, Servicio
  de Publicaciones, 2000.
The Fragmentary Latin histories of Late Antiquity (AD 300–620). Edited by Lieve Van Hoof
  and Peter Van Nuffelen. Cambridge: Cambridge University Press, 2020.
Garibay y Zamalloa, Esteban de. "Refranes vascongados, recogidos y ordenados por Esteban
  de Garibay y Zamalloa." In *Memorial histórico español: Colección de documentos,
  opusculos y antigüedades que publica la Real Academia de la Historia,* vol. 7, 630–659.
  Madrid: Imprenta de José Rodriguez, 1854.
General History of the Things of New Spain: The Florentine Codex. Translated by Arthur
  Anderson and Charles E. Dibble. 13 vols. Salt Lake City: University of Utah Press,
  1950–1982.
Gessner, Conrad. *Mithridates.* Translated by Bernard Colombat and Manfred Peters. Geneva:
  Droz, 2009.
Gramática de la lengua vulgar de España, impresa en Lovaina por Bartolomé Gravio en 1559.
  Zaragoza: La Derecha, 1892.
Guadix, Diego de. *Recopiliación de algunos nombres arábigos que los árabes pusieron a
  algunas ciudades, Y a otras muchas cosas* [1593]. Bibliotheca Arabo-Romanica et Islamica.
  Edited by Elena Bajo Pérez and Felipe Maíllo Salgado. Gijón: Ediciones Trea, S.L., 2005.
Guadix, Diego de. *Diccionario de arabismos: Recopilación de algunos nombres arábigos.*
  Edited by María Águeda Moreno Moreno and Ignacio Ahumada. Jáen: Universidad de
  Jáen, 2007.
Hernández, Esther. *Vocabulario en lengua castellana y Mexicana de Fray Alonso de Molina:
  Estudio de los indigenismos léxicos y registros de las voces españoles internas.* Madrid:
  Consejo Superior de Investigaciones Científica, 1996.

Hernández, Francisco. *Antigüedades de la Nueva España*. Edited by and Ascensión H. de León Portilla. Madrid: Historia 16, 1986.

Hernández, Francisco. *De antiquitatibus Novae Hispaniae*. Facsimile ed. México: Talleres Gráficos del Museo Nacional de Arqueología, Historia y Ethnografía, 1926.

Hernández, Francisco. "An Epistle to Arias Montano: An Epistle of a Poem by Francisco Hernández." Translated by Rafael Chabrán and Simon Varey. *Huntington Library Quarterly* 55, no. 4 (1992): 620–634.

Hernández, Francisco. *Obras completas de Francisco Hernández*. Edited by German Somolinos D'Ardois. 8 vols. México: Universidad Nacional de México, 1959–1985.

Hervás y Panduro, Lorenzo. *Catálogo de las lenguas de las naciones conocidas, y numeración, division, y clases de éstas, según la diversidad de sus idiomas y dialectos*. 6 vols. Madrid: Imprenta de la Administración del Real Arbitrio de Beneficiencia, 1800.

Horapollo. *The Hieroglyphics of Horapollo*. Translated by George Boas. Princeton, NJ: Princeton University Press, 1993.

Huehuehtlahtolli: Testimonios de la antigüa palabra [1600]. Facsimile ed. Translated by Librado Silva Galeana, introductory study by Miguel León-Portilla. México: Comisión Nacional Conmemorativa del V Centenario del Encuentro de dos Mundos, 1988.

Huerga, Cipriano de la. *Obras completas*. 10 vols. León: Secretariado de Publicaciones de la Universidad de León, 1990–2005.

Isidore of Seville. *The Etymologies of Isidore of Seville*. Translated by Stephen A. Barney, W. J. Lewis, J. A. Beach, and Oliver Berghof. Cambridge: Cambridge University Press, 2010.

Jiménez de la Espada, Marcos. *Relaciones geográficas de Indias, Peru*. 4 vols. Madrid: Tip. de M. G. Hernández, 1881.

Jiménez de la Espada, Marcos. *El codigo Ovandino*. Madrid: Imprenta de Manuel G. Hernández, 1891.

Jiménez de la Espada, Marcos. "Correspondencia del Doctor Benito Arias Montano con el Licenciado Juan De Ovando." *Boletín de la Real Academia de la Historia* 19 (1891): 476–498.

Juvenal and Persius. *The Satires of Persius / The Satires of Juvenal*. Translated by Susanna Morton Braund. Loeb Classical Library 91. Cambridge, MA: Harvard University Press, 2004.

Karttunen, Francis. *An Analytical Dictionary of Nahuatl*. Norman: University of Oklahoma, 1992.

La philosophía vulgar de Juan Mal Lara. Facsimile ed. Edited by José J. Labrador Herraiz and Ralph A. DiFranco. México: Frente de Afirmación Hispanista, A.C., 2012.

Las Casas, Bartolomé de. *Apologética historia sumaria*. Edited by Edmundo O'Gorman. 2 vols. México: Universidad Autónoma de México, 1967.

"Las Ordenanzas de 1571 del Real y Supremo Consejo de las Indias: Texto facsimilar de la edición de 1585; Notas de Antonio Muro Orejón." *Anuario de estudios americanos*, January 1957, 363–423.

León, Luis de. *Obras completas castellanas de Fray Luis de León*. Edited by Felix Garcia. 2 vols. Madrid: La Editorial Católica, S.A., 1957.

León, Luis de. *De los nombres de Cristo*. Edited by Cristóbal Cuevas. Madrid: Ediciones Cátedra, 1997.

León, Luis de. *Escritos sobre América*. Edited and translated by Andrés Moreno Mengíbar and Juan Martos Fernández. Madrid: Editorial Tecnos S.A. 1999.

Fradejas Lebrero, José. *Más de mil y un cuentos del Siglo de Oro*. Universidad de Navarra. Pamplona: Iberoamericana Vervuert, 2008.

Madariaga Orbea, Juan, and Juan Fornoff. *Anthology of Apologists and Detractors of the Basque Language*. Reno: Center for Basque Studies, University of Nevada, 2006.

Magnus, Olaus. *Historia de gentibus septentrionalibus: Romae 1555 = Description of the Northern Peoples: Rome 1555*. Translated by Peter Fisher and Humphrey Higgens. Edited by Peter Foote. Annotation derived from the commentary by John Granlund. London: Hakluyt Society, 1996.

Martínez de Bujanda, Jesús. *El índice de libros prohibidos y expurgados de la Inquisición española (1551–1819)*. Madrid: Biblioteca de Autores Cristianos, 2016.

Medicinas, drogas y alimentos vegetales del Nuevo Mundo: Textos e imágenes españolas que los introdujeron en Europa. Edited by José M. López Piñero, José Luis Fresquet Febrer, María Luz López Terrada, and José Pardo Tomás. Madrid: Ministerio de Sanidad y Consumo, 1992.

The Mexican Treasury: The Writings of Dr. Francisco Hernández. Edited by Simon Varey. Translated by Rafael Chabrán, Cynthia L. Chamberlin, and Simon Varey. Stanford, CA: Stanford University Press, 2000.

Motolinía, Toribio de Benavente. *Relaciones de la Nueva España*. Biblioteca del Estudiante Universitario. 2nd ed. México: Universidad Nacional Autónoma de México, 1964.

Motolinía, Toribio de Benavente. *Memoriales, o libro de las cosas de la Nueva España y los naturales a ella*. Edited by Edmundo O'Gorman. México: Universidad Nacional Autónoma, 1971.

Maxwell, Judith M., and Craig A. Hanson. *Of the Manners of Speaking that the Old Ones Had: The Metaphors of Andrés de Olmos in the Tulal Manuscript Arte para Aprender la Lengua Mexicana, 1547: With Nahuatl/English, English/Nahuatl Concordances*. Salt Lake City: University of Utah Press, 1992.

Mayans y Siscar, Gregorio. *Epistolario*. Edited by Antonio Mestre Sanchis. 25 vols. Vol. 6. Valencia: Publicaciones del Ayuntamiento de Oliva, 1977.

Mayans y Siscar, Gregorio. *Obras completas*. Edited by Antonio Mestre Sanchis. Valencia: Publicaciones del Ayuntamiento de Oliva, 1984.

Mendieta, Jerónimo de. *Historia eclesiástica Indiana*. México: Antigua librería Impr. F. Diaz de Leon y S. White, 1870.

Nebrija, Antonio de. *Gramatica de la lengua castellana (Salamanca, 1492); Muestra de la istoria de las antiguedades de España; Reglas de orthographia en la lengua castellana*. Edited by Ignacio Gonzalez-Llubera. Oxford: Oxford University Press, 1926.

Nebrija, Antonio de. *Gramática de la lengua castellana*. Facsimile ed. Madrid: Ediciones de Cultura Hispánica Instituto de Cooperación Iberoamericana, 1996.

Núñez, Hernán. *Refranes o proverbios en romance*. Edited by Louis Combet, Julia Sevilla Muñoz, Germán Conde Tarrío, and Josep Guia i Marín. Vol. 1. Madrid: Ediciones Guillermo Blázquez, 2001.

Núñez Muley, Francisco. *A Memorandum for the President of the Royal Audiencia and Chancery Court of the City and Kingdom of Granada*. Translated by Vincent Barletta. Chicago: University of Chicago Press, 2007.

Oration on the Dignity of Man: A New Translation and Commentary. Edited by Francesco Borghesi, Michael Papio, and Massimo Riva. Cambridge: Cambridge University Press, 2012.

Oviedo, Gonzalo Fernández de. *Historia general y natural de las Indias, islas y tierra-firme del mar océano*. Edited by D. José Amador de los Ríos. 4 vols. Madrid: Imprenta de la Real Academia de la Historia, 1851–1855.

Páez de Castro, Juan. "Memorial al Rey Don Felipe II, sobre las librerías por el Doctor Juan Páez de Castro." In *Carta del doctor Juan Páez de Castro al secretario Matheo Vázquez,*

sobre el precio de libros manuscritos. Digital ed. Valladolid: Junta de Castilla y León /
Consejería de Cultura y Turismo, 2009–2010.

Pané, Ramón. *Relación acerca de las antigüedades de los indios, las cuales, con diligencia,
como hombre que sabe el idioma de estos, recogió por mandato del Almirante*. Edited by
J. J. Arrom. México: Siglo XXI, 1988.

Perea, Juan Augusto. *Glosario etimólogico taíno-español histórico y etnográfico*. Mayagüez,
P.R.: Tipográfia Mayagüez, 194.

Plato. *Cratylus, Parmenides, Greater Hippias, Lesser Hippias*. Translated by Harold North
Fowler. Loeb Classical Library. Cambridge, MA: Harvard University Press, 1939.

Pliny. *Natural History*. Vols. 1–5. Translated by Horace Rackham. Loeb Classical Library 330,
352, 353, 370, and 371. Cambridge, MA: Harvard University Press, 1938–1950.

Pliny. *Natural History*. Vols. 6–8. Translated by William Henry Samuel Jones. Loeb Classical
Library 392–393, 418. Cambridge, MA: Harvard University Press, 1951–1963.

Pliny. *Natural History*. Vol. 9. Translated by Horace Rackham. Loeb Classical Library 394.
Cambridge, MA: Harvard University Press, 1952.

Pliny. *Natural History*. Vol. 10. Translated by David Edward Eichholz. Loeb Classical Library
419. Cambridge, MA: Harvard University Press, 1962.

*Pomar y Zurita: Pomar, Relación de Tezcoco; Zurita, Breve relación de los señores de la Nueva
España. Varias relaciones antiguas*. Edited by Joaquín García Icazbalceta. México:
Imprenta de Francisco Díaz de León, 1891.

Quintilian. *The Orator's Education*. Vol. 5: *Books 11–12*. Edited and translated by Donald A.
Russell. Loeb Classical Library 494. Cambridge, MA: Harvard University Press, 2002.

Relaciones geográficas de México siglo XVI. Edited by Rene de Acuña. 10 vols. México:
Universidad Autónoma de México, 1982–1986.

Relaciones topográficas de Felipe II, Madrid. Edited by Alfredo Alvar Ezquerra. Transcribed
by María Elena García Guerra and María de los Angeles Vicioso Rodríguez. 4 vols.
Madrid: Consejo Superior de Investigaciones Científicas, 1993.

Renaissance Invention: Stradanus's Nova Reperta. Edited by Lia Markey. Evanston, IL:
Northwestern University Press, 2020.

Reuchlin, Johannes. *Recommendation whether to Confiscate, Destroy and Burn All Jewish
Books*. Edited by Peter Wortsman and Elisheva Carlebach. New York: Paulist Press, 2000.

Recchi, Nardo Antonio de. *De materia medica Novae Hispaniae: Manuscrito de Recchi*.
Translated and edited by Florentino Fernández González and Raquel Álvarez Peláez. 2
vols. Madrid: Ediciones Dos Calles, 1998.

Sahagún, Bernardino de. *Historia general de las cosas de Nueva España*. Edited by Alfredo
López Agustín and Josefina García Quintana. México: Consejo Nacional Para la Cultura
y las Artes, 2000.

Sahagún, Bernardino de. *Historia general de las cosas de Nueva España*. Edited by Ángel
María Garibay Kintana. México: Editorial Porrúa, 2016.

Tubino, Francisco M. *Pablo de Céspedes: Obra premiada por voto unánime de la Academia de
Nobles Artes de San Fernando en el certamen de 1866*. Madrid: Manuel Tello, 1868.

Uztarroz, Juan Francisco A. de, and Diego J. Dormer. *Progresos de la historia en Aragon y
vidas de sus cronistas, desde que se instituyó este cargo hasta su extinción: Primera parte
que comprende la biografía este de Gerónimo Zurita*. Zaragoza: Impr. del Hospicio, 1878.

Valdés, Juan de. *Diálogo de la lengua*. Edited by Juan M. Lope Blanch. Madrid: Editorial
Castalia, 1969.

Valdés, Juan de. *Diálogo de lengua*. Edited by José Enrique Laplana. Barcelona: Crítica, 2010.

Vaquero Serrano, María del Carmen. *El maestro Álvar Gómez: Biografía y prosa inédita*.
Toledo: Impresión Torres, 1993.

Varro, Marcus. *De lingua latina*. Translated by Roland G. Kent. 2 vols. Cambridge, MA: Harvard University Press; London: Heinemann, 1977.

Viciana, Rafael Martí de. *Crónica de la ínclita y coronada ciudad de Valencia*. Valencia: Universidad, 1983.

Vives, Juan Luis. *Los comentarios de Juan Luis Vives a "La ciudad de Dios" de San Agustín*. Translated by Rafael Cabrera Petit. Valencia: Ajuntament de Valencia, 2000. Biblioteca Valenciana Digital, https://bivaldi.gva.es/va/corpus/unidad.do?idCorpus=1&idUnidad =10410.

Vives, Juan Luis. *Obras completas de Juan Luis Vives*. Translated by Llorenç Riber. 2 vols. Madrid: M. Aguilar, 1947–48.

Zorita, Alonso de. *Relación de la Nueva España*. Edited by Ethelia Ruiz Medrano and José Mariano Leyva. 2 vols. México: Consejo Nacional para la Cultura y Artes, 2011.

Index

relics, 4, 200n7
Reuchlin, Johann, 157
Reyneri, Cornelius, 167
Reyre, Dominique, 70, 190
Ricchieri, Lodovico, *Lectiones antiquae*, 138
Ríos Castaño, Victoria, 106
Rodríguez, Alónso, 97
Rodríguez Mediano, Fernando, 78, 79
Rodríquez de Francia, José Gaspar, 197
romance/Romance languages, 8, 29, 95, 101–04,
   107–8, 194–95, 210n63
romanceros, 43
romances, 35–36, 37, 43–44, 46, 47, 48, 49,
   194–95, 210n63, 213n113, 214n11
Romans, ancient, 18, 22, 38, 39, 44–45, 650, 51, 4,
   75, 76, 118, 129–130, 172, 177–78, 188, 233n25; of
Iberian Peninsula, 3, 17, 55, 56, 57, 60, 65, 67, 70,
   72, 75, 77, 97, 101, 121, 144
Román Zamora, Jerónimo, 58–59; *Republics of
   the World*, 81–82, 110–11, 178–79
Rondelet, Guillaume, *Books of Saltwater Fish,*
   133, 136, 138
Royal Academy of Language, 84
Royal Library of San Lorenzo de El Escorial,
   4–5, 17–19, 20–25, 26, 53–54, 163, 207n24

Sahagún, Berardino de, 89, 105, 147, 239n97;
   General History . . . , 106–8, 119, 120, 152, 153,
   228n94, 228n103
St. Ildefonso, 75, 164
San Juan (Puerto Rico), 39, 42
Santo Domingo, 13, 27–28, 42, 110, 145
Santo Tomás, Domingo de, 32, 93
Scaliger, Joseph Justus, 85, 196
Scolieri, Paul, 40, 42
Seville, Spain, intellectual elite, 46–47, 48, 101
Shalev, Zur, 176
Sidney, Philip, *Apology for Poetry*, 45
signification, in language, 11, 84–85, 93, 104, 136,
   137, 145, 181, 187; Arabic, 80–81; Basque, 70,
   71; in Castilian, 136; Hebrew, 84, 162, 165–66,
   173–75, 181; Indigenous languages, 1, 27, 29,
   32, 34, 80–81, 82, 86, 93, 185, 188–89, 193;
   metaphorical, 104, 107, 174–75; natural *versus*
   conventional, 11, 136–37, 162; of toponyms, 1,
   27, 70, 80, 82, 181

Siraisi, Nancy, 123
Sirleto, Guglielmo, 168–69
Sixtus of Sienna, *Bibliotheca* . . . , 165
Solomon, 70, 82, 116, 177, 180–81
Spanish conquest, of the Americas, 5, 8, 34, 40,
   42, 56, 96, 118–20, 179, 200n9, 229n128
Spanish Crown: Bourbon monarchy, 7, 60, 159,
   195, 196; linguistic diversity policies, 5–9, 195,
   198, 200n9
Spanish language, etymological histories, 9–10,
   55–85, 194–95; Hebrew lineage, 3, 58–59,
   60–61, 68–70, 71, 73, 82, 143–44, 161, 191,
   248n97; 85; Latin lineage, 73–74, 90, 161;
   linguistic change theories, 62–66, 83–84, 85;
   multilingualism, 4, 5–6, 8–9, 16, 58, 64, 67,
   84, 87; primordial language, 56, 58, 65–73,
   66–73, 78–79, 161, 194–95; relation to Spanish
   identity, 57–58, 68, 83–84, 198. *See also*
   toponym etymologies, of Spain
speech communities, 3–4, 9, 26, 66, 89, 121, 161,
   200n6
Strabo, 65, 67, 69, 76, 143
Stradanus (Jan van der Straet), 181–82
Suárez, Diego, 97

Tagalog language, 5
Taíno language, 7, 12, 19, 25, 37, 40, 42, 73,
   160–61, 210n60, 242n2
Tenochtitlan. *See* Mexico City
Tepehuan language, 105
Texcoco, Mexico, 105, 122–23, 186, 229n128,
   231n1
Tezconquense language, 151
Theophrastus, 23, 132, 156
Thomas Aquinas, 136
Tlacopan, Mexico, 105, 122, 231n1
Tlatelolco, Mexico, 104, 105, 106, 107, 229n117
Tlaxcalan language, 151
Toledo, Spain, 1, 3, 60, 65, 116, 130, 131
toponym etymologies, 1–3, 4, 11–12, 27, 55–56,
   81–83, 119, 143–45, 180
toponym etymologies, Indigenous American, 1,
   27, 28, 59, 70, 79–80, 81–83, 89, 90, 98–99, 110,
   111, 120–21, 157, 184, 191, 199n2, 225n37; Acosta's
   discussion of, 179, 180, 184; Arabic, 80–81;
   Basque, 71–72; Hebrew/biblical, 70, 81–82, 161,

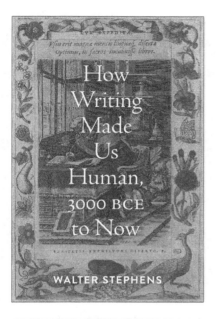
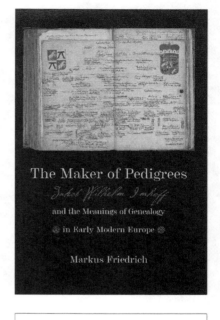
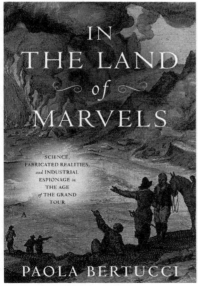
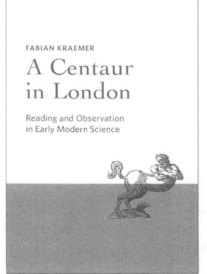